Diabetes insipidus in Man

Frontiers of Hormone Research

Vol. 13

Series Editor
Tj.B. van Wimersma Greidanus, Utrecht

S. Karger · Basel · München · Paris · London · New York · Tokyo · Sydney

International Symposium on Diabetes insipidus in Man, Paris, January 18–19, 1984

Diabetes insipidus in Man

Volume Editors
P. Czernichow, Paris
A.G. Robinson, Pittsburgh, Pa.

99 figures, 3 color plates, 30 tables, 1985

S. Karger · Basel · München · Paris · London · New York · Tokyo · Sydney

Frontiers of Hormone Research

National Library of Medicine, Cataloging in Publication
 International Symposium on Diabetes insipidus in Man (1984: Paris, France)
 Diabetes insipidus in man/International Symposium on Diabetes insipidus in Man, Paris, January 18–19, 1984
 Volume editors, P. Czernichow, A.G. Robinson. – Basel; New York: Karger, 1985. –
 (Frontiers of hormone research; vol. 13)
 Includes index.
 1. Diabetes insipidus – congresses   I. Czernichow, P.   II. Robinson, Alan G.
 W1 FR946F v. 13 [WK 550 I59d 1984]
 ISBN 3–8055–3921–5

Drug Dosage
 The authors and publisher have exerted every effort to ensure that drug selection and dosage set forth in this text are in accord with current recommendations and practice at the time of publication. However, in view of ongoing research, changes in government regulations, and the constant flow of information relating to drug therapy and drug reactions, the reader is urged to check the package insert for each drug for any change in indications and dosage and for added warnings and precautions. This is particularly important when the recommended agent is a new and/or infrequently employed drug.

All rights reserved.
 No part of this publication may be translated into other languages, reproduced or utilized in any form or by any means, electronic or mechanical, including photocopying, recording, microcopying, or by any information storage and retrieval system, without permission in writing from the publisher.

© Copyright 1985 by S. Karger AG, P.O. Box, CH-4009 Basel (Switzerland)
 Printed in Switzerland by Buchdruckerei 'Der Bund' Bern
 ISBN 3–8055–3921–5

Contents

Preface	VII
Zimmerman, E.A. (New York, N.Y.): Anatomy of Vasopressin-Producing Cells (with Color Plate I)	1
Robinson, A.G.; Verbalis, J.G. (Pittsburgh, Pa.): Biosynthesis Transport and Release of Vasopressin	22
Richter, D.; Schmale, H. (Hamburg): Molecular Aspects of the Expression of the Vasopressin Gene	37
Leake, R.D.; Fisher, D.A. (Torrance, Calif.): Ontogeny of Vasopressin in Man	42
Vincent, J.D.; Legendre, P.; Poulain, D.; Arnauld, E.; Theodosis, D. (Bordeaux): Electrophysiology of Vasopressin-Secreting Cells	52
Nicolaïdis, S. (Paris): Thirst Mechanisms and Antidiuretic Hormone	69
Jard, S. (Montpellier): Vasopressin Receptors	89
Valtin, H.; North, W.G.; Edwards, B.R.; Gellai, M. (Hanover, N.H.): Animal Models of Diabetes insipidus	105
Vokes, T.; Robertson, G.L. (Chicago, Ill.): Physiology of Secretion of Vasopressin	127
Moses, A.M. (Syracuse, N.Y.): Clinical and Laboratory Observations in the Adult with Diabetes insipidus and Related Syndromes	156
Robertson, G.L. (Chicago, Ill.): Diagnosis of Diabetes insipidus	176
Czernichow, P.; Pomarede, R.; Brauner, R.; Rappaport, R. (Paris): Neurogenic Diabetes insipidus in Children	190
Manelfe, C.; Balliana, M.O.; Louvet, J.P.; Sevely, A.; Prere, J.; Rochiccioli, P.; Bonafe, A. (Toulouse): Computed Tomography in Diabetes insipidus	210
Niaudet, P.; Dechaux, M.; Leroy, D.; Broyer, M. (Paris): Nephrogenic Diabetes insipidus in Children	224
Scherbaum, W.A. (Ulm); *Bottazzo, G.F.* (London); *Czernichow, P.* (Paris); *Wass, J.A.H.* (London); *Doniach, D.* (London): Role of Autoimmunity in Central Diabetes insipidus (with Color Plate II)	232
Pomarede, R.; Czernichow, P.; Brauner, R.; Rappaport, R. (Paris): Intracranial Germinoma in Children and Diabetes insipidus: Clinical Description and Search for Tumor Markers (with Color Plate III)	240

Contents

Verbalis, J.G.; Robinson, A.G. (Pittsburgh, Pa.); *Moses, A.M.* (Syracuse, N.Y.):
  Postoperative and Post-Traumatic Diabetes insipidus 247
Amico, J.A. (Pittsburgh, Pa.): Diabetes insipidus and Pregnancy 266
Pliška, V. (Zürich): Pharmacology of Deamino-*D*-Arginine Vasopressin 278
Robinson, A.G.; Verbalis, J.G. (Pittsburgh, Pa.): Treatment of Central Diabetes
  insipidus 292
Kauli, R.; Galatzer, A.; Laron, Z. (Tel Aviv): Treatment of Diabetes insipidus in
  Children and Adolescents 304
Vilhardt, H.; Hammer, M.; Bie, P. (Copenhagen): Peroral Administration of Antidiuretic Peptides to Conscious Dogs, Normal Humans and Patients with
  Diabetes insipidus 314

Subject Index 321

Preface

This book is the first attempt to provide a comprehensive clinical overview of the disease diabetes insipidus. It is remarkable that this disease which was distinguished from diabetes mellitus by the famous 'taste test' of *Willis* [1] as early as 1664, and has contributed greatly to advances in medical science, has not previously been the subject of an extensive review. A few historical landmarks illustrate the impact of basic research on understanding neurohypophyseal endocrinology. The studies of *Bernard* [2] that discrete lesions in the hypothalamus produced an effect on the kidney (1849) was an original demonstration of the role of the nervous system in control of peripheral organ function. It was subsequently shown that lesions in the hypothalamus, in the median eminence, or in the posterior pituitary might produce disorders of diabetes insipidus and the neurohypophysis was, for a time, thought to be a 3-part organ [3]. It was eventually shown the 3 parts were all a single gland originating in the hypothalamus and ending in the posterior lobe. The concept of neurosecretion and axon flow of neuropeptides was based on this organ as described in the classic studies of *Scharrer and Scharrer* [4] and the signal contribution of *Bargmann and Scharrer* [5] who proved conclusively that section of the pituitary stalk resulted in accumulation of neurosecretory material on the brain side of the section. *Bargmann's* studies used the chrome-hematoxylin-phloxine stain which *Gomori* [6] had used on pancreas and identified the classic neurohypophyseal system. More recently use of antibodies for immunohistology was described and again studies of the neurohypophysis became the model for identification of peptide hormones within nerve cells [7]. These studies have led to the identification of a vast innervation of the nervous system by peptidergic neurons and the physiologic significance of these multiple interventions is only beginning to be unraveled.

Possibly because of the large size of the magnocellular neurons, the discrete anatomic location, and the ease of demonstrating a physiologic effect (antidiuresis, pressor activity, uterine contraction, or milk ejection) the cells were a ready target for neurophysiologists and have provided the most detailed correlation of nerve cellular electrical activity with physiologic effect [8]. In studies of biosynthesis, vasopressin was hypothesized to be synthesized as a prohormone [9] before the more widely known description of synthesis of insulin as proinsulin. The prohormone has now been conclusively documented by the isolation of the gene which produces vasopressin, neurophysin and the glycopeptide of the posterior pituitary.

Therapy of diabetes insipidus has been equally illustrative. Extracts of the posterior pituitary were shown to have antidiuretic activity in patients with diabetes insipidus in 1913 [10, 11] and were further demonstrated to produce antidiuresis in normal subjects. These reports were one of the first uses of replacement hormone therapy which became known in the early 1900s as 'organotherapy'. In more recent years 1-desamino-8-*D*-arginine-vasopressin (DDAVP) has been used to treat this disorder. DDAVP is a model of the ability of molecular pharmacology to specifically tailor a drug which will enhance the desired effects and decrease the undesired effects [12].

These few highlights are not meant to be a comprehensive review of the history of the neurohypophysis and the references cited are of necessity highly selective. Nonetheless, they do illustrate the critical role that the study of the posterior pituitary has played in medical history. It is on this background that the editors convened an international congress in Paris, France, in January 1984, to provide a clinical and scientific update of the major disease of the posterior pituitary, diabetes insipidus.

We acknowledge the support of Ferring AB, Malmö, Sweden, and of USV Pharmaceuticals of Revlon Health Care, USA, who sponsored the symposium. The excellent organization by the secretary of the congress, *Sophie Battarel,* made the symposium possible.

Les Gets, January 1984 *P. Czernichow*
A.G. Robinson

References

1. Willis, T.: Cerebri anatome: cui accessit nervorum descriptio et usus (Flesher, London 1664).
2. Bernard, C.: Leçons sur les propriétés physiologiques et les altérations pathologiques des liquides de l'organisme (Paris 1859).
3. Motzfeld, K.: Diabetes insipidus. Endocrinology 2: 112 (1918).
4. Scharrer, E.; Scharrer, B.: Neurosekretion; in Bargmann, Handbuch der mikroskopischen Anatomie des Menschen, Bd 6, T1.5 (Springer, Berlin 1954).
5. Bargmann, W.; Scharrer, E.: The site of origin of the hormones of the posterior pituitary. Am. Scient. 39. 255–259 (1951).
6. Gomori, J.D.: Some aspects of the anatomy and function of the pituitary gland, with a special reference to the neurohypophysis. Alexander Blair Hosp. Bull. 6: 128–142 (1947).
7. Zimmerman, E.A.: Anatomy of vasopressin-producing cells (this volume).
8. Vincent, J.D.; Legendre, P.; Poulain, D.; Arnauld, E.; Theodosis, D.: Electrophysiology of vasopressin-secreting cells (this volume).
9. Sachs, H.; Takabatake, Y.: Evidence for a precursor in vasopressin biosynthesis. Endocrinology 75: 943–948 (1964).
10. Farini, F.: Diabete insipido ed opoterapia ipofisaria. Gaz. Osp. Clin. 34: 1135–1139 (1913).
11. Von den Velden, R.: Die Nierenwirkung von Hypophysenextrakten beim Menschen. Berl. klin. Wschr. 50: 2083–2086 (1913).
12. Robinson, A.G.; Verbalis, J.G.: Treatment of central diabetes insipidus (this volume).

Anatomy of Vasopressin-Producing Cells

E.A. Zimmerman

Department of Neurology, College of Physicians and Surgeons, Columbia University, New York, N.Y. USA

Introduction

The hypothalamic-neurohypophysial neuronal system, which produces oxytocin and vasopressin and secretes them into the general circulation, has served as the most widely studied model of the peptidergic neuron for more than 40 years [1]. It was shown from the earliest days using Gomori stains that the large neurons (magnocellular) concentrated in the supraoptic (SON) and paraventricular nuclei (PVN) are the source of fibers projecting in the hypothalamo-neurohypophysial tract carrying neurosecretory material to the posterior pituitary gland [2, 3]. Lesions in the nuclei, particularly the supraoptic, or involving the tract as far as the upper stalk, produce the loss of vasopressin secretion and clinical diabetes insipidus [2–6]. It is this system, the magnocellular neurosecretory system, that is the main subject of this symposium. Introduction of immunohistochemical techniques and other anatomical methods, including orthograde and retrograde tracing methods, to the study of vasopressin neurons in the last decade has led to further understanding of the organization of the magnocellular system in many mammalian species including man [see 7, 8 for recent reviews]. It has been shown, for example, that accessory magnocellular elements, at least in the rat, are as important in number as those concentrated in PVN and SON in terms of projections to posterior pituitary gland [9].

Immunohistochemical studies have also revealed a second vasopressin neurosecretory pathway to the hypophysial portal system which transports high concentrations of the hormone to the anterior pituitary gland [7]. Recent evidence from several laboratories indicates that vasopressin and corticotropin-releasing hormone (CRH) co-exist in smaller neurons (parvocellular) in adrenalectomized rats [10–12] in the medial portion of the paraventricular nucleus. Since neurons of this region are known to project to the portal capillary system, both CRH and vasopressin are most likely secreted together. Evidence continues to mount that vasopressin produced by these pathways participates in the secretion of adrenocorticotropin (ACTH) from the anterior pituitary gland.

Two additional parvocellular systems have been recently described [for reviews, see 7, 8]. One is located in cell bodies of the suprachiasmatic nucleus (SCN), the function of which is not yet known. The other is located in various cells of the PVN which projects fibers to many extra-hypothalamic sites of the brain and spinal cord. In the latter case, vasopressin and oxytocin are thought to be involved in neuronal communication between PVN and limbic structures and autonomic relay stations in brainstem and spinal cord. Some of the brainstem targets contain catecholaminergic neurons which in turn form regulatory relays to the PVN. The extrahypothalamic pathways suggest behavioral and autonomic functions for vasopressin and oxytocin in addition to more traditional neurosecretory functions.

Immunohistochemical Methods

Immunohistochemical methods (immunofluorescence, or immunoenzyme techniques such as those using horseradish peroxidase, HRP) became possible because of the identification, purification, and generation of antisera to vasopressin, oxytocin and their related neurophysin proteins. They are more sensitive and specific than were the earlier histochemical techniques of Gomori (aldehyde-fuchsin, chrome-alum-hematoxylin), which rely mainly on the sulfhydryl-rich neurophysins [3]. Although all the findings obtained with Gomori stains have been confirmed, the existence of vasopressin neurons in SCN was discovered using immunohistochemical methods [13, 14]. This probably reflects lower concentrations of vasopressin-neurophysin in SCN. Secondly, Gomori techniques could not differentiate vasopressin from oxytocin neurons. Many laboratories have preferred immunoperoxidase techniques to immunofluorescence because of the permanence of reaction products and the possibility of ultrastructural studies [15]. In the case of immunoelectron microscopy, however, problems remain in obtaining good preservation of the membranes and subcellular organelles while allowing penetration of antibodies.

Problems in specificity of antisera such as cross-reactivity of antiserum to vasopressin with oxytocin and vice versa, a problem in earlier studies, have been solved by solid-phase absorption with the cross-reacting antigen [16, 17] or more recently by application of monoclonal antibody to vasopressin [18]. Lack of specificity and inefficiency of liquid-phase absorptions led to erroneous conclusions, such as vasopressin and oxytocin may coexist in some neurons [19]. The homozygous Brattleboro rat with diabetes insipidus which lacks vasopressin and associated neurophysin (see chapter 6) also served as an excellent test for cross-reactivity of antisera to vasopressin with oxytocin and to vasopressin-neurophysin with oxytocin-neurophysin [16, 19]. Given the most specific antibody to vasopressin, however, there remains the problem inherent in immunocytochemistry that the antigenic determinants of the substance visualized is not vasopressin itself, but another peptide which shares the antigenic determinants of vasopressin. The solution at present is to co-localize vasopressin and other parts of the precursor with a number of different antibodies within the same cell. In this case the co-localization of specific neurophysins with oxytocin and vasopressin provided confidence that these cells have the biosynthetic mechanism to produce the hormones [15]. It also provided some evidence that the material identified was produced and not taken up from elsewhere [15]. Magnocellular neurons for example are known to take up albumin from the periphery. In the near future, application of libraries of monoclonal antibodies to various parts of vasopressin precursor and products may allow immunohistochemical studies of the processing of vasopressin in various vasopressin neurons and their processes.

False-negatives remain a significant problem in immunocytochemistry. Lack of sensitivity is often due to the antiserum used as well as the detection system. Application of the avidin-biotin detection method has recently improved results in some cases compared with the method of peroxidase-antiperoxidase complexes (PAP). Lack of visualization of vasopressin may also be due to inactivation of the antigenic site by formaldehyde fixation if it involves an aromatic amino acid [18], or possibly due to its inaccessibility in the precursor form. In some cases lack of visualization of perikarya due to a relatively low concentration of peptide compared with terminals can be obviated by pretreatment with intraventricular colchicine which increases cell body content by blocking axonal transport. Another improvement in recent years has been the use of 60- to 100-μm thick vibratome sections compared with 6- to 20-μm frozen or paraffin sections which provides a better appreciation of the distribution of cell bodies and fiber pathways. At the same time such thick sections provide problems for the study of two peptides in the same cell, such as CRH and vasopressin by serial sections, or by double staining elution techniques [10–12]. Finally, substances may not be visualized or the results altered by the endocrine or neurotransmitter status of the animal.

Classic Magnocellular Neurosecretory System to Posterior Pituitary

The distribution of oxytocin and vasopressin and their respective neurophysins have been described in a wide number of mammalian species [20–26] including monkey [27] and man [28–31]. Most studies have focussed on the distribution in the SON and PVN, although some have included accessory magnocellular groups, particularly in the rat (tab. I) [9]. These studies have been reviewed recently and only highlights and new concepts will be provided here [7, 8].

Table I. Magnocellular system

I   Supraoptic nucleus (SON)
II Paraventricular nucleus (PVN)
III Accessory nuclei
   1 The anterior commissural N.
   2 The perifornical N.: anterior
                          posterior
   3 N. circularis
   4 N. of the medial forebrain bundle
   5 The retrochiasmatic nucleus

Both hormones and their neurophysins are found in different neurons in the SON and PVN (fig. 1–3), in their dendrites, and in their axonal processes which converge in the zona interna of the rostral median eminence to traverse the pituitary stalk and end around systemic capillaries in the posterior pituitary gland (fig. 4). There are species differences in the relative number and distribution in these two nuclei [7]. In the rat equal numbers, or somewhat more vasopressin than oxytocin neurons [9, 13], are found in both SON and PVN, making the old concept that the SON is a vasopressin nucleus and PVN an oxytocin one [32] untenable. In fact the rat SON contains 3 times more magnocellular neurons than PVN. Since all SON neurons project to posterior pituitary, and all are reactive for one of the other hormones [13, 29], it is a major contributor of both hormones to the systemic circulation. In the human relatively few oxytocin neurons are found in SON compared with PVN, although vasopressin predominates in both [28, 29] (ratio of vasopressin/oxytocin neurons in SON = 4/1; PVN = 4/3). The number of magnocellular neurons in SON is only 36% greater than PVN in man [32].

The importance of the contribution of the accessory magnocellular [33] projections to posterior pituitary vasopressin projections was pointed out by the study of *Fisher* et al. [9] in the rat. By comparing retrograde tracing with HRP from posterior pituitary with immunohistochemical localization of vasopressin, it was found that the number of vasopressin neurons in the accessory groups contributing fibers to neural lobe equaled those in SON and PVN. The only exceptions were the very rostral magnocellular neurons in the anterior commissural and anterior perifornical groups which did not contain vasopressin. It is now known that these neurons primarily or solely contain oxytocin [26].

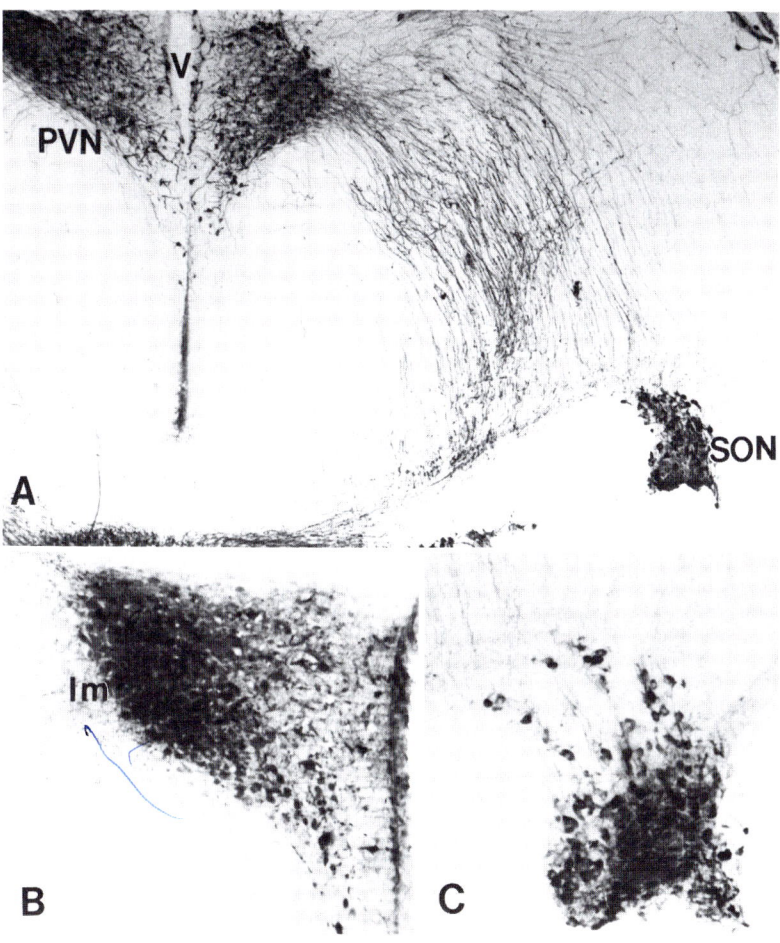

Fig. 1. Magnocellular neurosecretory system demonstrated in 100 μm sections of rat hypothalami by immunoperoxidase technique. *A* Low power photomicrograph of a section reacted with antiserum to both neurophysins. Magnocellular neurons are clustered in the paraventricular (PVN) and supraoptic (SON) nuclei. Axonal pathways passing laterally and then ventrally form the major projection of that nucleus to the posterior pituitary. V = Third ventricle. From *Sloviter and Zimmerman* [unpublished]. *B, C* Vasopressin neurons at slightly higher magnification clustered in the lateral magnocellular (lm) portion of the PVN *(B)*, and in the ventral portion of the SON *(C)*, both of which project to posterior pituitary gland. Reacted with monoclonal antibody. From *Hou-Yu* et al. [18], with permission.

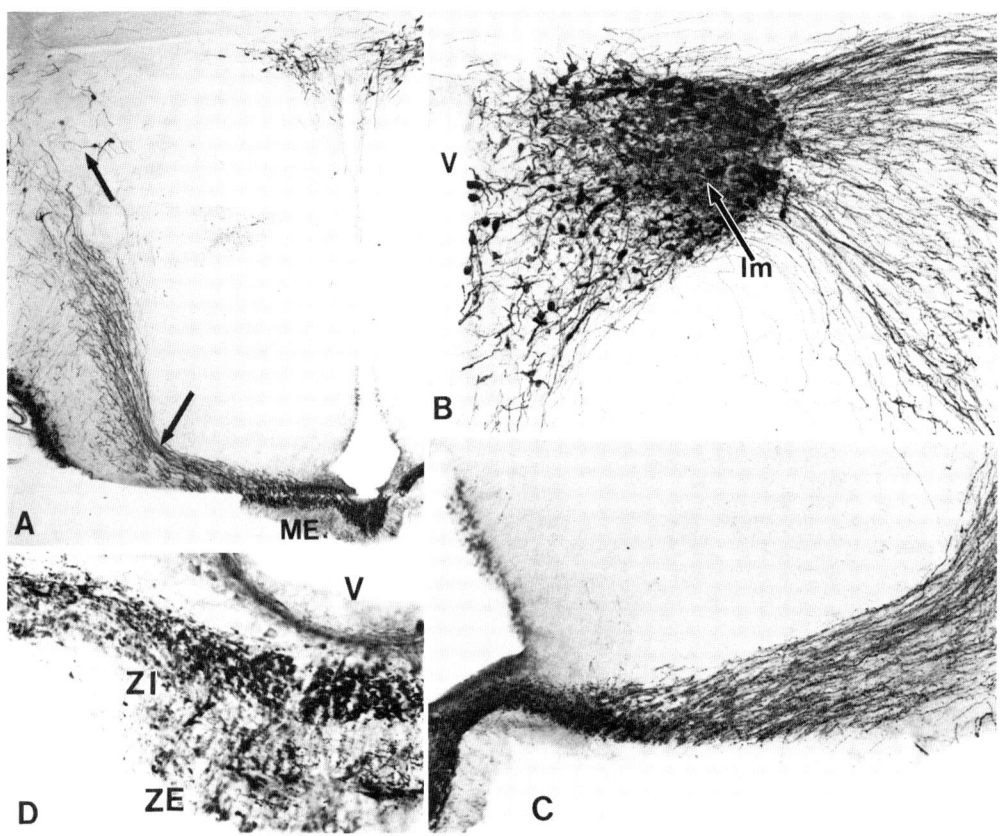

Fig. 2. A More caudal section to figure 1A reacted for neurophysins at the level of the posterior PVN and the median eminence (ME). Accessory posterior forniceal neurons (upper arrow) contribute to the axonal bundle (lower arrow) which passes into the ME. *B* PVN rostral to *(A)* showing extensive axonal projections extending laterally from the lateral magnocellular (1m) group. Some cells are found in the medial parvocellular portions including the periventricular zone along the third ventricle (V). *C* Higher magnification of the tract in *A* formed by the convergence of magnocellular projections as it enters the median eminence. *D* Vasopressin fibers in zona interna (ZI) of the median eminence which project to posterior pituitary, and in zona externa (ZE) where the hormone is secreted into portal blood. Normal rat reacted with monoclonal antibody and immunoperoxidase technique. V = Third ventricle. From *Hou-Yu* et al. [18], with permission.

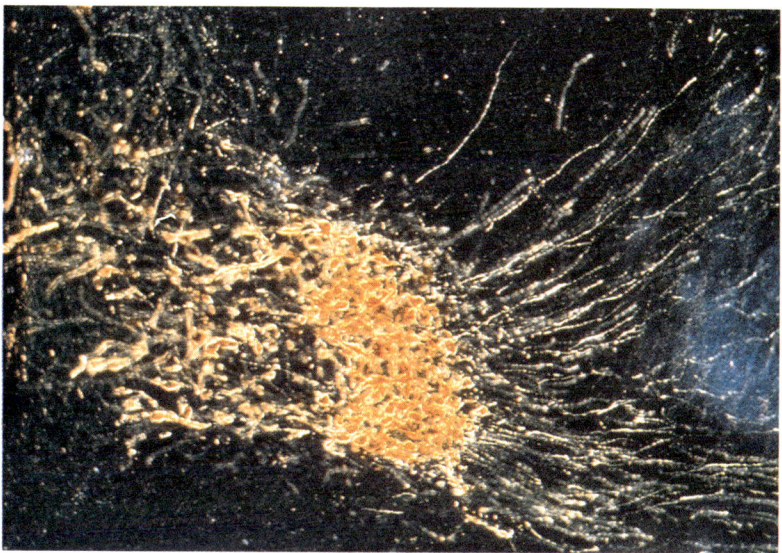

Fig. 3. Darkfield photomicrograph of vasopressin neurons in the PVN (left side at a similar level and magnification to fig. 2B) immunoreacted with monoclonal antibody and immunoperoxidase technique. Note the concentration of reactive perikarya in the lateral magnocellular portion and fibers exiting laterally. From *Hou-Yu and Zimmerman* [unpublished].

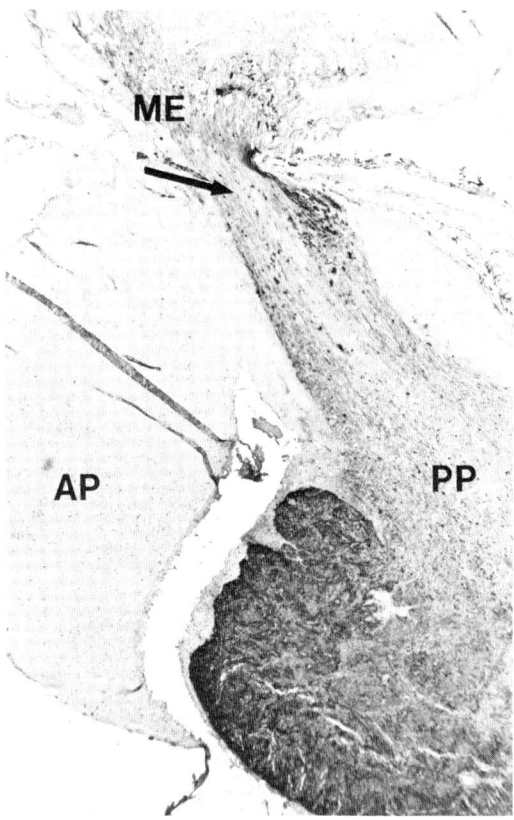

Fig. 4. Low power sagittal section of monkey median eminence (ME) and posterior pituitary gland (PP) stained by immunoperoxidase technique using antiserum to human estrogen-stimulated neurophysin. Note fibers descending through the infundibular stalk (arrow) to innervate PP. Anterior pituitary is unreactive. From *Antunes* et al. [27], photomicrography.

In their 3-dimensional reconstruction study, *Fisher* et al [9] also point out that the magnocellular system forms a cone encompassing all of the SON, PVN and accessory neurons described in table I. Cells in the bed nucleus of the stria terminalis or nucleus of the forebrain bundle, for example, can be included in the cones as part of one system. The rostral groups (anterior commissural and anterior perifornical nucleus) appear to form a miniature reduplication of the shape of the PVN in a nucleus separate from the PVN. It is of interest in this regard that *Dierick and Vandesande* [30] describe another group of neurons rostral and parallel to the PVN in human which contains more oxytocin than vasopressin cells.

Parvocellular Vasopressin/CRH Pathway to the Hypophysial Portal System

Soon after the introduction of immunocytochemical methods to study vasopressin pathways, a second neurophysin-containing neurosecretory field to the capillaries of the zona externa of the median eminence was described in sheep [34] and monkeys [35]. Concentrations of vasopressin in individual hypophysial portal vein blood of monkeys were 1,000 times higher than in peripheral blood, suggesting that the neurophysin was associated with vasopressin [35]. These observations activated the longstanding controversy as to whether vasopressin functioned as CRH or an assistant CRH [36–38]. This debate continued until the present when the structure of CRH was discovered [39]. At about the time of the neurophysin findings, it was shown that Gomorireactive material markedly increased after adrenalectomy in the zona externa of rats [40]. It was subsequently noted that this buildup was associated with increases in immunoreactive vasopressin and vasopressin neurophysin, and to a lesser extent if any with oxytocin or its neurophysin [41, 42]. Unlike the posterior pituitary projection, dehydration had little effect on these zona externa projections [42, 43]. Further data supported the concept that the median eminence vasopressin neurosecretory system differed from the one to posterior pituitary [for review, see 44]. Vasopressin granules in zona externa terminals were found to be smaller (100 nm) than those in the neural lobe (120–180 nm) [45].

It was demonstrated later that the vasopressin fibers to the zona externa (fig. 2) originated in the PVN and not the SON by lesions studies in the monkey [46] and rat [41, 47]. Anterograde tracing studies using tritiated amino acid precursor injections into PVN and SON also demonstrated that PVN rather than SON contributed to the zona externa pathway [48]. Differences in the two vasopressin systems were also suggested by preferential incorporation of tritiated cytidine in PVN after adrenalectomy [49], and in SON after dehydration [50]. The observation that the production of vasopressin precursor increased in PVN after adrenalectomy supported this and an additional hypothesis. The PVN-zona externa system produces more vasopressin in response to adrenal insufficiency. Increased content probably reflects increased secretion.

The next question was which cells in the PVN contributed to the median eminence pathway. Retrograde tracing studies suggested that median eminence afferents arise from mainly parvocellular [51–53] rather than magnocellular neurons which label from the posterior pituitary [9, 51, 54]. Median eminence projections arise from the anterior, dorsal and mainly from the medial parvocellular divisions of the PVN [51–53]. Some magnocellular elements are also labeled in most of these studies, which may be the true projections or which may reflect technical difficulties in preferentially injecting median eminence and posterior pituitary systems. It also remains possible, based on electrophysiological studies, that some neurons which project to the neural lobe may send collaterals to the zona externa [55]. In the normal rat very few vasopressin-containing neurons are found in these parvocellular regions of the PVN [12, 56]. However, a number of other peptides known to project to the median eminence reside in or near the medial parvocellular region: thyrotropin releasing hormone, somatostatin, neurotensin, and CRH [for review see 57].

Just prior to this publication three separate laboratories reported the co-existence of immunoreactive CRH and vasopressin in the same neurons in the medial and lateral parvocellular portion of the PVN after adrenalectomy [10–12] (fig. 5, 6). In animals treated with colchicine alone, only 1% of CRH neurons visualized in this region by immunohistochemistry contain vasopressin immunoreactivity. After adrenalectomy alone they contain 70% or more co-existence with immunoreactive vasopressin [10, 12]. Since CRH/vasopressin-reactive neurons also contained immunoreactive neurophysin, it seems certain that the vasopressin is indeed the peptide in question and not a cross-reacting substance. Further confirmation of this observation can be expected from future studies using antisera to the vasopressin precursor [58].

Although there are many other CRH-containing immunoreactive neurons in the brain [see 59, 60 for reviews], lesion studies suggest that it is the PVN system which provides the hormone to the median eminence capillary system [60, 61]. In view of the previous studies, these recent findings strongly suggest that vasopressin is produced by these particular CRH neurons, at least in conditions of adrenal insufficiency, in the rat, and is likely secreted along with CRH to assist in the release of ACTH from the anterior pituitary. CRH is found in the PVN of normal

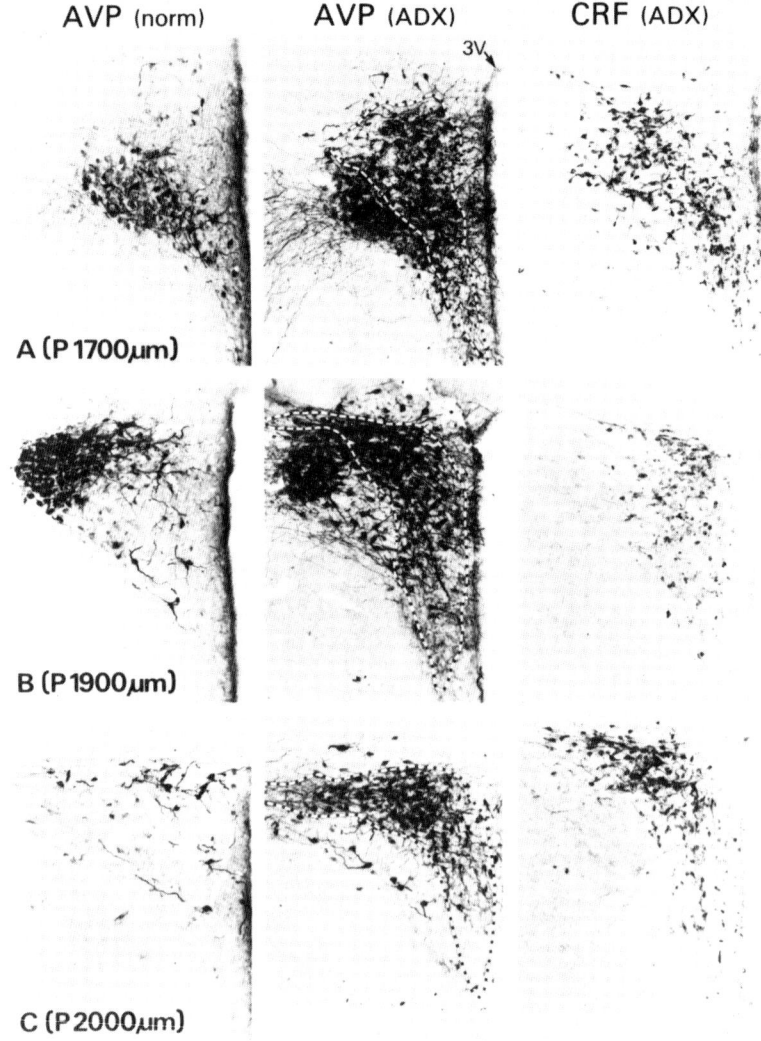

Fig. 5. Localization of vasopressin (AVP) and corticotropin releasing factor (CRF) by immunoperoxidase technique at various levels of the PVN of normal (norm) and adrenalectomized (ADX) rats. Coronal sections 1,700 μm *(A)*, 1,900 μm *(B)* and 2,000 μm *(C)* posterior (P) to the bregma. White dashed lines outline parvocellular subdivision of the PVN. Note that number of vasopressin-reactive neurons in the parvocellular zone increased markedly with adrenalectomy and overlaps with the distribution of CRF-reactive neurons in this region. 3V = Third ventricle. From *Kiss* et al. [11], with permission.

A. AVP (ADX)

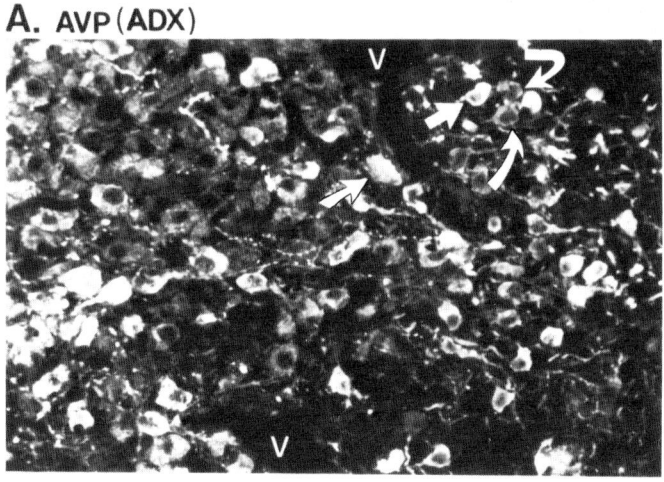

B. CRF (ADX)

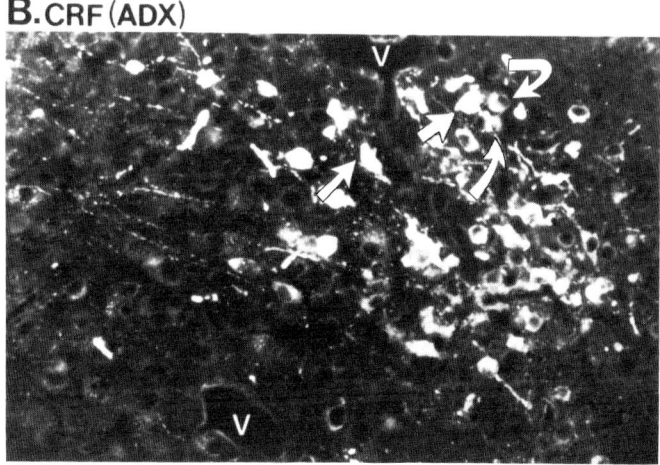

Fig. 6. Adjacent 5-μm sections of adrenalectomized (ADX) rat PVN reacted for *(A)* vasopressin (AVP) and *(B)* corticotropin releasing factor (CRF). Some of the same cells are reactive in both sections (arrows) indicating the co-existence of immunoreactivity for both hormones in the same cell. V = Vessels. Immunofluorescence technique. From *Kiss* et al. [11], with permission.

sheep [62]. Since a larger number of vasopressin/neurophysin terminals are found in the zona externa of the median eminence of normal sheep and monkeys than in rats [personal observation] further studies may reveal co-existence of CRH and vasopressin in these animals under normal conditions.

Since it was shown previously that replacement of glucocorticoid after adrenalectomy inhibited the increase in vasopressin in the zona externa [63], it seems likely that the adrenal cortex in turn has an inhibitory feedback on the vasopressin/CRH system. Whether the glucocorticoids act directly on the PVN neurons, or indirectly via a second neuron is not certain. Since reserpine administration inhibits the effects of adrenalectomy in the zona externa [43], it is possible that the effect of adrenalectomy may be mediated by an adrenergic or noradrenergic system arising from the brainstem.

Parvocellular Vasopressin Neurons in the Suprachiasmatic Nucleus

Vasopressin (fig. 7) and its neurophysin are found in the small parvocellular neurons of the SCN [13, 14, 30]. Contrary to earlier reports that it was not present in monkey [46] and human [29], vasopressin has been demonstrated in this nucleus in all mammals studied by immunohistochemistry [14, see 65, 66 for reviews]. The absence of vasopressin and neurophysin immunoreactivity in the suprachiasmatic nucleus of homozygous Brattleboro rats [19, 69] suggests that these neurons in normal animals synthesize the peptide from the vasopressin precursor.

The presence of the suprachiasmatic vasopressin system is intriguing because of the importance of the nucleus in diurnal rhythms. At present, however, such a role for vasopressin has not been demonstrated, since Brattleboro rats deficient in vasopressin exhibit normal adrenal, pineal, drinking and activity rhythms [see 65, 66 for reviews]. It has been suggested that the SCN is the source of the diurnal rhythm in vasopressin concentration in cerebrospinal fluid in cats [68] and monkeys [69] which is not reflected in the peripheral circulation. Vasopressin fibers arising from the SCN do project to the organum vasculosum [70], a circumventricular organ where the hormone might be secreted into the ventricular fluid system. However, a diurnal rhythm for oxytocin and its neurophysin, similar to the one for vasopressin, was also demonstrated [69], and oxytocin is not produced by suprachiasmatic neurons.

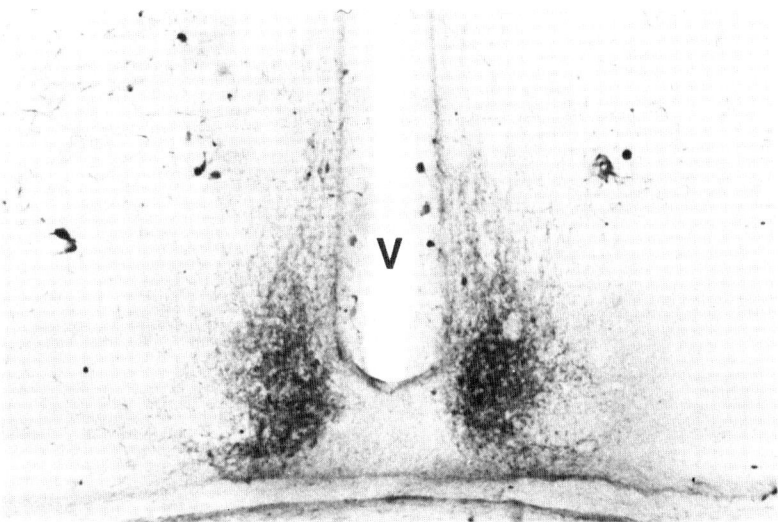

Fig. 7. Parvocellular vasopressin neurons in the rat SCN reacted with monoclonal antibody and immunoperoxidase technique. V = Third ventricle. From *Hou-Yu* et al. [18], with permission.

At present, the vasopressin axonal projections of the SCN do not appear to be fully established. A recent study in which bilateral lesions were made in the nucleus demonstrated the loss of immunoreactive projections to the organum vasculosum and the periventricular and dorsomedial nuclei of the hypothalamus [70]. There was no loss of fibers to extrahypothalamic sites such as lateral septum, habenula or amygdala, sites previously thought to be heavily innervated from the SCN. As for other suprachiasmatic neurons containing vasoactive intestinal polypeptide [71], many of the vasopressin-containing axons arising from cell bodies in the nucleus probably form dense synaptic plexuses with the dendrites of other neurons within the same and opposite side of the nucleus [65].

Parvocellular Extrahypothalamic Projections

A major new development in the study of neurohypophysial peptides in the last 5 years has been the discovery of fiber pathways containing vasopressin or oxytocin in many regions of forebrain, brainstem and spinal cord. This subject has been recently extensively reviewed [8, 66,

72–74]. It is well established that pathways to brainstem and spinal cord originate from primarily the parvocellular portions of the PVN [75]. However, the recent report of immunoreactive vasopressin and neurophysin in cells of the locus ceruleus in colchicine-treated rats raises the possibility that some fibers of the caudal system may have their origin there. Neither bilateral lesions of the SCN [70] nor the PVN [73] eliminate many of the forebrain vasopressin projection areas including the lateral septum. A new group of vasopressin-reactive cells discovered in the bed nucleus of the stria terminalis after colchicine [77], when lesioned in normal animals, produced a reduction in fibers to lateral septum [73]. Additional parvocellular extrahypothalamic vasopressin-reactive cells were found in the medial amygdaloid nucleus and dorsomedial hypothalamus in the colchicine-treated rats [77]. If confirmed, by application of antisera such as those to the precursor of vasopressin and probes to the vasopressin gene, these new findings would indicate that the vasopressin system in the brain does not originate solely in the hypothalamus and its environs [66].

Vasopressin Catecholamine Interactions

Catecholamine-containing fiber pathways ascending from the brainstem heavily innervate the SON and PVN [78–82]. Using paraformaldehyde-induced histofluorescence and immunohistochemistry on adjacent sections, vasopressin neurons in both the SON and PVN appear to be more heavily innervated than those containing oxytocin by light microscopy [82]. Whether these are mainly noradrenergic, and the nature of the contacts, awaits electron microscopic analysis with specific labels for catecholamines and the peptides. In the rat noradrenergic fibers originating in the lateral reticular nucleus (A1) innervate magnocellular subgroups in the PVN and SON [79, 80], and therefore serve to regulate the secretion of vasopressin projecting to the posterior pituitary. Fibers originating from cells in the locus ceruleus (A6) and nucleus of the solitary tract and adjacent dorsal motor nucleus (A2) project to various parvocellular divisions of the PVN including the medial region containing CRH/vasopressin neurons of the median eminence system. Since a dense network of adrenergic fibers is also found in medial PVN [83], either catecholamine could regulate ACTH secretion. Dopamine-containing cells are also found within the PVN [83].

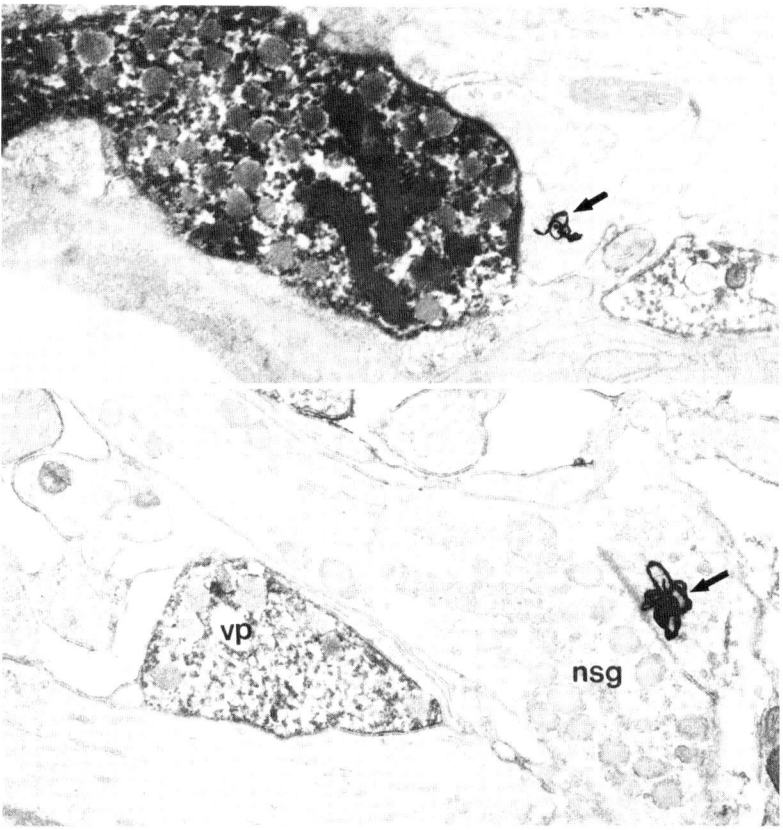

Fig. 8. Electron micrographs of the lateral magnocellular part of rat PVN. Combined immunoperoxidase technique using monoclonal antibody to localize vasopressin and autoradiography of ^{3}H-noradrenalin uptake to visualize noradrenalin terminals. Upper panel: A radioactive noradrenalin terminal (arrow) synapses with a vasopressin-reactive dendrite. Lower panel: A non-vasopressin-reactive dendrite containing neurosecretory granules (nsg) receives a noradrenalin (arrow) synapse. A vasopressin-reactive process is also seen in the plane of section. From *Silverman* et al. [85], with permission.

We have recently investigated synaptic terminals on PVN vasopressin neurons identified by an immunoperoxidase technique using a monoclonal antibody [84, 85] (fig. 8). Both somata and dendrites receive a diverse innervation. Axo-axonic contacts with vasopressin fibers were found along the PVN-neurohypophysial tract and in the median eminence. By simultaneously studying reuptake of ^{3}H-noradrenalin by autoradiography, well-defined synapses were found primarily on dendritic processes of non-vasopressin-reactive neurons in the medial periventricular zone. There were occasional terminals on vasopressin-reactive dendrites in this region as well, while terminals were rare on neurons of the lateral mangocellular groups. These results suggest that noradrenalin terminals in the PVN preferentially innervate the medial parvocellular system which projects to the median eminence. However, some of the dendrites in this region may originate from cells in the lateral magnocellular division. Extension of these studies to adrenalectomized rats may reveal more terminals on vasopressin neurons in the medial zone which may also contain CRH [10–12]. Similar studies of the neurohypophysis using ^{3}H-dopamine revealed a close association between dopamine and vasopressin terminals [86]. Dopamine fibers originating in the arcuate nucleus [86], and those containing norepinephrine originating in brainstem, appear to have inhibitory influences on vasopressin secretion [82].

References

1   Scharrer, E.; Scharrer, B.: Hormones produced in neurosecretory cells. Recent Prog. Horm. Res *10:* 183–240 (1954).
2   Rasmussen, A.T.: Effects of hypophysectomy and hypophysial stalk section of the hypothalamic nuclei of animals and man. Res. Publs Ass. Res. nerv. ment. Dis. *20:* 245–269 (1940)
3   Sloper, J.C.: The experimental and cytophathological investigation of neurosecretion in the hypothalamus and pituitary; in The pituitary gland, part III, pp. 131–180 (University of California Press, Berkeley 1966).
4   Fisher, C.; Ingram, W.R.; Ranson, S.W.: Diabetes insipidus and the neurohormonal control of water balance (Edwards, Ann Arbor 1938).
5   Magoun, H.W.; Ranson, S.W.: Retrograde degeneration of the supraoptic nuclei after section of the infundibular stalk in the monkey. Anat. Rec. *75:* 107–123 (1939).
6   Maccubin, D.A.; Van Buren, J.M.: A quantitative evaluation of hypothalamic degeneration and its relation to diabetes insipidus following interruption of the human hypophyseal stalk. Brain *86:* 443–469 (1963).
7   Silverman, A.-J.; Zimmerman, E.A.: Magnocellular neurosecretory system. A. Rev. Neurosci. *6:* 357–380 (1983).
8   Kozlowski, G.P.; Nilaver, G.; Zimmerman, E.A.: Distribution of neurophysial hormones in the brain. Pharmacol. Ther. *21:* 325–349 (1983).
9   Fisher, A.W.F.; Price, P.G.; Burford, G.D.; Leceris, K.: A 3-dimensional reconstruction of the hypothalamo-neurohypophysial system of the rat. Cell Tiss. Res. *204:* 343–354 (1979).
10 Tramu, G.; Croix, C.; Pillez, A.: Ability of the CRF immunoreactive neurons of the paraventricular nucleus to produce a vasopressin-like material. Immunohistochemical demonstration in adrenalectomize guinea pigs and rats. Neuroendocrinology *37:* 467–469 (1983).

11 Kiss, J.Z.; Mezey, E.; Skirboll, L.: Corticotropin releasing factor (CRF)-immunoreactive neurons of the paraventricular nucleus become vasopressin-positive following adrenalectomy. Proc. natn. Acad. Sci. USA (in press).
12 Sawchenko, P.E.; Swanson, L.W.; Vale, W.W.: Co-expression of CRF – and vasopressin immunoreactivity in parvo cellular neurosecretory neurons of the adrenalectomized rat. Proc. natn. Acad. Sci. (in press).
13 Swaab, D.F.; Pool, C.W.; Nijveldt, F.: Immunoflurescence of vasopressin and oxytocin in the rat hypothalamo-neurohypophysial system. J. neural Transm. *36:* 195–215 (1975)
14 Sofroniew, M.V.; Weindl, A.: Identification of parvocellular vasopressin and neurophysin neurons in the suprachiasmatic nucleus of a variety of mammals including primates. J. comp. Neurol. *193:* 659–675 (1980).
15 Zimmerman, E.A.; Krupp, L.; Hoffman, D.L.; Matthew, E.; Nilaver, G.: Exploration of peptidergic pathways in brain by immunocytochemistry: a ten-year perspective. Peptides *1:* suppl. 1, pp. 3–10 (1980).
16 Swaab, D.F.; Pool, C.W.: Specificity of oxytocin and vasopressin immunofluorescence. J. Endocr. *66:* 263–272 (1975).
17 Vandesande, F.; Dierickx, K.: Identification of vasopressin-producing and/or the oxytocin-producing neurons in the hypothalamic magnocellular neurosecretory system of the rat. Cell Tiss. Res. *164:* 153 (1975).
18 You-Yu, A.; Ehrlich, P.; Valiquette, G.; Engelhardt, D.L.; Sawyer, W.H.; Nilaver, G.; Zimmerman, E.A.: A monoclonal antibody to vasopressin. Preparation, characterization and application in immunohistochemistry. J. Histochem. Cytochem. *30:* 1249–1260 (1982).
19 Sokol, H.W.; Zimmerman, E.A.; Sawyer, W.H.; Robinson, A.C.: The hypothalamoneurohypophysial system of the rat: localization and quantification of neurophysin by light microscopic immunocytochemistry in normal rat and in Brattleboro rats deficient in vasopressin and a neurophysin. Endocrinology *98:* 1176–1188 (1976).
20 Vandesande, F.; Dierickx, K.: Identification of the vasopressin producing and oxytocin producing neurons in the hypothalamic magnocellular neurosecretory system of the rat. Cell. Tiss. Res *164:* 153–162 (1975).
21 Vandesande, F.; Dierickx, K.; DeMey, J.: Identification of the vasopressin-neurophysin II and the oxytocin-neurophysin I producing neurons in the bovine hypothalamus. Cell Tiss. Res. *156:* 189–200 (1975).
22 Watkins, W.B.: Immunocytochemical study of the hypothalamo-neurohypophysial system. I. Localization of neurosecretory neurons containing neurophysin I and neurophysin II in the domestic pig. Cell Tiss. Res. *175:* 165–181 (1976).
23 Zimmerman, E.A.: Localization of hypothalamic hormones by immunocytochemical techniques: in Martini, Ganong, Frontiers in neuroendocrinology, vol. 4, pp. 25–62 (Raven Press, New York 1976).
24 Sokol, H.W.; Zimmerman, E.A.; Sawyer, W.H.; Robinson, A.G.: The hypothalamoneurohypophysial system of the rat. Localization and quantification of neurophysin by light microscopic immunocytochemistry in normal rat and in Brattleboro rats deficient in vasopressin and a neurophysin. Endocrinology *98:* 1176–1188 (1976).
25 Sofroniew, M.W.; Weindl, A.; Schinko, I.; Wetzstein, R.: The distribution of vaso-

pressin, oxytocin and neurophysin producing neurons in the guinea pig brain. I. The classical hypothalamoneurohypophysial system. Cell Tiss. Res. *196:* 367–384 (1979).

26 Rhodes, C.H.; Morrell, J.L.; Pfaff, D.W.: Immunohistochemical analysis of magnocellular elements in rat hypothalamus: distribution and numbers of neurophysin, oxytocin and vasopressin-containing cells. J. comp. Neurol. *198:* 45–64 (1981).

27 Antunes, J.L.; Zimmerman, E.A.: The hypothalamic magnocellular system of the Rhesus monkey: an immunocytochemical study. J. comp. Neurol. *181:* 539–566 (1978).

28 Dierickx, K.; Vandesande, F.: Immunocytochemical localization of the vasopressinergic and the oxytocinergic neurons in the human hypothalamus. Cell Tiss. Res. *184:* 15–27 (1977).

29 Defendini, R.; Zimmerman, E.A.: in Reichlin, Baldessarini, Martin, The hypothalamus, pp. 137–154 (Raven Press, New York 1978).

30 Dierickx, K.; Vandesande, F.: Immunocytochemical demonstration of separate vasopressin-neurophysin and oxytocin-neurophysin neurons in the human hypothalamus. Cell Tiss. Res. *196:* 203–212 (1979).

31 Fellman, D.; Block, B.; Bugnon, C.; Lenys, D.: Etude immunocytologique de la maturation des axes neuroglandulaires hypothalamo-neurohypophysaires chez le foetus humain. J. Physiol., Paris *75:* 37–43 (1979).

32 Morton, A.: A quantitative analysis of the normal neuron population of the hypothalamic magnocellular nuclei in man and of their projections to the neurohypophysis. J. comp. Neurol. *136:* 143–158 (1969).

33 Peterson, R.P.: Magnocellular neurosecretory centers in the rat hypothalamus. J. comp. Neurol. *128:* 181–190 (1966).

34 Parry, H.B.; Livett, B.G.: A new hypothalamic pathway to the median eminence containing neurophysin and its hyptertrophy in sheep with natural scrapie. Nature, Lond. *242:* 63–95 (1973).

35 Zimmerman, E.A.; Carmel, P.W.; Husain, M.K.; Ferin, M.; Tannenbaum, M; Frantz, A.G.; Robinson, A.G.: Vasopressin and neurophysin: high concentrations in monkey hypophyseal portal blood. Science *198:* 925–927 (1973).

36 Yates, F.E.; Russell, S.M.; Dallman, M.F.; Hedge, G.A.; McCann, S.M.; Dhariwal, A.P.S.: Potentiation of vasopressin of corticotropin release induced by corticotropin-releasing factor. Endocrinology *88:* 3–15 (1971).

37 Krieger, D.T.; Zimmerman, E.A.: The nature of CRF and its relationship to vasopressin; in Besser, Martini, Clinical neuroendocrinology, pp. 363–391 (Academic Press, New York 1977).

38 Gillies, G.E.; Linton, E.A.; Lowry, P.J.: Corticotropin releasing activity of the new CRF is potentiated several times by vasopressin. Nature, Lond *299:* 355–357 (1982).

39 Vale, W.; Spiess, J.; Rivier, C.; Rivier, J.: Characteristics of a 41-residue ovine hypothalamic peptide that stimulates the secretion of corticotropin and B-endorphin. Science *213:* 1394–1397 (1981).

40 Wittowski, W.; Bock, R.: Electron microscopical studies of the median eminence following interference with the feedback system anterior pituitary-adrenal cortex; in Knigge, Scott, Weindl, Brain endocrine interaction: median eminence structure and function, pp. 171–180 (Karger, Basel 1972).

41 Vandesande, F.; Dierickx, K.; DeMey, J.: The origin of the vasopressin and oxytocinergic fibers of the external region of the median eminence of the rat hypophysis. Cell Tiss. Res. *180:* 443–452 (1977).

42 Zimmerman, E.A.; Stillman, M.A.; Recht, L.D.; Antunes, J.L.; Carmel, P.W.; Goldsmith, P.C.: Vasopressin and corticotropin-releasing factor: an axonal pathway to portal capillaries in the zona externa of the median eminence containing vasopressin and its interaction with adrenal corticoids. Ann. N.Y. Acad. Sci. *279:* 405–419 (1977).

43 Seybold, V.; Elde, R.; Hokfelt, T.: Terminals of reserpine-sensitive vasopressin neurophysin neurons in the external layer of the rat median eminence. Endocrinology *108:* 1803–1809 (1981).

44 Zimmerman, E.A.; Silverman, A.J.: Vasopressin and adrenal cortical interactions; in Cross, Leng, The neurohypophysis: structure, function and control, vol. 60, pp. 493–504 (Elsevier, New York 1983).

45 Silverman, A.J.; Zimmerman, E.A.: Ultrastructural localization of neurophysin and vasopressin in the median eminence and posterior pituitary of the guinea pig. Cell Tiss. Res. *159:* 291–301 (1975).

46 Antunes, J.L.; Carmel, P.W.; Zimmerman, E.A.: Projections from the paraventricular nucleus to the zona externa of the median eminence of the rhesus monkey: an immunohistochemical study. Brain Res. *137:* 1–10 (1977).

47 Silverman, A.J.; Zimmerman, E.A.: Adrenalectomy increases sprouting in a peptidergic neurosecretory system. Neuroscience *7:* 2705–2714 (1982).

48 Alonso, G.; Assenmacher, I.: Radioauthographic studies on the neurophysial projections of the supraoptic and paraventricular nuclei in the rat. Cell Tiss. Res. *219:* 525–534 (1981).

49 Silverman, A.J.; Gadde, C.A.; Zimmerman, E.A.: The effects of adrenalectomy on the incorporation of ^{3}H-cytidine into RNA in neurophysin and vasopressin containing neurons of the rat hypothalamus. Neuroendocrinology *30:* 285–290 (1980).

50 George, J.M.: Localization in hypothalamus of increased incorporation of ^{3}H cytidine into RNA in response to oral hypertonic saline. Endocrinology *92:* 1550–1555 (1973).

51 Wiegand, S.J.; Price, J.L.: Cells of origin of the afferent fibers to the median eminence in the rat. J. comp. Neurol. *192:* 1019 (1980)

52 Swanson, L.W.; Sawchenko, P.E.; Wiegand, S.J.; Price, J.L.: Separate neurons project to the median eminence and to the medulla or spinal cord. Brain Res. *198:* 190–195 (1980)

53 Lechan, R.M.; Nestler, J.L.; Jacobson, S.: Immunohistochemical localization of retrogradely and anteriorgradely transported wheat germ agglutinin (WGA) within the central nervous system of the rat: application to immunostaining of a second antigen within the same neuron. J. Histochem. Cytochem. *29:* 1255–1262 (1981).

54 Armstrong, W.E.; Hatton, G.I.: The localization of projection neurons in the rat hypothalamic paraventricular nucleus following vascular and neurohypophysical injections of HRP. Brain Res. Bull. *5:* 473–477 (1980).

55 Pittman, Q.J.; Blume, H.W.; Renaud, L.P.: Electrophysiological indications that individual hypothalamic neurons innervate both median eminence and neurohypophysis. Brain Res. *157:* 364–368 (1978).

56 Swanson, L.W.; Sawchenko, P.E.: Paraventricular nucleus: a site for the integration of neuroendocrine and autonomic mechanisms. Neuroendocrinology *31:* 410–417 (1980).

57 Zimmerman, E.A.; Nilaver, G.: The organization of neurosecretory pathways to the hypophysial portal system; in Molinatti, Camanni, Müller, Pituitary hyperfunction: pathophysiology and clinical aspects (Raven Press, New York, in press).

58 Watson, S.J.; Seidah, N.G.; Chretien, M.: The carboxy terminus of the precursor to vasopressin and neurophysin: immunocytochemistry in rat brain. Science *217:* 853–855 (1982).

59 Swanson, L.W.; Sawchenko, P.E.; Rivier, J.; Vale, W.W.: Organization of ovine corticotropin-releasing factor immunoreactive cells and fibers in the rat brain: an immunohistochemical study. Neuroendocrinology *36:* 165–186 (1983).

60 Antoni, F.A.; Palkovits, M.; Makara, G.B.; Linton, E.A.; Lowry, P.J.; Kiss, J.Z.: Immunoreactive corticotropin-releasing hormone in the hypothalamoinfundibular tract. Neuroendocrinology *36:* 415–423 (1983).

61 Bruhn, T.O.; Plotsky, P.M.; Vale, W.W.: Effect of paraventricular lesions on corticotropin-releasing factor (CRF)-like immunoreactivity in the stalk-median eminence: studies on the adrenocorticotropin response to ether stress and exogenous CRF. Endocrinology *114:* 57–62 (1984).

62 Paull, W.K.; Schöler, J.; Arimura, A.; Meyers, C.A.; Chang, J.K.; Chang, D.; Shimizu, M.: Immunocytochemical localization and CRF in the ovine hypothalamus. Peptides *3:* 183–191 (1982).

63 Silverman, A.-J.; Hoffman, D.; Gadde, C.A.; Krey, L.C.; Zimmerman, E.A.: Adrenal steroid inhibition of the vasopressin-neurophysin neurosecretory system to the median eminence of the rat. Neuroendocrinology *32:* 129–133 (1981).

65 Silverman, A.-J.; Pickard, G.E.: The hypothalamus; in Emson, Chemical neuroanatomy, pp. 295–336 (Raven Press, New York 1983).

66 Zimmerman, E.A.; Hou-Yu, A.; Nilaver, G.; Silverman, A.-J.: Anatomy of pituitary and extrapituitary vasopressin secretory systems; in Reichlin, Neurohypophysis, vasopressin and vasopressin analogues: basic and clinical aspects (Plenum Publishing, New York, in press).

67 Swaab, D.F.; Pool, C.W.: Specificity of oxytocin and vasopressin immunofluorescence. J. Endocr. *66:* 263–272 (1975).

68 Reppert, S.M.; Artman, A.; Swaminathan, S.; Fisher, D.: Vasopressin exhibits a daily pattern in cerebrospinal fluid but not blood. Science *213:* 1256–1257 (1981).

69 Perlow, M.J.; Reppert, S.M.; Artman, H.A.; Fisher, D.A.; Seif, S.M.; Robinson, A.G.: Oxytocin, vasopressin, and estrogen-stimulated neurophysin: daily patterns of concentration in cerebrospinal fluid. Science *216:* 1416–1418 (1982).

70 Hoorneman, E.M.D.; Buijs, R.M.: Vasopressin fiber pathways in the rat brain following suprachiasmatic nucleus lesions. Brain Res. *243:* 235–241 (1982).

71 Card, J.P.; Brecha, N.; Karten, H.J.; Moore, R.Y.: Immuno-cytochemical localization of vasoactive intestinal polypeptide-containing cells and processes in the suprachiasmatic nucleus of the rat: light and electron microscopic analysis. J. Neurosci. *1:* 1289–1303 (1981).

72 Sofroniew, M.V.: Morphology of vasopressin and oxytocin neurones and their central and vascular projections; in Cross, Leng, The neurohypophysis: structure, function, and control, vol. 60, pp. 101–114 (Elsevier, New York 1983).

73 Buijs, R.M.; De Vries, J.; VanLeeuwen, F.W.; Swaab, D.F.: Vasopressin and oxytocin: distribution and putative functions in the brain; in Cross, Leng, The neurohypophysis: structure, function and control, vol. 60, pp. 115–122 (Elsevier, New York 1983).
74 Swanson, L.W.; Sawchenko, P.E.: Hypothalamic integration: organization of the paraventricular and supraoptic nuclei. A. Rev. Neurosci. *6:* 269–324 (1983).
75 Swanson, L.W.; Sawchenko, P.E.: Paraventricular nucleus: a site for the integration of neuroendocrine and autonomic mechanisms. Neuroendocrinology *31:* 410–417 (1980).
76 Caffe, A.R.; Van Leeuwen, F.W.: Vasopressin-immunoreactive cells in the dorsomedial hypothalamic region, medial amygdaloid nucleus and locus coeruleus of the rat. Cell Tiss. Res. *233:* 23–33 (1983).
77 Van Leeuwen, F.W.; Caffe, R.: Immunoreactive vasopressin cell bodies in the bed nucleus of the stria terminalis. Cell Tiss. Res. *228:* 525–534 (1983)
78 Berk, M.L.; Finkelstein, J.A.: Afferent projections of the preoptic area and hypothalamic regions in the rat brain. Neuroscience *6:* 1601–1624 (1981).
79 Sawchenko, P.E.; Swanson, L.W.: Central noradrenergic pathways for the integration of hypothalamic neuroendocrine and automatic responses. Science *214:* 685–687 (1981).
80 Sawchenko, P.E.; Swanson, L.W.: The organization of noradrenergic pathways from the brainstem to the paraventricular and supraoptic nuclei in the rat. Brain Res. Rev. *4:* 275–325 (1982).
81 Tribollet, E.; Dreifuss, J.J.: Localization of neurons projecting to the hypothalamic paraventricular nucleus area of the rat: a horseradish peroxidase study. Neuroscience *6:* 1315–1328 (1981).
82 Sladek, J.R., Jr.; Zimmerman, E.A.: Simultaneous monoamine histofluorescence and neuropeptide immunocytochemistry. VI. Catecholamine innervation of vasopressin and oxytocin neurons in the rhesus monkey hypothalamus. Brain Res. Bull. *9:* 431–440 (1982).
83 Swanson, L.W.; Sawchenko, P.E.; Berod, A.; Hartman, B.K.; Heile, K.B.; vanOrden, D.E.: An immunohistochemical study of the organization of catecholamine cells and terminal fields in the paraventricular and supraoptic nuclei of the hypothalamus. J. comp. Neurol. *196:* 271–285 (1981).
84 Silverman, A.J.; Hou-Yu, A.; Zimmerman, E.A.: Ultrastructural studies of vasopressin neurons of the paraventricular nucleus of the hypothalamus using monoclonal antibody to vasopressin: analysis of synaptic input. Neuroscience *9:* 141–155 (1983).
85 Silverman, A.-J.; Oldfield, B.; Hou-Yu, A.; Zimmerman, E.A.: The noradrenergic innervation of vasopressin neurons in the paraventricular nucleus of the hypothalamus: an ultrastructural study using radioautography and immunocytochemistry. Brain Res. (submitted).
86 Pelletier, G.: Identification of endings containing dopamine and vasopressin in the rat posterior pituitary by a combination of radioautography and immunocytochemistry at the ultrastructural level. J. Histochem. Cytochem. *31:* 562–564 (1983).

E.A. Zimmerman, MD, Department of Neurology, College of Physicians and Surgeons, Columbia University, 630 West 168th Street, New York, NY 10032 (USA)

Biosynthesis Transport and Release of Vasopressin[1]

Alan G. Robinson, Joseph G. Verbalis

Department of Medicine, School of Medicine, University of Pittsburgh, Pittsburgh, Pa., USA

Introduction

Vasopressin is synthesized within clusters of magnocellular neurons which are identified as the paired supraoptic nuclei and paraventricular nuclei of the hypothalamus. The hormone is then transported in neurosecretory granules along the long axons to the posterior pituitary where the hormone is stored prior to release [1–3]. Much of our knowledge about the biosynthesis of vasopressin has derived from the study of neurophysins [4]. Neurophysins have been considered intragranular carrier proteins for the hormone since the studies of *Acher* [5] which documented that the neurophysins formed a non-covalent pH-dependent reversible bond with the hormone. The neurophysins, like the hormones, are cysteine-rich peptides. Of the 95 amino acid residues in neurophysins, positions 10 through 74 are highly conserved between all neurophysins in all species studied. Both the N-terminal region (1–9) and the C-terminal region (75–95) are variable between individual neurophysins [6, 7]. In every species which has been carefully studied there is a single neurophysin for each neurohypophyseal hormone, i.e. a vasopressin-neurophysin, an oxytocin-neurophysin, a vasotocin neurophysin, etc. [6–10]. Furthermore, between species there is for the N-terminal residues a characteristic pattern of amino acid residues which have been described as the MSEL neurophysins for vasopressin (and the VLDV neurophysins for oxytocin) according to the amino acid residues in positions 2,

---

[1] Supported by NIH Grant MO1 RR00056, NIH Grant AM 16166, NIH Grant NS 17138, The Veterans Administration Research Career Development Program.

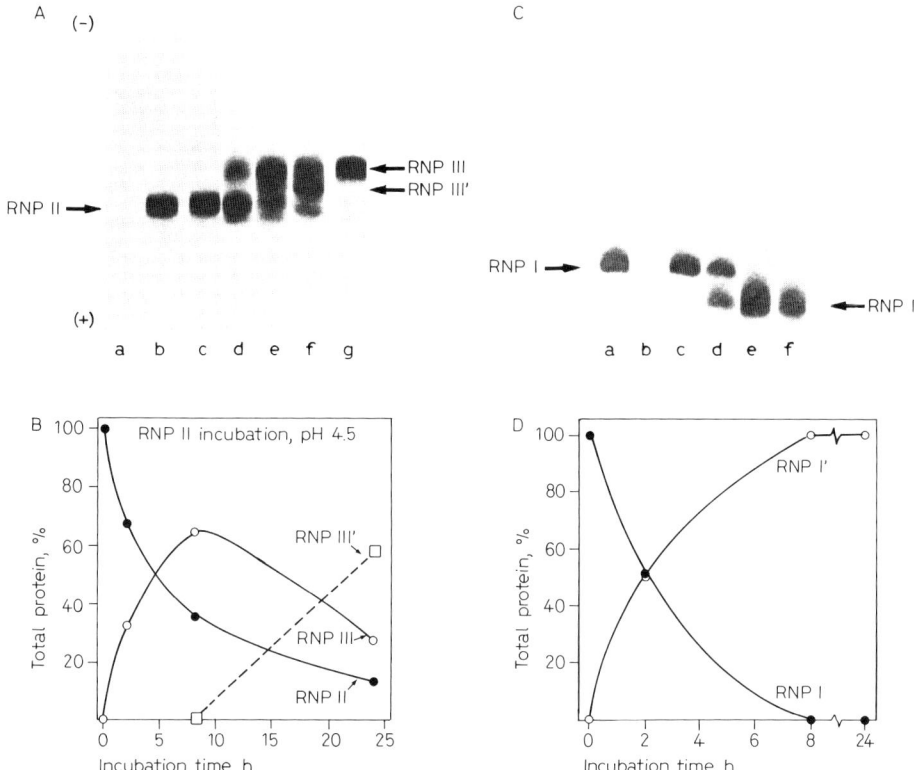

Fig. 1. Electrophoretic analysis of neurophysins isolated in the rat. In *A* rat neurophysin (RNP) II was incubated with extract of rat neural lobe at pH 4.5. *A* illustrates the conversion of RNP II to RNP III and RNP III': *a* control substrate blank; *b* purified RNP to reference; *c–f* products after 0, 1, 2 and 4 h, respectively; *g* purified RNP 3 reference. *B* is the quantitative densitometry of data in *A*. In *C* RNP I was incubated with extract of rat neural lobe at pH 4.5 and was converted to RNP I': *a* purified RNP I; *b* control substrate blank; *c–f* products after 0, 4, 8 and 24 h. *D* quantitative densitometry of the data presented in *B*.

3, 6 and 7 [6, 7]. In many species more than two neurophysins have been identified depending upon the method of extraction and isolation of the neurophysins [8, 11, 12]. It is now clear (as discussed below) that the heterogeneity of neurophysins is probably due to intragranular conversion. This has been well documented in the rat where intragranular conversion of the oxytocin neurophysin was described based upon pulse-chase labeling studies [13], and was studied in detail by *North* et al. [14] (fig. 1).

He isolated specific neurophysins from the rat and incubated labeled individual neurophysins with extracts of rat neural lobe. RNpII was converted to RNpIII and III' and RNpI was converted to RNpI'. In the human and pig two vasopressin neurophysins were isolated which have been proven to be identical with the exception of an Arg-Ala which was clipped from the C-terminus of one of the neurophysins [7, 12]. Another observation which was unexplained until the identification of the vasopressin precursor was the presence of a glycopeptide which had been isolated from bovine, sheep, pig and human neural lobes [15]. This 39 amino acid glycopeptide was shown to be present in neurosecretory granules but its function was unknown.

Biosynthesis

A 'common precursor' hypothesis for the synthesis of vasopressin and neurophysin was first proposed by *Sachs and Takabatake* [16] in 1964. This was prior to the discovery of proinsulin and represents a landmark in neuroscience. Figure 2 is an illustration from *Sachs* et al. [17] which illustrates the proposed intragranular conversion of the precursor. *Sachs* et al. [17] used pulse-labeling studies to demonstrate that vasopressin was initially synthesized as part of a larger molecule and also showed a striking parallel between the synthesis of neurophysin and vasopressin. However, because precursors are rapidly broken down in the intact cell, isolation and identification of the common precursor awaited more sophisticated techniques. Cell-free translation studies using messenger RNA (mRNA) from neurohypophyseal tissue incubated with heterologous cell-free translation systems allowed synthesis of and isolation of prohormone forms in vitro in media in which the prohormones were not further catabolized [18, 19]. *Schmale* et al. [20, 21] and *Richter and Schmale* [19] isolated messenger RNA from bovine hypothalamus and translated the message in a reticulocyte lysate system which has low endogenous levels of mRNA. Since both the neurophysins and vasopressin are rich in cysteine, ^{35}S-cysteine was used as a radioactive amino acid for labeling of translated peptides. The translated peptides were then precipitated using specific antibodies for either individual neurophysins or for arginine vasopressin. It is important to recognize that most neurophysin antisera will react with precursor because of the long and exposed 'constant' region of neurophysin in the precursor. Only

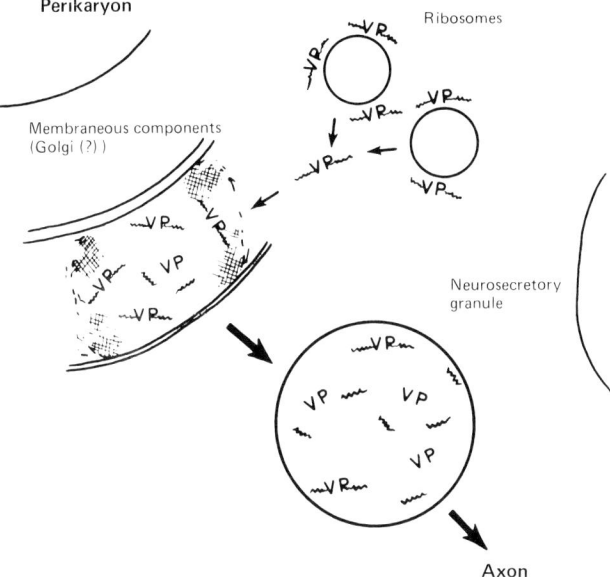

Fig. 2. The original 'precursor model' of Sachs demonstrating the biosynthesis of vasopressin as part of a large protein including vasopressin and neurophysin. Conversion of precursor to hormone is demonstrated to occur in neurosecretory granules. Reprinted from [16].

selected vasopressin antisera will react because they must be tail specific and not discriminate between the acid or deaminated form of vasopressin [22]. Using selected antibodies with these techniques vasopressin and bovine neurophysin II were found to be synthesized as part of a common 21,000 molecular weight precursor. It was subsequently shown that the precursor for vasopressin was glycosylated at the C-terminal end. Messenger RNA was also used to clone cDNA and the nucleotide sequences of cDNA confirmed the predicted prohormone sequence of 166 amino acids.

A summary of the proposed biosynthetic pathway for the rat is shown in figure 3. Almost certainly human precursor synthesis is similar. The gene for vasopressin (fig. 3 a, b) is proposed to be 1.85 kilobase pairs consisting of three exons [23]. The first exon, Exon A, encodes the 19 amino acid signal peptide, arginine vasopressin, a spacer sequence between vasopressin and neurophysin consisting of Gly-Lys-Arg, and the variable N-terminal sequence of neurophysin. Exon B encodes the internal constant portion of the neurophysin (which is the conserved

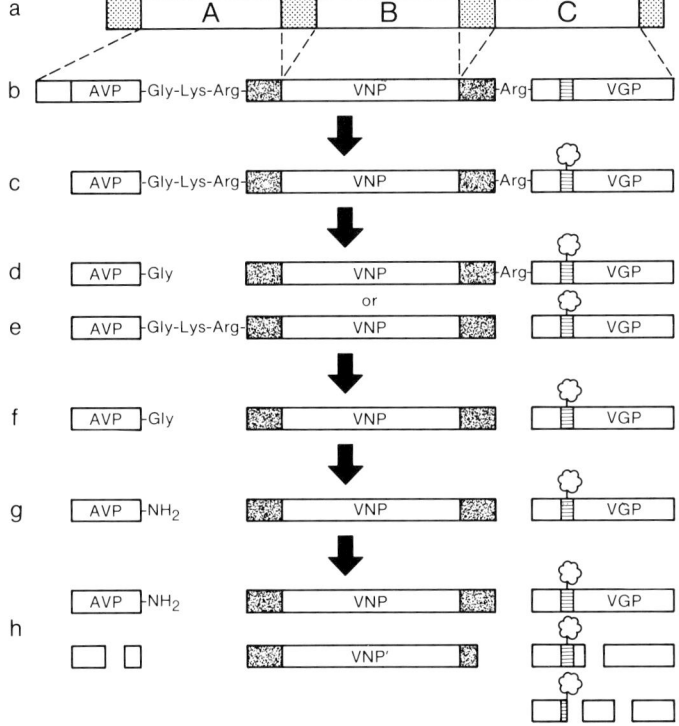

Fig. 3. Biosynthetic pathway for vasopressin via a common precursor including arginine vasopressin (AVP), vasopressin neurophysin (VNp), and vasopressin glycopeptide (VGP). *a* The three exons, A, B, and C of the vasopressin gene. *b* Original peptide chain including leader sequence. *c* Pro-vasopressin which is glycosylated on the glycopeptide. *d, e* Cleavage of vasopressin, neurophysin, and glycopeptide as the three products of the prohormone. *f* The three products of the prohormone. *g* Terminal amidation of AVP. *h* Further metabolic breakdown of the three peptide products which may occur within neurosecretory granules or after release.

sequence between neurophysins within a species and between neurophysins in different species). Exon C encodes the variable C-terminus of neurophysin, a single Arg, and then a 39 amino acid polypeptide which contains a typical glycosylation site, Asn-X-Thr at residues 114–116. Within the endoplasmic reticulum the signal peptide is cleaved and mannose-rich side chains are attached to the special glycosylation site to form the 23,000-dalton prohormone (fig. 3c). The glycosylated prohormone is transported to the Golgi apparatus where it is packaged into neurosecretory granules. Within the neurosecretory granules the 23,000-dalton glycosylated prohormone is cleaved. Tissue-specific enzymes cleave vasopressin and neurophysin and produce amidation of the C-terminal glycine of vasopressin (fig. 3d). The glycine which is part of the bridge between vasopressin and neurophysin is probably necessary for the oxidative transamidation of the C-terminus of vasopressin (fig. 3f). A relatively non-specific proteolytic enzyme cleaves the glycopeptide from the prohormone (fig. 3e). This was demonstrated experimentally by microinjection of mRNA from the bovine hypothalamus into oocytes from the *Xenopus laevis*. In this species a 14,000-dalton vasopressin and neurophysin intermediate is found, but mature neurophysin and vasopressin are never isolated, indicating that more specific enzymes are necessary to cleave the vasopressin from the neurophysin and are involved in proteolysis and amidation of vasopressin [18–23].

The three peptides, vasopressin, neurophysin and glycopeptide (fig. 3g), represent three individual functional units. Only vasopressin has been shown to have biologic activity outside the neuron. The only known function of neurophysin is as an intragranular binding protein for vasopressin and no physiologic role has been identified for the glycoprotein. Nonetheless, the high sequence conservation which is present beween species for each of the three units has been interpreted by many to imply that each has an independent function similar to the multiple biologically active hormones which are all derived from the single propiomelanocortin prohormone [24]. There is some evidence that each of the neurohypophyseal peptides is further cleaved. Within the neurosecretory granule, subfragments of the glycoprotein have been isolated from different species (fig. 3h) [25]. In several species there is heterogeneity of neurophysins, and in the human we found that one human neurophysin (hNp IV) was identical to the vasopressin neurophysin of man but with an Arg-Ala extension at the C-terminus (fig. 3h) [7]. Within the central nervous system a fragment of vasopressin has recently been identified which has enhanced vasopressin-like memory consolidation actions (fig. 3h) [26]. The extent of further processing within granules and after secretion is not known. Nor is it proven that any of the fragments will have specific biologic actions.

Documentation that similar vasopressin precursors are present in humans was provided by *Verbalis and Robinson* [27–29] and by *Yamaji* et al. [30]. *Verbalis* extracted acetone-dried human posterior pituitary and performed gel filtration on Sephadex G-75 of an acid extract. In the presence of 6 M urea four distinct peaks of neurophysin immunoreactivity were found. In the 10,000-dalton area 80–90% of the neurophysin was eluted but 10–15% was also eluted in the 20,000-dalton area and was proven to be 20,000-daltons by sodium dodecyl sulfate (SDS) gel electrophoresis. Following short incubation with chymotrypsin the 20,000-dalton neurophysin was partially converted to a 10,000-dalton neurophysin with generation of increased AVP immunoreactivity. *Yamaji* et al. [30] isolated a similar 20,000-dalton neurophysin immunoreactive protein from human oat cell carcinoma tissue and also showed the biosynthesis of this material using pulse chase studies with labeled amino acids by carcinoma tissue in organ culture. The 20,000-dalton neurophysin was converted to 10,000-dalton neurophysin in the pulse chase studies, and isolated 20,000-dalton neurophysin could be converted enzymatically to 10,000-dalton neurophysin and additionally vasopressin. Therefore, it is highly likely that human precursors are similar to those described in other animals and are similarly packaged and transported in the neurohypophysis as well as being present in tissues capable of ectopic production of vasopressin.

Transport

While the elegant pulse-chase labeling studies of neurophysin precursors in the rat by *Brownstein* and co-workers [31–33] were the first clear identification of the common precursor of neurophysin and vasopressin, they are equally important for their contribution to our understanding of the transport of vasopressin. These workers injected ^{35}S-cysteine into the hypothalamus adjacent to the supraoptic nucleus (SON) and studied the incorporation of the labeled amino acid into neurohypophyseal peptides. In figure 4 are labeled proteins separated on acid urea gels [32]. Peaks a and b are 19,000- to 20,000-dalton precursors, c and d intermediate forms and e mature 10,000-dalton neurophysin. The studies confirmed rapid incorporation and transport of labeled amino acids. Within 20 min of injection neurophysin precursors could be isolated from the SON, within 1 h from the median eminence, and within 1.5 h from the posterior pituitary. This rapid appearance of labeled neurophysin in the posterior pituitary was probably secondary to fast axonal transport because there was virtual complete inhibition of the appearance of labeled peptides in the pituitary if the rats were pretreated

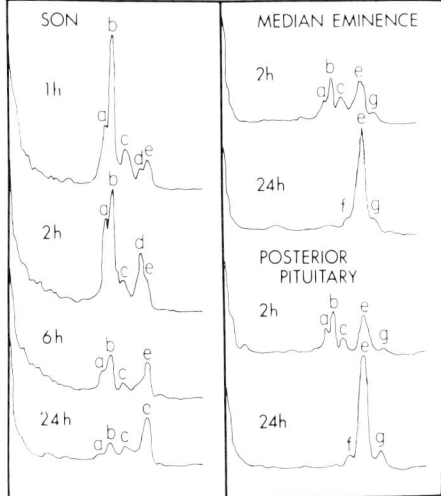

Fig. 4. Autoradiograph of acid urea gels illustrating ^{35}S-labeled proteins from the supraoptic nucleus (SON), the median eminence, and the posterior pituitary of normal rats 1–24 h after injection of ^{35}S-cysteine into the SON. Peaks a and b are 19,000- to 20,000-dalton precursors of vasopressin, c and d are intermediate forms, and e is the 12,000- to 13,000-dalton neurophysin. f, g Unexplained. Reprinted from [32].

with colchicine. There was both a temporal (fig. 4a) and spatial (fig. 4a–c) progression from the precursor neurophysin of approximately 20,000 daltons to an intermediate form of about 15,000 daltons and finally to a mature neurophysin of approximately 10,000 daltons. Of importance, the first labeled product isolated from the median eminence and from the posterior pituitary contained at least 50% 20,000-dalton precursor neurophysins (fig. 4b). The only machinery for synthesis of precursor is in the perikarya in the hypothalamus and all of the synthesis of vasopressin took place in the SON. As all of the neurophysin present within the median eminence and the posterior pituitary is within neurosecretory granules, the studies confirmed conclusively that precursor neurophysins were packaged into neurosecretory granules. It is reasonable to assume that all of the neurophysin (and vasopressin) which is packaged within the SON is in the form of precursor. By 1 h when neurophysin was detected in the median eminence, no more than 50% of the neurophysin was precursor. As the time for incorporation may require as much as 20 min, we can approximate the time for conversion to the first intermediate requires something under 40 min. When hormone is stored in the neural lobe it is virtually all converted to mature neurophysin, vasopressin, and glycopeptide (fig. 4c, 24 h).

Release Control

With the development of radioimmunossays for neurophysins and the measurement of release of vasopressin-neurophysin in response to stimulated release of the hormone vasopressin, it was clear that the entire contents of the neurosecretory granule were released [3, 8, 11, 34]. This has been amply confirmed by in vitro studies of isolated neurohypophyses [36] and by studies with electron microscopy which show increased exocytosis [36, 37] during stimulated release of hormone. As reviewed by *Douglas* [36] and *Nordmann* [38], calcium plays a key role in this process. Impulses originating in the SON or paraventricular nucleus (PVN) depolarize the axon terminals and promote influx of calcium. 'Free' calcium in the nerve ending causes exocytosis and release of hormone from neurosecretory granules. In the human we found specific release of one neurophysin with vasopressin and another neurophysin with oxytocin [33]. Similar results were obtained in several species. Based on the findings one would also predict that the glycopeptide is released. Indeed, a radioimmunoassay has recently been developed for the glycopeptide of rat and it was documented that this peptide was released in response to stimulation of vasopressin release [34].

Regulation

The factors which regulate biosynthesis, transport and release are not well understood from a mechanistic point of view. The physiologic release of vasopressin by osmotic and non-osmotic stimuli is becoming better defined with the availability of sensitive radioimmunoassays for vasopressin and has been described elsewhere in this book (chapter 9, 10). The neurotransmitters and neuropeptides which may be involved in the stimulation and/or inhibition of release of vasopressin are only beginning to be clarified. Multiple neuropathways have been described to project to the magnocellular neurons both of the PVN and the SON. Catecholamines, especially norepinephrine given both peripherally and centrally affect the release of vasopressin. These may act as neurotransmitters within the chain of neurons which project to the SON as well as on the neurons themselves [see *Zimmerman*, this volume]. GABA (γ-aminobutyric acid), histamine and substance P have all been described to influence the release of vasopressin although the exact role of these transmitters is not known [37]. As each of the pathways to the SON and PVN may have different neurotransmitters, it is certain that in the intact animal various transmitters will have either stimulatory or inhibitory action which may vary from species to species, with route of injection, and with the physiologic state of the animal, i.e. the relative activity of competing pathways in that animal. Use of neurohypophyseal organ explants has allowed some understanding of the more proximal (to the vasopressin neuron) neurotransmitter regulation of vasopressin release. Acetyl choline acting on nicotinic receptors is an important final common pathway in the SON, and this is the final neurotransmitter which is involved in the osmotic regulation of release of vasopressin [39]. Angiotensin II receptors which are separate from the nicotinic cholinergic receptors are also present on the magnocellular neurons of the SON and studies with the explant demonstrate that angiotensin II antagonists will decrease osmotically stimulated release of vasopressin [40, 41].

Opiate peptides are of special interest because these are the only neurotransmitters which have been described to act at the level of the posterior pituitary [42, 43]. Both Met- and Leu-enkephalin have been localized to the posterior pituitary by radioimmunoassay and by immunohistochemistry [44, 45]. Stereo-specific opiate binding sites have also been described. There is debate as to whether the enkephalins are co-localized with hormones in neurohypophyseal neurons, are separate nerve terminals which terminate upon the vasopressin (and oxytocin) axons, or are nerve terminals which terminate on pituicytes and/or on capillaries. Recent data using immunoelectron microscopy would favor the interpretation that the enkephalin stained fibers abut on pituicytes [46, 47]. Certainly opiate receptors in the posterior pituitary exist on cells other than axon terminals, because after section of the neurohypophyseal stalk when all of the nerve fibers disappear and only pituicytes remain opiate receptors can still be found in the neural lobe [47]. Most studies would agree that opiate peptides inhibit the release of vasopressin (and oxytocin) from the neural lobe. It is not clear how the pituicyte might accomplish the decreased release of the hormone, but it has been proposed that the pituicytes enclose the nerve endings and thereby regulate the immediate environment of the nerve terminal. It is postulated that the pituicytes may form a 'potassium buffer' and that the opiate peptide may influence the pituicytes to enclose more or less of the distal axons. The axons would then take up more or less potassium to regulate release of vasopressin [46]. At this time evidence to support the hypothesis for the role of pituicytes in secretion is preliminary, however, it seems clear that opiate peptides do act at the distal pituitary to inhibit the release of vasopressin and it seems likely that pituicytes play some role in regulating the release of vasopressin.

Central (or hypothalamic) diabetes insipidus has only been thought of as a lack of the hormone vasopressin, and this has been interpreted to mean a lack of biosynthesis of vasopressin. The Brattleboro rat which does not synthesize vasopressin nor vasopressin neurophysin [31] has recently been reported to have an abnormal gene for the vasopressin precursor. A single nucleotide deletion results in a phase shift of the remaining bases and probably removes a stop codon so that mRNA is not translated and precursor not transcribed. The present review, however, demonstrates multiple sites in which there may be abnormalities of vasopressin biosynthesis, post-translational processing, transport or release, any of which may produce disorders of diabetes insipidus. Some trigger for biosynthesis of vasopressin occurs when hormone release is increased, but virtually nothing is known about the signal which stimulates transcription and translation and how this signal might be disturbed. A rudimentary understanding of the enzymes which are necessary for post-translational processing was described above and constitutes another potential site(s) of disturbed synthesis.

Because the prohormone is packaged within neurosecretory granules and because the entire contents of the neurosecretory granules are known to be released by exocytosis, it is reasonable to postulate that disordered enzymatic processing of the vasopressin prohormone would result in release of precursor forms of neurophysin and vasopressin into the circulation. These have not yet been identified, but neither have investigators known how to test for them. In at least one family of familial diabetes insipidus no vasopressin neurophysin was found, indicating there was probably no prohormone secreted [48]. (As described above most neurophysin antisera would be expected to react with vasopressin prohormone, whereas only selected vasopressin antisera would react.) Therefore, assay of neurophysin provides a potential marker for diabetes insipidus in which the abnormality is due to the processing of the prohormone.

Although rate of transport of hormone is a potential source of control, studies in hemorrhaged and in adrenalectomized rats indicated that there was no increase in the rate of flow of neurosecretory granules during increased delivery of vasopressin to the posterior pituitary [49, 50]. Rather, the bulk of neurosecretory granules delivered accounted for the increased total delivery of product. The role of nonneuronal elements in the posterior pituitary on release of vasopressin is only beginning to be understood and is yet another site of potential regulation of vasopressin release.

References

1 Bargmann, W.; Scharrer, E.: The site of origin of the hormones of the posterior pituitary. Am. Sci. *39:* 255–259 (1951).
2 Zimmerman, E.A.: Localization of hypothalamic hormones by immunocytochemical techniques; in Martini, Ganong, Frontiers in neuroendocrinology, vol. 4, p. 25 (Raven Press, New York 1976).
3 Zimmerman, E.A.; Robinson, A.G.: Hypothalamic neurons secreting vasopressin and neurophysin. Kidney int. *10:* 12–24 (1976).
4 Robinson, A.G.: The contribution of measured secretion of neurophysins to our understanding of neurohypophyseal function; in Reichlin, The neurohypophysis, p. 65 (Plenum Publishing, New York 1984).
5 Acher, R.: Chimie des hormones neurohypophysaires. Ergebn. Physiol. *48:* 286 (1955).
6 Chauvet, M.T.; Codogno, R.; Chauvet, J.; Acher, R.: Comparison between MSEL- and VLDV-neurophysins; complete amino acid sequences of porcine and bovine VLDV-neurophysins. FEBS Lett. *98:* 37 (1979).
7 Chauvet, M.T.; Chauvet, J.; Acher, R.; Robinson, A.G.: Identification of the two types of neurophysins, MSEL- and VLDV-neurophysins, in human pituitary gland, FEBS Lett. *101:* 391 (1979).
8 Robinson, A.G.: Isolation, assay and secretion of individual human neurophysins. J. clin. Invest. *55:* 360 (1975).
9 Burford, G.D.; Jones, C.W.; Pickering, B.T.: Tentative identification of a vasopressin-neurophysin and an oxytocin-neurophysin in the rat. Biochem. J. *124:* 809 (1971).
10 North, W.G.; Naurer, L.H.; Valtin, H.; O'Donnell, J.F.: Human neurophysins as potential tumor markers for small cell carcinoma of the lung: application of specific radioimmunoassays. J. clin. Endocr. Metab. *51:* 884–891 (1980).
11 Seif, S.M.; Huellmantel, A.B.; Platia, M.P.; Haluszczak, C.; Robinson, A.G.: Isolation, radioimmunoassay and physiologic secretion of rat neurophysins. Endocrinology *100:* 1317 (1977).
12 Cheng, K.W.; Friesen, H.G.: Isolation and characterization of a third component of porcine neurophysin. J. biol. Chem. *246:* 7656–7665 (1971).

13 Pickering, B.T.; Jones, C.W.; Burford, G.D.; McPherson, M.; Swann, R.W.; Heap, P.F.; Morris, J.F.: The role of neurophysin proteins: suggestions from the study of their transport and turnover. Ann. N.Y. Acad. Sci. *248:* 15 (1975).

14 North, W.G.; Morris, J.F.; LaRochelle, F.T.; Valtin, H.: Enzymatic interconversions of neurophysins: the nature of an enzyme(s) within neurosecretory granules of the neurohypophysis; in Moses, Share, Neurohypophysis, p. 43 (Karger, Basel 1977).

15 Seidah, N.G.; Benjannet, S.; Chretian, M.: The complete sequence of a novel human pituitary glycopeptide homologous to pig posterior pituitary glycopeptide. Biochem. biophys. Res. Commun. *100:* 901–907 (1981).

16 Sachs, H.; Takabatake, V.: Evidence for a precursor in vasopressin biosynthesis. Endocrinology *75:* 943–948 (1964).

17 Sachs, H.; Fawcett, P.; Takabatake, Y.; Portanova, R.: Biosynthesis and release of vasopressin and neurophysin, Recent Prog. Horm. Res. *25:* 447–491 (1969).

18 Lin, C.; Joseph-Bravo, P.; Sherman, T.; Chan, L.; McKelvy, J.F.: Cell-free synthesis of putative neurophysin precursors from rat and mouse hypothalamic poly (A)-RNA. Biochem. biophys. Res. Commun. *89:* 943–959 (1979).

19 Richter, D.; Schmale, H.: The structure of the precursor to arginine-vasopressin: a model preprohormone; in Cross, Leng, The neurohypophysis: structure, function and control. Progr. Brain Res., vol. 60, pp. 223–227 (Elsevier, Amsterdam 1983).

20 Schmale, H.; Leipold, B.; Richter, D.: Cell-free translation of bovine hypothalamic mRNA, synthesis and processing of the prepro-neurophysin I and II, FEBS Lett. *108:* 311–316 (1979).

21 Schmale, H.; Richter, D.: In vitro biosynthesis and processing of composite common precursors containing amino acid sequences identified immunologically as neurophysin I/oxytocin and neurophysin II/arginine vasopressin. FEBS Lett. *121:* 358–362 (1980).

22 Ivell, R.; Schmale, H.; Richter, D.: Vasopressin and oxytocin precursors as model preprohormones. Neuroendocrinology *37:* 235–239 (1983).

23 Richter, D.: Vasopressin and oxytocin are expressed as polyproteins. Trends biochem. Sci. *8:* 278–281 (1983).

24 Eipper, B.A.; Mains, R.E.: Structure and biosynthesis of proadrenocorticotropin/endorphin and related peptides. Endocr. Rev. *1:* 1–27 (1980).

25 Smyth, D.G.; Massey, D.E.: A new glycopeptide in pig, ox, and sheep pituitary. Biochem. biophys. Res. Commun. *87:* 1006–1010 (1979).

26 Burbank, J.P.; Kovacs, G.L.; deWied, D.; Nispen, J.W. van; Greven, H.M.: A major metabolite of arginine vasopressin in the brain is a highly potent neuropeptide. Science *221:* 1310–1312 (1983).

27 Verbalis, J.G.; Robinson, A.G.: Identification of high molecular weight neurophysins in extracts of human neurohypophyseal tissue. Brain Res. *237:* 504–509 (1982).

28 Verbalis, J.G.; Robinson, A.G.: Characterization of neurophysin-vasopressin prohormones in human posterior pituitary tissue. J. clin. Endocr. Metab. *57:* 115–123 (1983).

29 Verbalis, J.G.; Robinson, A.G.: Human pituitary neurophysin precursors; in Cross, Leng, The neurohypophysis: structure, function and control. Progr. Brain Res., vol. 60, pp. 247–251 (Elsevier, Amsterdam 1983).

30 Yamaji, T.; Ishibashi, M.; Katayama, S.: Nature of the immunoreactive neuro-

physins in ectopic vasopressin-producing oat cell carcinomas of the lung. J. clin. Invest. *68:* 388–398 (1981).

31 Brownstein, M.J.; Gainer, H.: Neurophysin biosynthesis in normal rats and in rats with hereditary diabetes insipidus. Proc. natn. Acad. Sci. USA *74:* 4046–4049 (1977).

32 Gainer, H.; Sarne, V.; Brownstein, M.J.: Biosynthesis and axonal transport of rat neurohypophyseal proteins and peptides. J. Cell Biol. *73:* 366–381 (1977).

33 Brownstein, M.J.; Russell, J.T.; Gainer, H.: Biosynthesis of posterior pituitary hormones; in Ganong, Martini, Frontiers in neuroendocrinology, vol. 7, pp. 31–43 (Raven Press, New York 1982).

34 Groesbeck, M.D.; Shome, B.; Parlow, A.F.: The isolated carboxy terminal glycopeptide of rat vasopressin-neurophysin precursor: factors affecting its secretion. Endocr. Soc. Abstr. *259:* 145 (1983).

35 Thorn, N.A.: In vitro studies of the release mechanism for vasopressin in rats. Acta endocr., Copenh. *53:* 644 (1966).

36 Douglass, W.W.: Mechanism of release of neurohypophyseal hormones: stimulus secretion coupling; in Knobil, Sawyer, Handbook of physiology, vol. IV, pp. 191–224 (Waverly Press, Baltimore 1974).

37 Theodosis, D.T.; Dreifuss, J.J.: Ultrastructural evidence for exoendocytosis in the neurohypophysis; in Moses, Share, Neurohypophysis, pp. 84-94 (Karger, Basel 1976).

38 Nordmann, J.J.: Stimulus-secretion coupling; in Cross, Leng, The neurohypophysis, function and control. Progr. Brain Res., vol. 60 (Elsevier, Amsterdam 1983).

39 Robinson, A.G.; Verbalis, J.G.; Amico, J.A.; Seif, S.M.: Recent advances in neurohypophyseal research; in McCann, Endocrine physiology. III. International review of physiology, vol. 24, pp. 1–40 (University Park Press, Baltimore 1981).

40 Sladek, C.D.; Joynt, R.J.: Angiotensin stimulation of vasopressin release from the rat hypothalmo-neurohypophyseal system in organ culture. Endocrinology *104:* 148–163 (1979).

41 Sladek, C.D.; Joynt, R.J.: Cholinergic involvment in osmotic control of vasopressin release by the organ-cultured rat hypothalamus-neurohypophyseal system. Endocrinology *105:* 367–371 (1979).

42 Iversen, L.L.; Iverson, S.D.; Bloom, F.E.: Opiate receptors influence vasopressin release from nerve terminals in the rat neurohypophysis. Nature, Lond. *284:* 350–351 (1980).

43 Lightman, S.L.; Iversen, L.L.; Forsling, M.L.: Dopamine and (D-Ala^2, D-Leu^5) enkephalin inhibit the electrically stimulated neurohypophyseal release of vasopressin in vitro: evidence for calcium-dependent opiate action. J. Neurosci. *2:* 78–81 (1982).

44 Rossier, J.; Vargo, T.M.; Minick, S.; Ling, N.; Bloom, F.E.; Guillemin, R.: Regional dissociation of beta-endorphin and enkephalin contents in rat brain and pituitary. Proc. natn. Acad. Sci. USA *74:* 5162–5165 (1977).

45 Rossier, J.; Battenberg, E.; Pittmann, Q.; Bayon, A.; Koda, L.; Miller, R.; Guillemin, R.; Bloom, F.: Hypothalamic enkephalin neurones may regulate the neurohypophysis. Nature, Lond. *277:* 653–655 (1979).

46 Van Leeuwen, F.W.; De Vries, G.J.: Enkephalin-glial interaction and its consequence for vasopressin and oxytocin release from the rat neural lobe; in Cross, Leng,

The neurohypophysis: structure, function and control. Progr. Brain Res. vol. 60, pp. 343–350 (Elsevier, Amsterdam 1983).

47 Lightman, S.L.; Ninkovic, M.; Hunt, S.P.: Neurohypophysial opiate receptors: Are they on pituicytes?; in Cross, Leng, The neurohypophysis: structure, function and control. Progr. Brain Res. vol. 60, pp. 353–356 (Elsevier, Amsterdam 1983).

48 Blackett, P.R.; Seif, S.M.; Altmiller, D.H.; Robinson, A.G.: Familial dominant diabetes insipidus, a defect with vasopressin and nicotine stimulated neurophysin (NSN) deficiency and subnormal oxytocin and estrogen-stimulated neurophysin (ESN). Am. J. med. Sci. (1983).

49 Norstrom, A.; Sjostrand, J.: Effect of salt-loading, thirst and water-loading on transport and turnover of neurohypophysial proteins of the rat. J. Endocr. *52:* 87–105 (1972).

50 Norstrom, A.; Sjostrand, J.: Effect of haemorrhage on the rapid axonal transport of neurohypophysial proteins of the rat. J. Neurochem. *18:* 2017–2026 (1971).

A.G. Robinson, MD, Department of Medicine, School of Medicine, University of Pittsburgh, Pittsburgh, PA 15261 (USA)

Molecular Aspects of the Expression of the Vasopressin Gene

Dietmar Richter, Hartwig Schmale

Institut für Physiologische Chemie, Abteilung Zellbiochemie,
Universität Hamburg, Hamburg, FRG

Vasopressin and its corresponding carrier protein are synthesized as a composite precursor by neurones in both the supraoptic and paraventricular nuclei of the hypothalamus. The precursor is packaged into neurosecretory granules and transported axonally in the stalk to the posterior pituitary. En route to the neurohypophysis the precursor is processed into the active hormone (fig. 1).

Recombinant DNA methodology was used to determine the structure of the vasopressin precursor and its gene [*Richter*, 1983]. The data showed that the precursor not only contained the hormone and its neurophysin but additionally a third unit, a glycoprotein of yet unknown function (fig. 2).

Expression of Vasopressin in Diabetes insipidus (Brattleboro) Rats

Gene Structure

It is well established that mutant (Brattleboro) rats with hereditary hypothalamic diabetes insipidus lack biologically active vasopressin and its neurophysin, while the synthesis of the structurally related hormone oxytocin is not affected by the mutation [*Sokol and Valtin*, 1982]. In these animals control of water retention in the distal kidney tubuli is greatly impaired. Hence, the mutants show a high water uptake accompanied by excretion of an excessive amount of dilute urine.

By isolating and sequencing the vasopressin gene from homozygous Brattleboro rats it was possible to locate the genetic defect [*Schmale and Richter*, 1983]. It is due to a single nucleotide deletion within the second

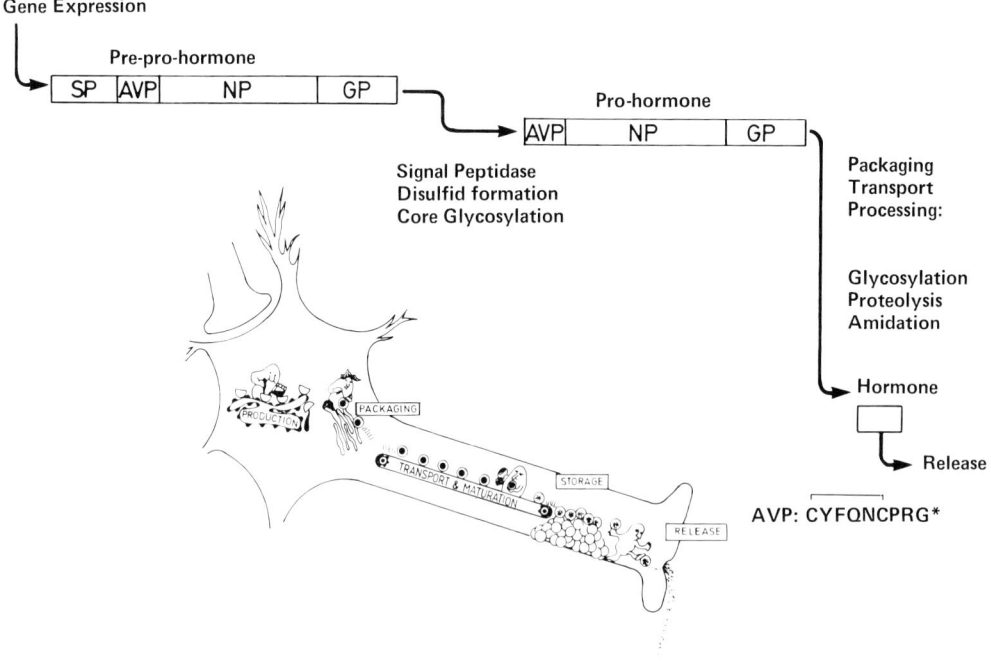

Fig. 1. Cascade of vasopressin biosynthesis. SP = Signal peptide; AVP = arginine vasopressin; NP = neurophysin; GP = glycoprotein.

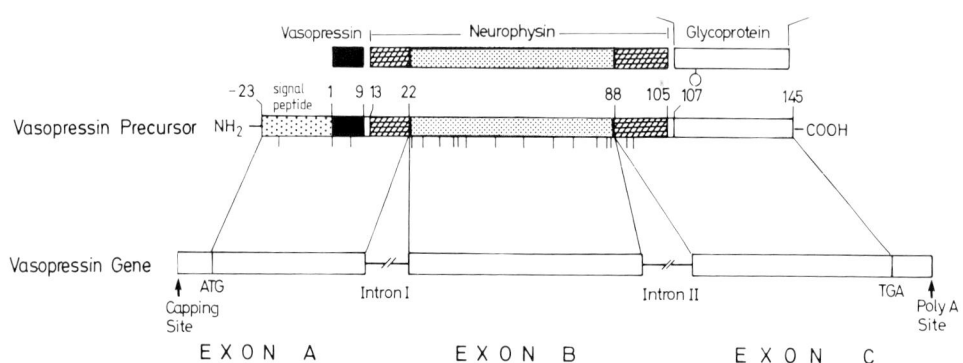

Fig. 2. Structural organization of the vasopressin gene and the hormone precursor. For details see *Richter* [1983].

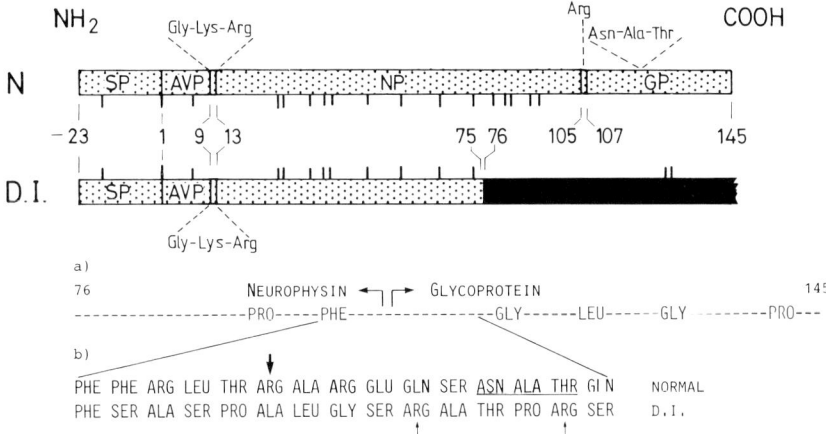

Fig. 3. Comparison of the vasopressin precursors from normal (N) and diabetes insipidus (DI) rats deduced from the gene structures. From *Schmale and Richter* [1984].

exon encoding the highly conserved region of the carrier protein. As a result of this deletion, the reading frame is shifted predicting a precursor with an entirely different C-terminus (fig. 3).

Comparison of the precursor from normal and mutant rats showed that the amino acid sequences are identical from the N-terminus to the deletion site (roughly two-thirds of the normal precursor). The sequence following the deletion site is almost entirely different. The most drastic changes found are: (1) Because of the new reading frame the Brattleboro vasopressin mRNA lacks the signal (stop codon) which normally terminates protein synthesis; (2) the reading frame predicts that the mutated precursor no longer contains a glycosylation site nor the basic amino acid separating the neurophysin from the C-terminal peptide; (3) the amino acid sequence of the C-terminal neurophysin part is drastically changed lacking 5 cysteine residues, probably of great importance for the secondary folding of the molecule (fig. 3).

Transcription and Translation

Analysis of hypothalamic mRNA from Brattleboro rats indicated that their mRNA is of similar size as the normal one suggesting that it has been correctly transcribed and spliced. However, when mRNA from the Brattleboro rat was translated in a cell-free reticulocyte lysate system, little or no precursor could be identified using antibodies raised against vasopressin.

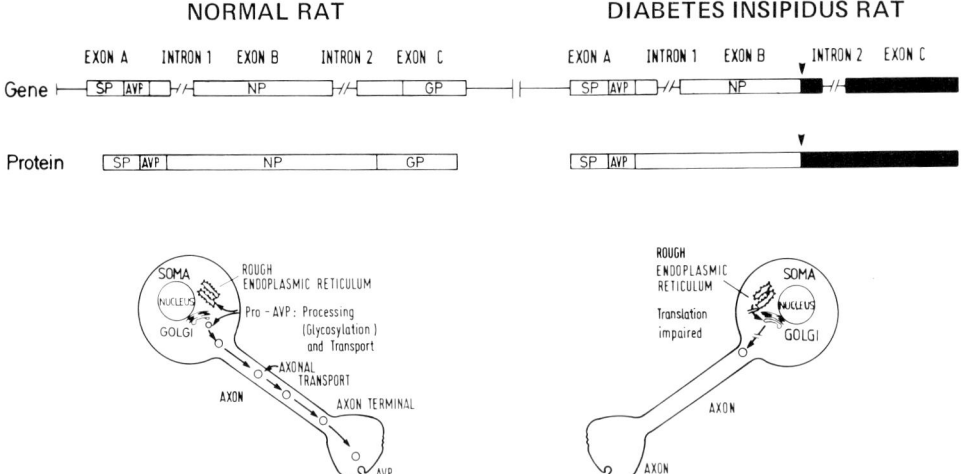

Fig. 4. Pathway of the expression of vasopressin in normal and diabetes insipidus rats. For abbreviations see figure 1. Black bars = gene sequence which leads to the altered C terminus of the predicted precursor. The left scheme was redrawn from *Russell* et al. [1980].

The observed translation defect might be due to the lacking stop codon which normally signals termination of protein synthesis. Theoretically the mutation predicts a precursor with a C-terminal poly-lysine tail. Alternatively, the absence of a vasopressin precursor in a cell-free translation system may be due to the antibodies used (anti-vasopressin) which no longer recognize the mutated product with possible altered secondary folding.

The situation in the Brattleboro rat itself suggests that – like the in vitro system – translation stops at the rough endoplasmic membrane where the ribosomes may come to a halt in the 3'-untranslated region of the vasopressin mRNA. Again lack of a stop codon would predict that ribosomes cannot leave the mRNA thus jamming the protein synthesizing machinery at the membrane level (fig. 4). Uncompleted precursors may be the result of the defective translational termination event. Whether the 'peptide X' isolated from Brattleboro hypothalami is a candidate for such a mutated precursor remains to be seen [*Russell* et al., 1980]. 'Peptide X' appears to have antigenic sites for vasopressin and neurophysin and lacks the carbohydrate chain which would fit with the structure of the predicted precursor.

In summary, Brattleboro rats have, in theory at least, the capacity to produce vasopressin – the part of the mRNA encoding the hormone is completely intact. However, most likely because of the lacking stop codon, the mRNa is inefficiently translated.

References

Richter, D.: Vasopressin and oxytocin are expressed as polyproteins. Trends biochem. Sci. *8:* 278–281 (1983).
Russell, J.T.; Brownstein, M.J.; Gainer, H.: Biosynthesis of vasopressin, oxytocin, and neurophysins. Endocrinology *107:* 1880–1891 (1980).
Sokol, H.W.; Valtin, H. (eds.): The Brattleboro rat. Ann. N.Y. Acad. Sci. *394:* 1–828 (1982).
Schmale, H.; Richter, D.: Single base deletion in the vasopressin gene is the cause of diabetes insipidus in Brattleboro rats. Nature, Lond. *308:* 705–709 (1984).

Dietmar Richter, Institut für Physiologische Chemie, Abteilung Zellbiochemie, Universität Hamburg, D-2000 Hamburg 20 (FRG)

Ontogeny of Vasopressin in Man

Rosemary D. Leake[a], Delbert A. Fisher[b]

Departments of [a] Pediatrics and [b] Medicine, University of California, Los Angeles, School of Medicine and The Harbor/UCLA Medical Center, Torrance, Calif., USA

Appearance of Arginine Vasopressin

Arginine vasopressin (AVP) has been identified in the human pituitary by radioimmunoassay (RIA) at 11–12 weeks of gestation, and its presence confirmed by differential bioassay and by high-pressure liquid chromatography (HPLC) [1, 2]. By 16 weeks gestation stainable neurosecretory granules are visualized in the hypothalamic nuclei [3]. At term, fetal pituitary AVP content is 20% of the adult human value. Infusion of hypertonic saline to the human infant with indirect measurement of AVP secretion (increase in urine osmolality) as well as direct measurements of cord blood AVP concentrations have indicated that the newborn has the capability to secrete and modulate secretion of AVP [4, 5]. Additionally, arginine vasotocin (AVT), an ancestral precursor of AVP having only one amino acid substitution, has been shown by HPLC and RIA to be present in ovine fetal plasma (and human cord blood) [6].

Fetal AVP Secretory Control, Production Rates and Metabolic Clearance

Since there is little information regarding the ontogenesis of AVP secretory control (and none concerning AVT secretory control) in the human fetus, we and others utilized the fetal sheep model to characterize fetal AVP metabolism [7, 8], applying a variety of stimuli in the third trimester ovine fetus to study the control of AVP secretion. The ovine (and human) placenta is impermeable to both AVP and AVT so that fetal plasma levels reflect fetal production [9]. We and others found that baseline fetal plasma AVP concentrations are similar to maternal levels from the time of earliest measurement (88 days' gestation); both approximate 1–1.5 µU/ml [10–12]. The fetal lamb responds with a vigorous and prompt increase in plasma AVP to hypertonic saline infusion (10 mEq/kg) [10, 13], to hemorrhage (6–30% of estimated fetal/placental blood volume) [14–16], to furosemide (2 mg/kg, which stimulates AVP secretion by increasing renin-angiotensin production and angiotensin II stimulation of AVP release) [10], to hypoxia (induced by maternal ewe breathing 10% oxygen), and to fetal dehydration [21, 22] (produced by evoking a fetal-to-maternal water transfer during the infusion of hypertonic saline or mannitol to the ewe or by maternal water deprivation) [22, 23]. The AVP response (at least to furosemide, hypertonic saline and dehydration) matures with increasing gestational age. *Robillard* et al. [15] and *Bell* et al. [23] demonstrated maturation of the renal response to endogenous AVP in fetal lambs approaching term and *Wintour* et al. [24] demonstrated a similar renal response to exogenous AVP.

An earlier report from our laboratory suggests that the baseline AVP production rate in the ovine fetus exceeded the maternal rate (on a mU/kg/h basis) by 100- to 200-fold [7]. In retrospect, these high 'basal' values probably were due to stress in the chronically catheterized animals. In recent studies in carefully monitored, unstressed, chronically catheterized ewes we reported that fetal, maternal and nonpregnant ewe metabolic clearance rates and production rates are similar for both AVP and AVT [12].

Additionally, urinary AVP clearance rates in the ovine fetus are nonsaturable, and it appears that the fetal kidney plays only a minor role in AVP clearance, perhaps reflecting the relatively low blood flow and limited glomerular filtration rate characteristic of the fetus [25].

Effects of Fetal AVP on Transplacental Water Exchange

The ovine fetal neurohypophysis appears to maintain some degree of autonomy over fetal osmolar homeostasis. Under basal conditions fetal and maternal plasma AVP levels are significantly correlated. When a maternal to fetal osmolar gradient is created (by infusion of hypertonic saline or mannitol into the mother), water is rapidly transferred from the fetal to maternal circulation [22, 26]. If an identical osmolar gradient is created during fetal AVP (or AVT) infusion, the fetal-to-maternal water flux is markedly diminished [26]. Thus, fetal plasma AVP and AVT levels modulate transplacental water transfer. When AVT is infused into the fetal circulation with cortisol, the fetal osmolar increase following maternal mannitol infusion is augmented suggesting that cortisol facilitates fetal to maternal water transfer under these circumstances. In contrast, water flow from amniotic fluid to the maternal circulation (following the same stimulus) is not affected by amniotic fluid AVP or AVT levels [27].

Effects of Fetal Hormones on Lung Fluid Production

During fetal AVP infusion lung fluid production, like renal free water clearance, is significantly decreased [28]. Moreover, studies in the acutely exteriorized fetal goat and chronically catheterized fetal lamb show that lung fluid production is reduced and there is some net lung water absorption when fetal plasma AVP is elevated. Additionally, catecholamines and cortisol infused into the fetus also decrease tracheal fluid production in the chronically tracheotomized fetal lamb. Thus, lung fluid kinetics seem to be influenced by several hormones as is the case with fetal/maternal/amniotic fluid water exchange.

Elevated levels of plasma AVP may play a critical role in the clearance of alveolar and tracheal fluid during parturition. Additionally, infants born by cesarean section delivery have a diminished AVP surge at birth and this may lead to less efficient lung water removal, contributing, at least in part, to the increased prevalence of the syndrome of transient tachypnea of the newborn after cesarean delivery.

These studies suggest that the fetus is capable of diminishing, on a short-term basis, water loss via the placenta, fetal lung and kidney by secreting AVP or AVT. Other studies support the view that AVP plays an important physiologic role in blood pressure regulation in fetal lambs during periods of hypovolemia [29, 30]. AVP or AVT receptors have not yet been identified in the placenta, fetal lung or kidney, but their presence in these sites as well as in smooth muscle is evidenced by the observed renal and pulmonary responses as well as the increased fetal blood pressure and associated bradycardia noted during AVP or AVT infusion.

Table I. AVP concentrations in newborn lambs [10, 31]

	Plasma AVP, μU/ml	
	baseline values	peak values
Hypertonic saline (10 mEq/kg, n=11)	2.9±0.7	22.2±9*
Phlebotomy (10 ml/kg, n=10)	1.9±0.4	72 ±40*
Dehydration (18 h, n=9)	0.6±0.1	4.8±1.8*
Water loading (100 ml/kg, n=7)	3.4±1.2	1.1±0.3*
Furosemide (2 mg/kg, n=10)	2.1±0.4	13.8±2.1*

* p = <0.05 baseline versus peak concentration.

Newborn Secretion, Metabolism and Effects

Studies of AVP responsiveness to provocative stimuli in the newborn sheep also were conducted in the chronically catheterized (jugular vein and carotid artery) newborn lamb. As seen in table I, plasma AVP levels change significantly and appropriately in the 1- to 7-week-old newborn lamb following stimulation with hypertonic saline [31], furosemide [10], hemorrhage, dehydration or water loading [31]. There are no comparable data for the human newborn.

Urine concentrating capacity in the newborn human kidney as well as the kidney of the developing lamb is limited even in the presence of AVP. AVP acts on the kidney to increase water permeability of the collecting duct and to increase tubular to interstitial NaCl transfer by the ascending limb of the loop of Henle. This creates and maintains the osmotic gradient to facilitate water reabsorption from the collecting duct. As discussed by *Wintour* et al [24], it is unlikely that immaturity of kidney anatomy or morphology accounts for the reduced AVP responsiveness of the immature kidney. A more likely explanation is a reduced population and/or responsiveness of renal AVP receptors or the post-receptor effect systems. In the piglet and rat a reduced responsiveness of the medullary cAMP system to AVP has been demonstrated [32, 33], but maturation of AVP receptors, cAMP activation and post-receptor responses have not been systematically explored in the loop of Henle or in the collecting ducts.

AVP in Pregnancy and Parturition

Beginning as early as 6 weeks of human gestation, there appears to be a resetting of maternal AVP osmostat control which persists throughout pregnancy. Basal and stimulated plasma osmolality decreases to values 5–10 mosm below those of nonpregnant women, despite normal basal and stimulated (dehydrated and water-loaded) levels of plasma AVP. The mechanism of the osmostat resetting is not clear, nor is the purpose. It may be related (as cause or effect) to the increase in extracellular fluid volume characteristic of human pregnancy. Interestingly, the osmotic threshold stimulus for thirst also is reduced [34]. Maternal plasma AVP levels do not increase during either labor or infant suckling [35, 36].

A cystine aminopeptidase (CAP) which degrades AVP, AVT and oxytocin appears in primate and human maternal and cord blood as well as amniotic fluid during pregnancy [37]. CAP is produced by the syncytiotrophoblastic cells of the placenta and circulating concentrations increase with advancing gestational age. Although CAP rapidly degrades plasma AVP and AVT in vitro, it has little effect on the metabolic clearance in vivo; AVP and AVT metabolic clearance rates are identical in nonpregnant and pregnant women [37].

Basal amniotic fluid AVP and AVT levels approximate 1.5 and 10 µU/ml, respectively [38], and there is a significant, sustained increase in amniotic fluid AVP levels following fetal hypoxia (half-life = 8 h) [39]. These elevated values appear not to affect amniotic-to-maternal water transfer [27] but may prove to be a useful marker for the presence of fetal hypoxia.

AVP in CSF and Urine

AVP and AVT have been identified in cerebrospinal fluid (CSF) of laboratory animals and human subjects (including infants) [38]. AVP appears to be secreted into CSF in association with its neurophysin [40, 41]. Moreover, there is a diurnal rhythm for levels in CSF whereas no such rhythm has been observed in blood levels [42]. The time of appearance of the CSF AVP diurnal rhythm in the fetus or newborn has not been defined nor has the physiological role of CSF AVP. It is known that AVP has corticotropin-releasing factor-like activity and that AVP potentiates learning and memory in animals. However, any role for either of these functions in the neonate is unknown.

Godard et al. [43], measuring urine AVP from infancy to young adulthood, reported an increase in absolute values with age; however, AVP excretion in infancy was proportional to creatinine clearance, and there is no change with maturation when values are corrected for surface area. Using a specific RIA for urinary AVP [44], we measure mean basal urinary AVP levels of 8 μU/ml in well-hydrated infants and levels of 48 μU/ml following chronic furosemide use for congestive heart disease in infancy.

AVP Physiology in Healthy Infants

Elevated plasma AVP levels have been reported by bioassay and radioimmunoassay in the immediate newborn period in both preterm and full-term infants [5, 45–47]. Umbilical arterial AVP levels are greater than umbilical venous levels, implying fetal production. AVT levels in both umbilical arterial and venous plasma approach 5 pg/ml. There have been no careful studies measuring fetal plasma AVP levels during labor in sheep, but cord plasma AVP concentrations in vaginally delivered human infants exceed those of infants delivered by cesarean section. Cord plasma AVP levels are positively correlated with the length of labor and degree of cervical dilatation at the time of C-section, suggesting that head compression during delivery may be an important stimulus to fetal AVP secretion [46]. Hypoxia, acidosis, hypercarbia and hypotension may contribute additionally, as suggested by studies in the fetal sheep and the human neonate [17–21, 47–49]. All of these results suggest the fetal AVP levels are elevated during the process of parturition. The exact function of the strikingly elevated plasma AVP levels in cord blood at birth is not understood. A role for such levels in vascular homeostasis has been proposed [29, 30]; both AVP and catecholamines may play important roles in blood flow redistribution during hypoxic episodes and during birth itself.

The human newborn responds to dextran, hypertonic saline and cold exposure by appropriate changes in serum and urine osmolality, implying normal function of both osmolar and volume AVP secretion control systems as in the newborn sheep [4, 50]. Moreover, renal responsiveness to AVP seems normal within the limits of the relatively reduced concentrating capacity of the newborn kidney [51]. The mechanism of the limited concentrating capability is not entirely clear as discussed earlier.

In summary, AVP secretory dynamics are well developed in the ovine model by the third trimester; the fetus is responsive to all appropriate stimuli examined. AVP may play a role in water conservation in the fetus by means of its effects on transplacental, and amniotic-to-maternal water transfer as well as on lung fluid and fetal urine production. Chronic fetal secretion of AVP may produce oligohydramnios as lung fluid and urinary contributions to amniotic fluid volume decrease. Fetal AVP secretion during labor may facilitate lung fluid reabsorption. AVP secretion also is well developed in the human newborn and AVP secretion responses to hypoxia and volume depletion are relatively augmented. A role of AVP in vascular homeostasis in these conditions has been proposed. It is clear that the human neonate is not predisposed to AVP deficiency and diabetes insipidus is rare during this period of life in the absence of hypothalamic anomaly or damage. Rather the tendency is for AVP hypersecretion in response to metabolic stress, hypoxia, central nervous system disease or pulmonary disorders.

References

1 Skowsky, W.R.; Fisher, D.A.: Fetal neurohypophyseal arginine vasotocin in man and sheep. Pediat. Res. *11:* 627 (1977).
2 Fisher, L.A.; Fernstrom, J.D.: Measurement of nonapeptides in pineal and pituitary using reversed-phase, ion-pair liquid chromatography with post columnar detection by radioimmunoassay. Life Sci. *28:* 1471 (1981).
3 Rinne, U.K.; Kivalo, E.; Talanti, S.: Maturation of human hypothalamic neurosecretion. Biol. Neonate *4:* 351 (1982).
4 Fisher, D.A.; Pyle, H.R.; Porter, J.C.; Panos, T.C.: Studies of control water balance in the newborn. Am. J. Dis. Child. *106:* 137 (1963).
5 Chard, Y.; Hudson, C.V.; Edwards, C.R.W.; Boyd, N.R.H.: Release of oxytocin and vasopressin by the human foetus during labor. Nature, Lond. *234:* 352 (1971).
6 Leake, R.D.; Artman, H.G.; Fisher, D.A.: Plasma arginine vasotocin (AVT) in newborn infants. Pediat. Res. *16:* 114A (1982).
7 Skowsky, W.R.; Bashore, R.G.; Smith, F.G.; Fisher, D.A.: Vasopressin metabolism in the fetus and newborn; in Foetal and neonatal physiology, pp. 439–447 Cambridge University Press, London (1973).
8 Czernichow, P.: Vasopressin in fetal sheep: a review. J. Physiol., Paris *75:* 33 (1979).

9   Stegner, H.; Leake, R.D.; Palmer, S.M.; Fisher, D.A.: Permeability of the sheep placenta to ^{125}I arginine vasopressin. Dev. Pharmacol. Ther. *7:* 140 (1984).
10 Siegel, S.R.; Leake, R.D.; Weitzman, R.E.; Fisher, D.A.: Effects of furosemide and acute salt loading on vasopressin and renin secretion in the fetal lamb. Pediat. Res. *14:* 869 (1980).
11 Alexander, D.P.; Bashore, R.A.; Britton, H.G.; Forsling, M.L.: Maternal and fetal arginine vasopressin in the chronically catheterized sheep. Biol. Neonate *25:* 242 (1974).
12 Stegner, H.; Leake, R.D.; Palmer, S.M.; Morris, A.M.; Fisher, D.A.: Arginine vasopressin metabolic clearance and production rates in fetal sheep, lambs, maternal and nonpregnant adult sheep. Dev. Pharmacol. Ther. *7:* 87 (1984).
13 Weitzman, R.E.; Fisher, D.A.; Robillard, J.; Erenberg, A.; Kennedy, R.; Smith, F.G.: Arginine vasopressin response to an osmotic stimulus in the fetal sheep. Pediat. Res. *12:* 35 (1978).
14 Rurak, D.W.: Plasma vasopressin levels during hemorrhage in mature and immature fetal sheep. J. Dev. Physiol. *1:* 91 (1979).
15 Robillard, J.E.; Weitzman, R.E.; Fisher, D.A.; Smith, F.G.: The dynamics of vasopressin release and blood volume regulation during fetal hemorrhage in the lamb fetus. Pediat. Res. *13:* 606 (1979).
16 Drummond, W.H.; Rudolph, A.M.; Keill, L.C.; Gluckman, P.D.; MacDonald, A.A.; Heymann, M.A.: Arginine vasopressin and prolactin after hemorrhage in the fetal lamb. Am. J. Physiol. *238:* E214 (1980).
17 Alexander, D.P.; Forsling, M.L.; Martin, M.J.; Britton, H.G.: The effect of maternal hypoxia on fetal pituitary hormone release in the sheep. Biol. Neonate *21:* 219 (1972).
18 Stark, R.I.; Wardlow, S.L.; Daniel, S.S.; Husain, M.K.; Sanocka, V.M.; James, L.S.; Van de Weile, R.L.: Vasopressin secretion induced by hypoxia in sheep: developmental changes and relationship to B endorphin release. Am. J. Obstet. Gynec. *143:* 204 (1982).
19 Rurak, D.W.: Plasma vasopressin levels during hypoxemia and the cardiovascular effects of exogenous vasopressin in foetal and adult sheep. J. Physiol., Lond. *277:* 341 (1978).
20 Robillard, J.E.; Weitzman, R.E.; Burmeister, L.; Smith, F.G.: Developmental aspects of the renal response to hypoxia in the lamb fetus. Circulation Res. *48:* 128 (1981).
21 Stegner, H.; Artman, H.G.; Leake, R.D.; Oakes, G.; Fisher, D.A.: The effect of hypoxia on neurohypophyseal hormone release in fetal and maternal sheep. Pediat. Res. *18:* 188 (1984).
22 Leake, R.D.; Weitzman, R.E.; Effros, R.M.; Siegel, S.R.; Fisher, D.A.: Maternal fetal osmolar homeostasis: fetal posterior pituitary autonomy. Pediat. Res. *13:* 841 (1979).
23 Bell, R.J.; Congiu, M.; Hardy, K.J.; Wintour, E.M.: Gestation-dependent aspects of the response of the ovine fetus to the osmotic stress induced by maternal water deprivation. Q. Jl. exp. Physiol. (in press).
24 Wintour, E.M.; Congiu, M.; Hardy, K.J.; Hennessy, D.P.: Regulation of urine osmolality in fetal sheep. Q. Jl. exp. Physiol. *67:* 427 (1982).
25 Assali N.S.; Bekey, G.A.; Morrison, L.W.: Fetal and neonatal circulation; in Assali, Biology of gestation, vol. 2, chap. 2, pp. 48–62 (Academic Press, New York 1978).

26 Leake, R.D.; Stegner, H.; Palmer, S.M.; Oakes, G.K.; Fisher, D.A.: Arginine vasopressin and arginine vasotocin inhibit ovine fetal to maternal water transfer. Pediat. Res. *17:* 583 (1983).
27 Ross, M.G.; Ervin, M.G.; Leake, R.D.; Oakes, G.K.; Hobel, C.J.; Fisher, D.A.: Bulk flow of amniotic fluid water in response to maternal osmotic challenge. Am J. Obstet. Gynec. *147:* 697 (1983).
28 Ross, M.G.; Leake, R.D.; Ervin, M.G.; Fisher, D.A.: Ovine fetal lung liquid regulation by neuropeptides. Proc. 31st Annu. Meet. Society for Gynecologic Investigation, San Francisco 1984.
29 Kelley, R.T.; Rose, J.C.; Meis, P.J.; Hargrave, P.Y.; Morris, M.: Vasopressin is important for restoring cardiovascular homeostasis in fetal lambs subjected to hemorrhage. Am. J. Obstet. Gynec. *146:* 807 (1983).
30 Iwamoto, H.S.; Rudolph, A.M.; Keil, L.C.; Heymann, M.A.: Hemodynamic responses of the sheep fetus to vasopressin infusion. Circulation Res. *44:* 430 (1979).
31 Leake, R.D.; Weitzman, R.E.; Weinberg, J.A.; Fisher, D.A.: Control of vasopressin secretion in the newborn lamb. Pediat. Res. *13:* 257 (1979).
32 Schlondorff, D.; Weber, H.; Trizna, W.; Fine, L.G.: Vasopressin responsiveness of renal adenylate cyclase in newborn rats and rabbits. Am J. Physiol. *234:* F16 (1978).
33 Rajerison, R.; Butlen, D.; Jard, S.: Ontogenic development of antidiuretic hormone receptors in rat kidney: comparison of hormonal binding and adenylcyclase activity. Mol. cell. Endocrinol. *4:* 271 (1976).
34 Lindheimer, M.D.; Weston, P.V.: Effect of hypotonic expansion on sodium, water and urea excretion in late pregnancy: the influence of posture on these results. J. clin. Invest. *48:* 947 (1969).
35 Weitzman, R.E.; Leake, R.D.; Rubin, R.T.; Fisher, D.A.: The effect of nursing on neurohypophyseal hormones and prolactin in human subjects. J. clin. Endocr. Metab. *51:* 836 (1980).
36 Leake, R.D.; Weitzman, R.E.; Glatz, T.; Fisher, D.A.: Plasma oxytocin concentrations in men, nonpregnant women and pregnant women before and during spontaneous labor. J. clin. Endocr. Metab. *53:* 730 (1981).
37 Rosenbloom, A.A.; Sack, J.; Fisher, D.A.: The circulating vasopressinase of pregnancy: species comparison using radioimmunoassay. Am. J. Obstet. Gynec. *121:* 316 (1975).
38 Artman, H.G.; Leake, R.D.; Weitzman, R.E.; Sawyer, W.H.; Fisher, D.A.: Radioimmunoassay of arginine vasotocin, vasopressin and oxytocin in human neonatal cerebrospinal fluid and amniotic fluid. Dev. Pharmacol. Ther. *7:* 39–49 (1984).
39 Stark, R.I.; Wardlow, S.L.; Daniel, S.S.; Husain, M.K.; Sanocka, V.M.; James, L.S.; Van de Weile, R.: Vasopressin secretion induced by hypoxia in sheep. Am. J. Obstet. Gynec. *143:* 204 (1982).
40 Schrier, R.W.; Leaf, A.: Effect of hormones on water, sodium chloride and potassium metabolism; in Williams, Textbook of endocrinology; 6th ed, pp. 1032–1046 (Saunders, Philadelphia 1981).
41 Perlow, M.J.; Reppert, S.M.; Artman, H.A.: Oxytocin, vasopressin and estrogen-stimulated neurophysin: daily patterns of concentration in cerebrospinal fluid. Science *216:* 1416 (1982).
42 Reppert, S.M.; Artman, H.A.; Swaminathan, S.; Fisher, D.A.: Vasopressin exhibits a

rhythmic daily pattern in cerebrospinal fluid but not in blood. Science *213* 1256 (1981).
43 Godard, C.; Vallotton, M.B.; Favre, L.: Urinary prostaglandins, vasopressin and kallikrein excretion from birth to adolescence. J. Pediat. *100:* 898 (1982).
44 Tausch, A.; Stegner, H.; Leake, R.D.; Artman, H.G.; Fisher, D.A.: Radioimmunoassay of arginine vasopressin in urine: development and application. J. clin. Endocr. Metab. *57:* 777 (1983).
45 Hoppenstein, J.M.; Miltenberger, F.W.; Moran, W.H.: The increase in blood levels of vasopressin in infants during birth and surgical procedures. Surgery Gynec. Obstet. *123:* 966 (1968).
46 Hadeed, A.J.; Leake, R.D.; Weitzman, R.E.; Fisher, D.A.: Possible mechanisms of high blood levels of vasopressin during the neonatal period. J. Pediat. *94:* 805 (1979).
47 DeVane, G.W.; Porter, J.C.: An apparent stress-induced release of arginine vasopressin by human neonates. J. clin. Endocr. Metab. *51:* 1412 (1980).
48 Leung, A.K.C.; McArthur, R.G.; McMillan, D.D.: Circulating antidiuretic hormone during labor and in the newborn. Acta paediat. scand. *69:* 505 (1980)
49 DeVane, G.W.; Naden, R.P.; Porter, J.C.: Mechanism of arginine vasopressin release in the sheep fetus. Pediatrics, Springfield *16:* 504 (1982).
50 Fisher, D.A.: Cold diuresis in the newborn. Pediatrics, Springfield *40:* 636 (1967).
51 Svennigsen, M.W.; Aronson, A.S.: Postnatal development of renal concentrating capacity as estimated by DDAVP-test in normal and asphyxiated neonates. Biol. Neonate *25:* 230 (1974).

Rosemary D. Leake, MD, Department of Pediatrics, Harbor/UCLA Medical Center, 1000 West Carson Street, Torrance, CA 90509 (USA)

Electrophysiology of Vasopressin-Secreting Cells[1]

J.D. Vincent, P. Legendre, D. Poulain, E. Arnauld, D. Theodosis[2]

INSERM – U.176, Domaine de Carreire, Bordeaux, France

The vasopressin-secreting cell is one of the few neurons of the mammalian brain which can be characterized by a specific pattern of electrical activity. During the active release of its neurosecretory product, vasopressinergic neurons of the hypothalamic magnocellular system display a bursting pattern of firing. This has been called phasic activity and has been described in all the mammalian species so far studied [2, 15, 16, 25]. In this review we shall consider the functional meaning of phasic patterns both at the physiological and cellular level and then the mechanism by which such an activity is generated in the cell.

Electrophysiological Characteristics of the Phasic Pattern

In the absence of stimulation to evoke neurohypophysial hormone release, magnocellular neurons display firing activity which can be assigned to one of three categories (fig. 1). (1) *The slow irregular pattern* is characterized by a very low mean firing rate ≤2 spikes/s) with no regular distribution. (2) *The fast continuous pattern* shows a continuous discharge with a mean firing rate generally above 5 spikes/s and no regular

---

[1] This work was supported by INSERM (U.176) PRC 120052, CNRS ERA 493, Université de Bordeaux II and DGRST Grant 80.7.0106.

[2] We thank Dr. *J. Seal* for his help with the manuscript, *R. Bonhomme* and *B. Dupouy* for technical assistance, and *M.A. Delaage* for the sera against vasopressin. We are also grateful to *Viviane Robert* for typing the manuscript.

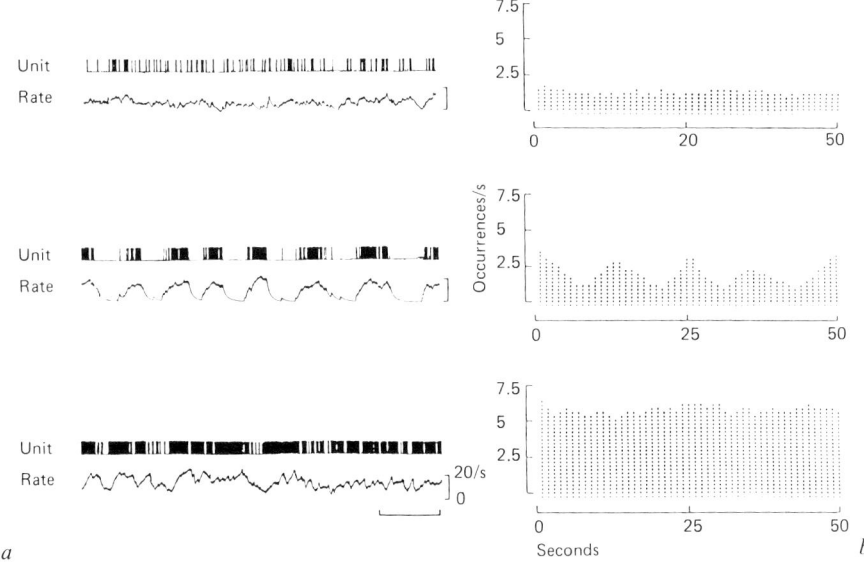

Fig. 1. The three types of firing pattern for supraoptic neurosecretory cells. *a* Unit represents the pulse output from a window discriminator, and Rate the analogue output proportional to the rate of unit discharge (time calibration, 10 s). *b* Expectation density correlograms [from ref. 2].

distribution. (3) *The phasic pattern* first described in the rat consists in successive periods of electrical activity and electrical silence occurring in a more or less regular manner. The analysis of periodicity by expectation density function shows a sinusoidal curve. This pattern has been recognized in a variety of mammalian species and presents similar characteristics in the anaesthetized rat (fig. 2a) and in the unanaesthetized monkey (fig. 2b).

The neuron being recorded can be identified as a magnocellular neuron by means of antidromic stimulation from the posterior pituitary. Nevertheless, this technique does distinguish the type of the hormone secreted by the neuron since it has been demonstrated by immunocytological techniques that vasopressinergic and oxytocinergic neurons coexist in the same nucleus (supraoptic and paraventricular). However, *Poulain* et. al. [21] have clearly demonstrated that the phasic pattern only appears in vasopressin neurons. In a lactating rat the oxytocin neurons can be identified by their specific response to suckling. Since on the one

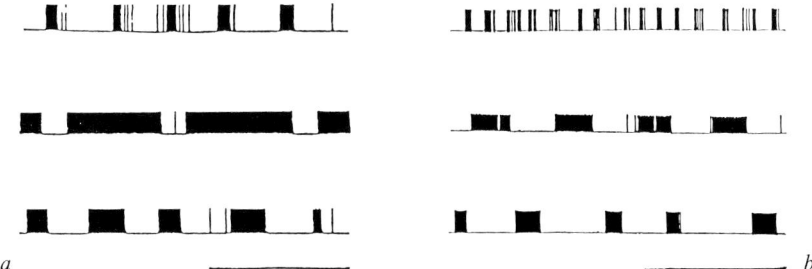

Fig. 2. Examples of phasic cells in the anaesthetized rat (*a*) and in the unanaesthetized monkey (*b*). Time calibration: 1 min.

hand the phasic pattern cannot be evoked in oxytocinergic neurons and on the other hand the phasically firing neurons do not respond to suckling, it is possible to conclude that the phasic pattern is specific of the vasopressin neurons.

The parameters of phasic neurons can vary considerably within the same neuron and from one neuron to another in the same animal (fig. 2). The range of burst and silence duration is rather wide: burst 4–100 s; silence 4–240 s. The neuron intraburst firing rates range from 3 to 15 spikes/s. From analysis of their response to stimulation it seems that vasopressin neurons display a phasic activity when the mean firing rate reaches a certain threshold (ca. 2–5 spikes/s). In contrast, the changes in phasic parameters related to increased stimulation are rather subtle and will be considered in the next section.

What Are the Functional Meanings of the Phasic Pattern?

After its initial description the significance of the phasic pattern has remained mysterious and phasic cells were considered as an electrophysiological curiosity. *Arnauld* et. al. [4] first showed that the percentage of cells with phasic activity in the total population of magnocellular neurons increased with the level of plasma osmolality in dehydrated monkeys; they concluded therefore that phasic firing might represent a state of activation of the magnocellular neurosecretory system. Other evidence has also led to the conclusion that the phasic pattern was correlated with an active release of vasopressin.

Functional Significance of the Phasic Pattern at the Physiological Level

The phasic pattern of neurosecretory cells is involved whenever vasopressin release occurs in response to physiological stimuli. If the phasic pattern is involved in vasopressin release, it will be interesting to study how vasopressin cells react in various physiological regulations.

Osmoregulation and Water Balance. Since the original work of *Verney* [24], which demonstrated the importance of osmotic stimulation in the regulation of vasopressin release, changes in osmolality have become the stimulus of choice for electrophysiologists studying the electrical responses of vasopressin neurons. In the unanaesthetized monkey, five days of water deprivation increased plasma osmotic pressure from 290 to 340 mosm. Under these conditions, the percentage of supraoptic neurosecretory cells displaying a phasic pattern rose from 10% on the first day up to 50–60% on the fifth day [2]. The proportion of neurosecretory cells firing phasically after dehydration corresponds well with the proportion (70%) of vasopressin neurons described immunohistochemically in the supraoptic (SO) nucleus of the same species [*Burlet,* personal communication] (fig. 3a). In the anaesthetized lactating rat, water deprivation caused a very fast rise in plasma osmotic pressure (from 290 to 340 mosm in 24 h). After 6 h or more of dehydration, 94% of the neurons which did not react to suckling displayed a phasic activity (fig. 3b). From statistical analysis of their activity during dehydration, it seems that vasopressin neurons display a phasic activity when the mean firing rate reaches a certain threshold. Most of the neurons with a mean firing rate under 1 spike/s showed a slow irregular pattern with no definite periodicity in firing. All vasopressin neurons firing above 2.5 spikes fired phasically. When the osmolality and the vasopressin release increased, the mean overall firing rate of the vasopressin neurons increased proportionally. This augmentation of the frequency corresponded to an increase of the burst duration and of the intraburst frequency [26]. Long-term and short-term electrophysiological responses are also observed during experimental procedures likely to inhibit the release of vasopressin. With progressive decrease of osmotic stimulation by rehydration after a period of dehydration in the monkey there was a reduction of the mean firing rates of neurosecretory cells, as well as a decrease in the proportion of units firing phasically.

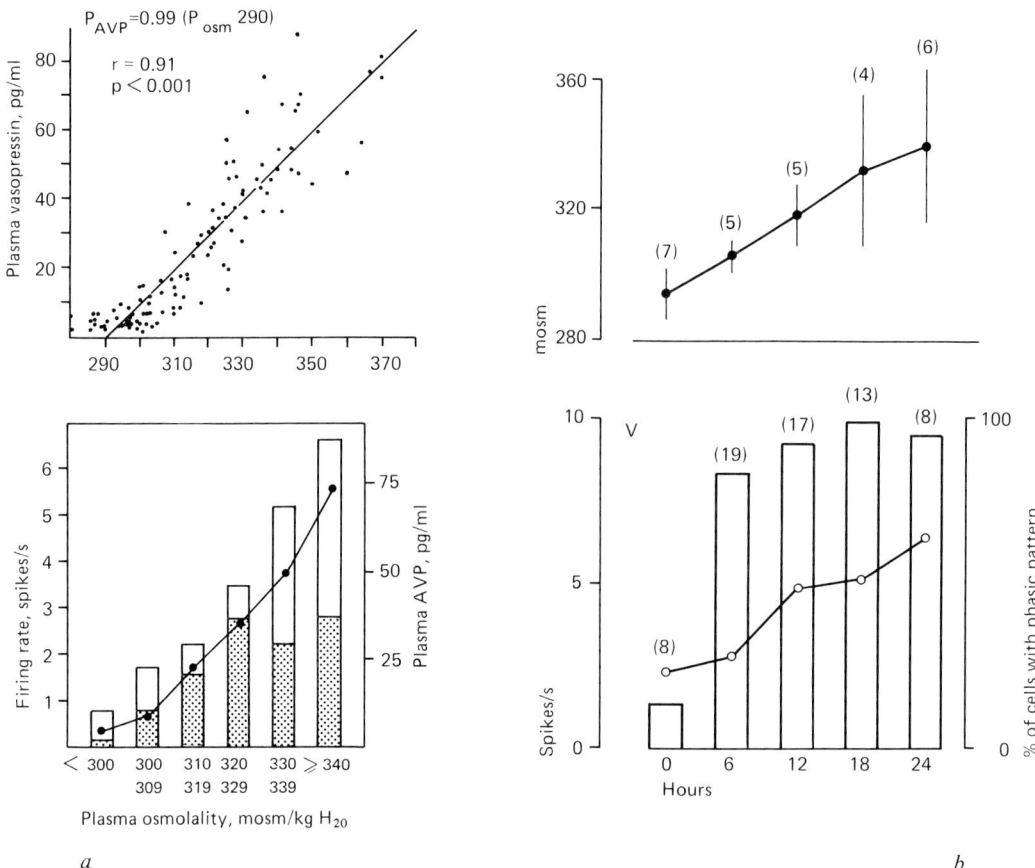

Fig. 3. Mean data observed during dehydration in the unanaesthetized monkey (*a*) and in the lactating rat (*b*). *a* Top: Values of plasma vasopressin plotted against value of plasma osmolality. Bottom: Mean firing rate of vasopressin neurons recorded for different values of plasma osmolality. The height of the block represents the mean firing rate of neurons; the relative proportion of phasic cells is represented by the stippled part of the blocks. *b* Effect of water deprivation on plasma osmolality (mosm) and on the electrical activity of vasopressin neurons in lactating rat. Top: Plasma osmolality (mean + standard deviation) in relation to the hours of water deprivation; the number of animal is given in parentheses. Bottom: Mean firing rates (circles) and proportion of cells displaying a phasic pattern [adapted from ref. 21].

An interesting reaction to drinking has been observed in the conscious monkey. In dehydrated animals, SO magnocellular neurons, including phasically firing neurons, showed a transient inhibition during the time of water intake [3]. This transient inhibition of phasic firing is correlated with a drop in vasopressin release. As plasma osmotic pressure was still very high, this favours the hypothesis of inhibitory influences of orogastric origin.

Cardiovascular Regulation. Vasopressin release is also affected by a number of cardiovascular reflexes [13]. Haemorrhage produced by slow withdrawal of blood from the right atrium causes a change in the frequency and the pattern of firing of vasopressin neurons. If before haemorrhage the neuron displays a slow irregular pattern, blood withdrawal leads to a decrease in blood pressure and concomitantly to an increase in the firing rate of the cell. For 10 min the neuron fires continuously and then gradually evolves a phasic pattern. On replacement of the blood, the rapid increase in blood pressure is accompanied by a total inhibition of firing (fig. 4). When the neuron is previously phasic, the haemorrhage induces an increase in burst duration and intraburst frequency.

Acute stimuli have permitted study of the role of various peripheral cardiovascular receptors in the control of the activity of magnocellular neurons. Carotid occlusion, which stimulates vasopressin release, precipitates the appearance of a burst of activity in phasically firing SO neurons. In non-phasic neurons, the effect of the stimulus seems negligible [8]. However, carotid occlusion can activate both baroreceptors and chemoreceptors. Stimulation of baroreceptors alone, in 'isolated carotid sinus' preparations, by increasing carotid sinus pressure with inflated balloons [27] or by increasing systemic blood pressure with i.v. injection of phenyl-ephrine [14] inhibited the activity of SO cells in the cat and the rat. In the rat, the inhibitory effect was particularly evident on phasically firing neurons whose bursts became shorter as the stimulus was applied. In contrast, chemoreceptor stimulation with solutions saturated with CO_2 caused excitation of magnocellular neurons. In phasically firing cells, the stimulus systematically precipitated a burst of action potentials, whereas local anaesthesia of the carotid bifurcation decreased their activity [14].

To conclude this section, one may consider that the phasic pattern of firing in vasopressin neurons is associated with the release of the neuro-

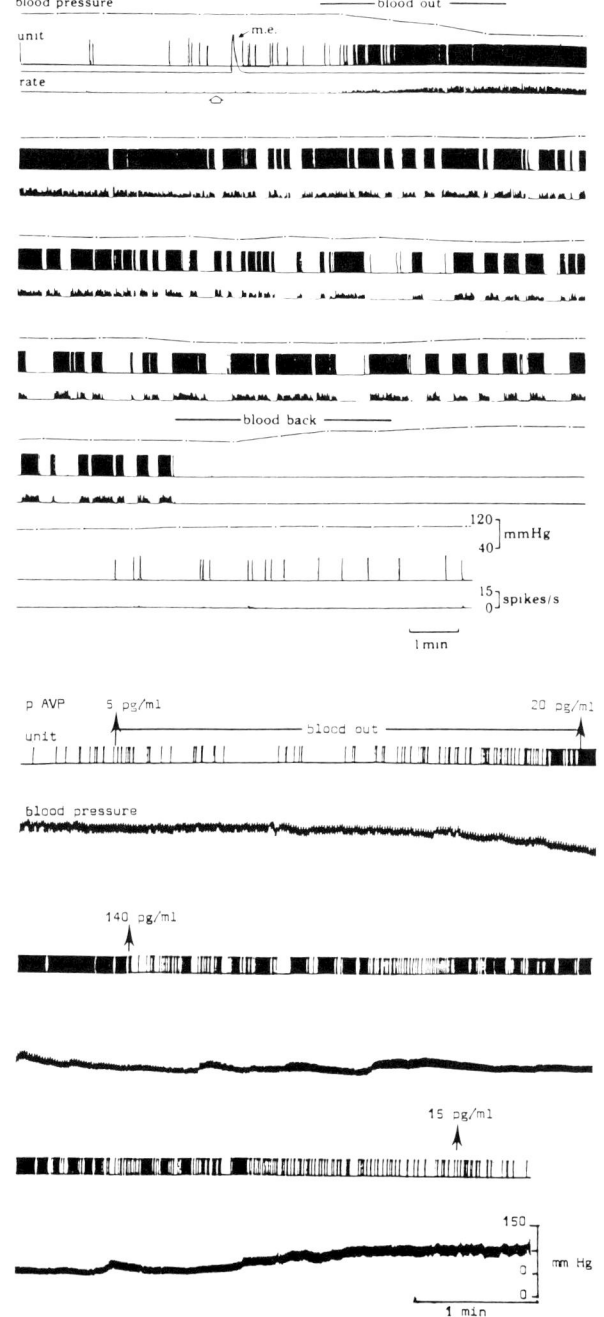

secretory product in response to various physiological stimulations. It is interesting now to consider the relevance of this pattern to the hormone release.

Functional Significance of Phasic Pattern at the Cellular Level

In isolated neural lobes stimulated electrically in a medium in which action potentials can propagate, the release of hormone is maximal at a frequency of about 35 Hz [9]. Less release is observed at higher frequencies. The most likely explanation is that the axons of the pituitary stalk fail to conduct action potential at high frequencies. It has been shown recently that the response of units in the neural lobe was abolished after a 10-second stimulation period at 50 Hz. This phasic organization of the firing in vivo may permit the conduction of high-frequency impulses along neurosecretory axons.

Recent experiments by *Dutton and Dyball* [10] have given some clue to the firing pattern observed in the hypothalamo-neurohypophysial tract. They used isolated neural lobes and compared the amount of hormone released by pulses given at a constant frequency with the output of hormones induced by a phasic pattern of stimuli identical to that of a vasopressin neuron. Although the mean firing frequencies were identical, the phasic pattern was more powerful in releasing vasopressin (fig. 5a). In other words, not only the frequency but also the pattern of firing is important for inducing hormone release.

To determine the mechanism by which phasic pattern enhances the release of hormone. *Nordmann* [19] has studied the Ca uptake into electrically stimulated neural lobes. The neurohypophyses were stimulated via the stalk using a stimulator triggered by a train of spikes previously recorded in vivo (fig. 5b). One burst of stimuli (mean firing rate, 13.3 Hz)

Fig. 4. a Patterns of electrical activity of one supraoptic vasopressin neuron recorded continuously during suckling and haemorrhage. Before haemorrhage (top), the neuron was displaying a slow irregular pattern, and was not activated (open arrow) by suckling prior to the reflex milk ejection (me). Blood withdrawal (5 ml) led to a decrease in blood pressure and concomitantly to an increase in the firing rate of the cell. For 10 min, the neuron fired continuously, then gradually evolved a phasic pattern. On replacement of the blood, the rapid increase in blood pressure was accompanied by a total inhibition of firing. A few minutes later, the cell again displayed its original slow irregular pattern [from ref. 20]. *b* Haemorrhage of 15% of the total blood volume in an unanaesthetized monkey. Values of the plasma vasopressin are indicated on the upper line.

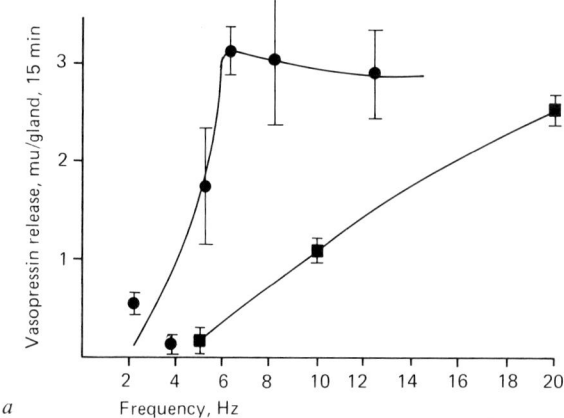

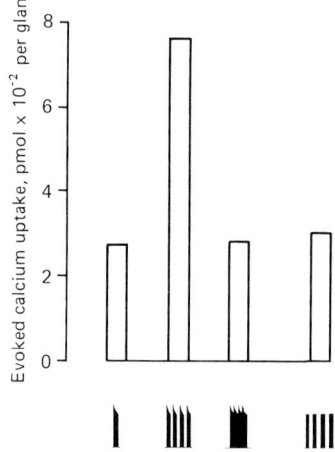

Fig. 5. a The effect of the frequency of pulses in vasopressin release. The closed circles represent the amount of hormone released when glands were stimulated with trains of spikes previously recorded in vivo from identified vasopressin cells. The squares represent the values obtained with a regular pattern [from ref. 10]. *b* Evoked Ca uptake induced by electrical stimulation of isolated neural lobes. The stimulator was triggered by train of spikes previously recorded in vivo. The bottom indicates the shape and the duration of the trains [from ref. 19].

gives rise to an evoked Ca uptake of 272 pmol/gland. Moreover, whereas four bursts with intervals of 21.1 s between each increase the evoked Ca uptake to a value of 751 pmol, the same number of bursts, but without silent periods, enhances Ca uptake to a value of 278 pmol. This is very similar to that observed after stimulation with a single burst. Also, four bursts of stimuli given at a constant frequencey of 13.3 Hz with 21.1-second intervals increased the Ca uptake by only 306 pmol.

These data show that the discharge pattern of the magnocellular neurons has some facilitatory effect on Ca entry and hence on hormone release. We do not know exactly at which step of the stimulus-secretion coupling mechanism facilitation occurs, but the data above suggests that the phasic pattern of discharge has some effects on the late Ca channel. Plausible explanations are that it increases either the number of channels activated at a given time or the time for a channel to inactivate.

In experiments involving electrical stimulation of the gland with a stimulation triggered by trains of spikes previously recorded in vivo, one only looks at the effect of the frequency of pulses of the same duration. This might not be the case in vivo. In invertebrate neurons there is a considerable increase in the duration of action potentials during a burst of activity. If this is also the case at the neurohypophysial terminals, one would expect an even greater effect of the pattern of discharge on Ca entry and hence on hormone release.

Mechanisms Controlling the Phasic Activity

The mechanisms by which the phasic pattern is generated has remained during the past decade a puzzle. Whether this periodicity is synaptically driven or is an intrinsic property of neurosecretory cells was unknown. The only method available for answering this question was to record intracellularly from vasopressin cells. Intracellular recordings of neurosecretory cells in mammals are rare [6, 7, 11, 12] and most have been performed on explants or slices (preparations whose complexity leaves unresolved the relative contribution of endogenous properties and neuronal interactions on pattern generation). In the hope of finding a less complex model more amenable to analysis of endogenous neuronal properties and their separation from the influence of neuronal interactions, we have examined cultures of dispersed tissue from fetal mouse hypothalamus. It quickly became evident that differentiation of the

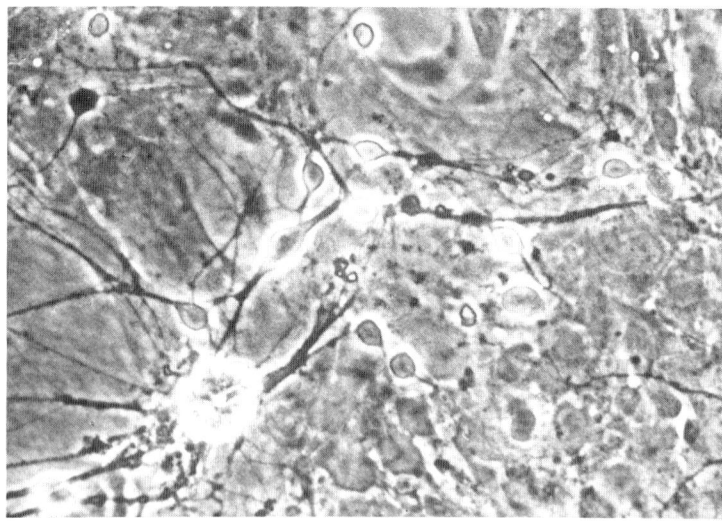

Fig. 6. Photomicrograph of neurons in dispersed cultures of fetal mouse hypothalamus: phase contrast of living neurons, 37 days in culture.

neurons required a minimum density of cells in the cultures. Differentiation paralleled the formation of contacts among the neuronal processes.

Our cultures were derived from 13- to 15-day-old mouse (IOPS/OFI) fetuses, by a procedure adapted from *Benda* et. al. [5]. Phase microscopy showed that after 4 weeks, the cultures consisted of a continuous basal layer of flat cells, on which birefringent cells were growing singly or in clusters (fig. 6). Intracellular recordings identified the overlying cells as neurons since they displayed overshooting action potentials and post-synaptic potentials (psps) similar to those recorded from other cultured neurons derived from various areas of the central nervous system. They exhibited resting potentials of –40 to –60 mV and input resistances of 50–200 MΩ.

Certain of the largest neurons exhibited, in addition to their usual neuronal activity, a particular pattern of long duration (30–90 s) depolarizing phenomena (plateau potentials) [18]. These plateau potentials arose in response to spontaneous or evoked depolarizing events. Plateau potentials recurred periodically. Following each plateau there was a refractory period during which another plateau could not be evoked by depolarizing current (fig. 7).

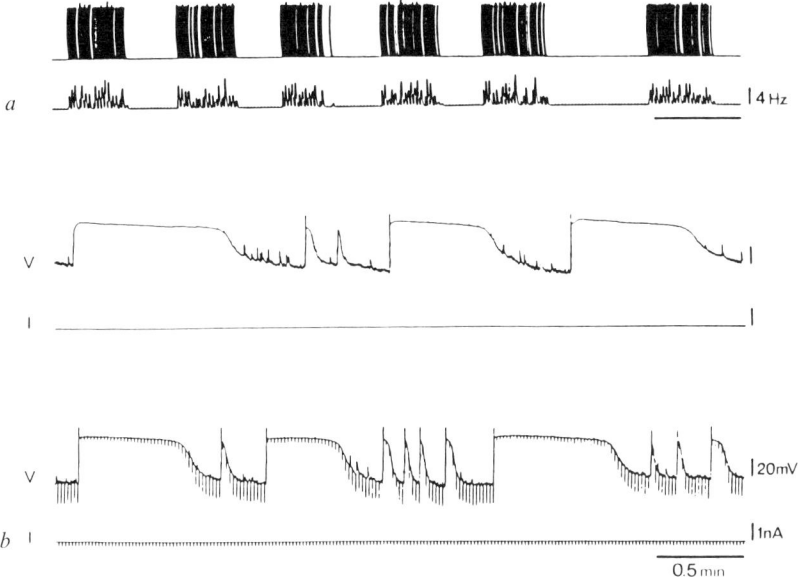

Fig. 7. a Extracellular recording from a vasopressin neuron in a normal rat. *b* Intracellular recording from a vasopressin neuron in vitro. Upper: Spontaneous activity showing plateau potentials and between them, slow potentials in response to synaptic bombardment. Lower: Spontaneous activity with hyperpolarizing pulses to show decreased input resistance (from 80 to 12 M) during plateau potentials and its recovery with repolarization. (Same time scale in *a* and *b*) [from ref. 23].

Plateau potentials are resistant to TTX (a blocker of Na channel) and can be suppressed by Cd^{2+} and Co^{2+}. They can be interrupted by a hyperpolarizing pulse applied during their ascending and fast repolarization phases. Simultaneous TEA (a blocker of K channel) application inhibits the effects of the hyperpolarizing pulses (fig. 8). TEA also increases the amplitude of the plateau depolarization, prolongs its duration and shortens the refractory period. By contrast, Ca^{2+} does not change the level of the plateau depolarization but increases its duration, prolongs the time course of fast repolarization and gives rises to superimposed action potentials on the plateau. We conclude that plateau potentials involve at least three conductances, one to Ca^{2+} and two to K^+, one of which is Ca^{2+}-dependent.

In order to study the morphological characteristics of the cells which exhibited plateau potentials, after recording, they were injected with lucifer yellow or horseradish peroxidase (HRP). Neurons which exhibit-

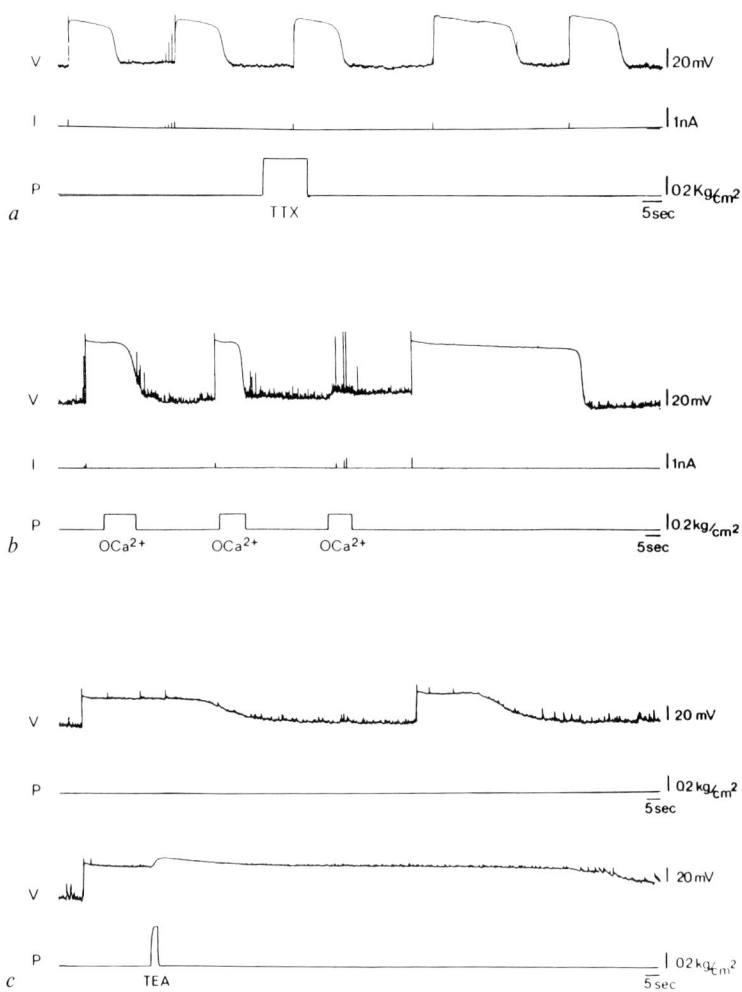

Fig. 8. Plateau potentials are resistant to TTX application (*a*) but suppressed in zero calcium medium (*b*). On the contrary TEA prolongs the duration of the plateau and increases the level of depolarization (*c*).

ed plateau potentials were distinguishable from others by their size and their morphology (fig. 9). Electron microscopy of these neurons, after injection with HRP, showed that they possess characteristics of neurosecretory cells. By immunocytochemical means, at the light and electron microscopic levels, it was possible to see that such cells contain vaso-

Electrophysiology of Vasopressin-Secreting Cells

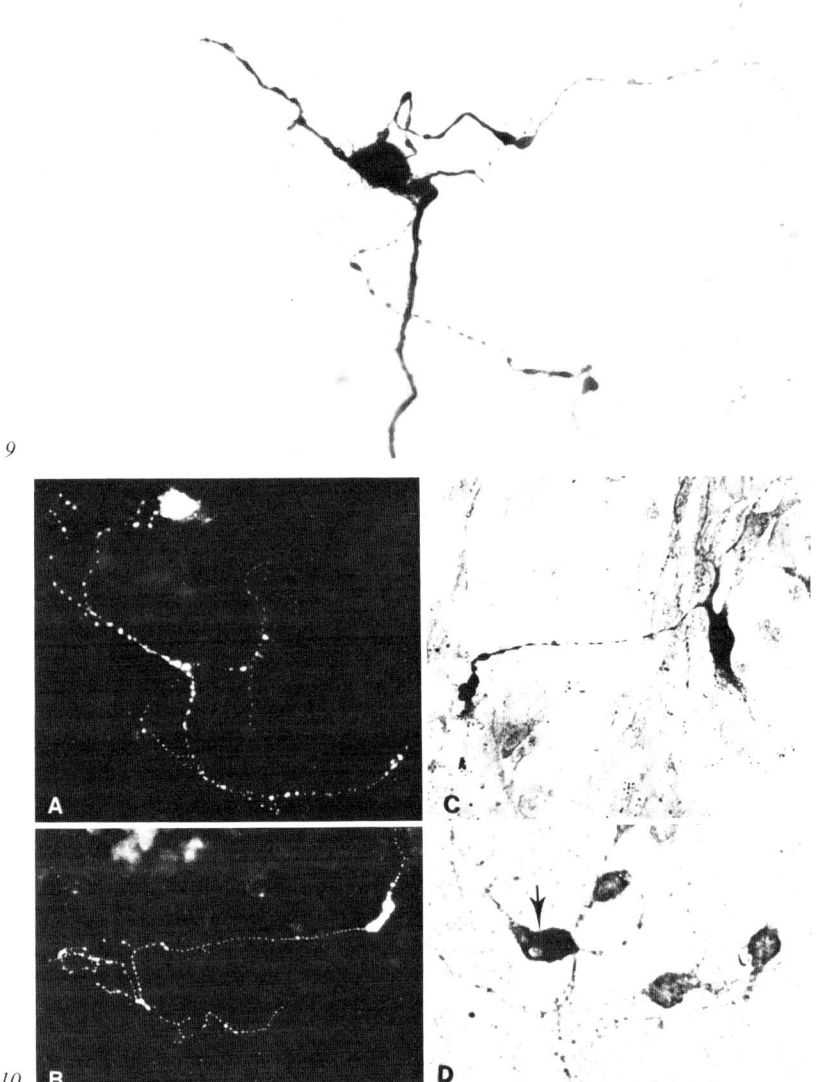

Fig. 9. Scale neuron injected with horseradish peroxydase.

Fig. 10. Cultures treated with rabbit sera against vasopressin. The immunoprecipitate was made evident by fluorescein in *A* and *B* and by peroxidase in *C* and *D*. It was localized in axonal-like processes and their dilatations and in cell bodies. The stained neuron in *D* (arrow) had been recorded intracellularly before immunohistochemical staining and had displayed plateau potential electrical activity.

Table I

Age of culture days	AVP fmol/10⁶ cells	Precursor pmol/10⁶ cells
2	0.45	437
8	3.6	449
14	10.7	450
24	26.0	447
28	30.7	445
40	68	449
48	111	443

pressin-like material [23]. We thus conclude that the periodic, calcium-dependent plateau potentials observed in such cells may represent mechanisms, perhaps in the process of maturation, underlying the phasic activity described in vasopressin neurons in vivo (fig. 10).

In order to determine whether these cells are functional, radioimmunoassay was performed on cultures of various ages. Both vasopressin and its precursor were detected in the cells from the second day of culture. The amount of vasopressin was low after a few days and then began to increase exponentially after 18 days. Our results indicate that the maturation of vasopressin secretion can take place under in vivo conditions and parallels the morphological and electrophysiological differentiation of the cultured neurons (table I).

Recent data has recently confirmed that slow depolarizing events which are calcium-dependent also exist in vasopressin cells recorded from slice preparations [1]. Finally, it is possible to conclude that the vasopressin cell is characterized by a phasic electrophysiological activity, the periodicity of which is endogenously driven. The physiological factors which control this endogenous property of the vasopressin neuron remains to be determined.

References

1 Andrew, R.D.; Dudek, F.E.: Burst discharge in mammalian neuroendocrine cells involves an intrinsic regenerative mechanism. Science *221:* 1050–1052 (1982).
2 Arnauld, E.; Dufy, B.; Vincent, J.D.: Hypothalamic supraoptic neurons: rates and pattern of action potentials firing during water deprivation in the unanaesthetized monkey. Brain Res. *100:* 315–325 (1975).

3   Arnauld, E.; Du Pont, J.: Vasopressin release and firing of supraoptic neurosecretory neurones during drinking in the dehydrated monkey. Pflügers Arch. *394:* 199–201 (1982).
4   Arnauld, E.; Vincent, J.D.; Dreifuss, J.J.: Firing patterns of hypothalamic supraoptic neuron during water deprivation in monkeys. Science *250:* 535–537 (1974).
5   Benda, P.; De Vitry, F.; Picart, R.; Tixier-Vidal, A.: Dissociated cells cultures from fetal mouse hypothalamic: patterns of organization and ultrastructural features. Exp. Brain Res. *23:* 29–47 (1975).
6   Canick, J.A.; Vaccaro, D.E.; Ryan, K.J.; Leeman, S.E.: The aromatization of androgens by primary monolayer cultures of fetal rat hypothalamus. Endocrinology *100:* 250–253 (1977).
7   Cooke, I.M.: Electrical activity of neurosecretory terminals and control of peptide hormone release; in Gainer, Peptides in neurobiology, pp. 345–374 (Plenum Publishing, New York 1977).
8   Dreifuss, J.J.; Harris, M.C.; Tribollet, E.: Excitation of physically firing hypothalamic supraoptic neurones by carotid occlusion in rats. J. Physiol., Lond. *257:* 337–354 (1976).
9   Dreifuss, J.J.; Kalmins, I.; Kelly, J.S.; Ruf, K.B.: Action potentials and release of neurohypophysial hormones in vitro. J. Physiol., Lond. *215:* 805–817 (1971).
10 Dutton, A.; Dyball, R.E.J.: Phasic firing enhances vasopressin release from the rat neurohypophysis. J. Physiol., Lond. *290:* 433–440 (1979).
11 Gähwiller, B.H.; Dreifuss, J.J.: Hypothalamic neurones in culture. II. A progress report. J. Physiol., Paris *75:* 23–26 (1979).
12 Gähwiller, B.H.; Sandoz, P.; Dreifuss, J.J.: Neurones with synchronous bursting discharges in organ cultures of the hypothalamic supraoptic nucleus area. Brain Res. *151:* 245–253 (1978).
13 Harris, M.C.: The concept of the neuroendocrine reflex; in Vincent, Kordon, Cell biology of hypothalamic neurosecretion, pp. 47–61 (CNRS, Paris 1978).
14 Harris, M.C.: Effects of chemoreceptor and baroreceptor stimulation on the discharge of hypothalamic supraoptic neurones in rats. J. Endocr. *82:* 115–125 (1979).
15 Haskins, J.T.; Jennings, D.P.; Rogers, J.M.: Response of neuroendocrine cell firing pattern types to measured changes in plasma osmolality. J. Physiol., Lond. *18:* 240 (1975).
16 Hayward, J.N.: Functional and morphological aspects of hypothalamic neurones. Physiol. Rev. *57:* 574–658 (1977).
17 Hayward, J.N.; Jennings, D.P.: Activity of magnocellular neuroendocricells in the hypothalamus of unanaesthetized monkeys. I. Functional cell types and their anatomical distribution in the supraoptic nucleus and the internuclear zone. J. Physiol., Lond. *232:* 515–543 (1973).
18 Legendre, P.; Cooke, I.M.; Vincent, J.D.: Regenerative responses of long duration recorded intracellularly from dispersed cell cultures of fetal mouse hypothalamus. J. Neurophysiol. *48:* 1121–1141 (1982).
19 Nordmann, J.J.: Stimulus-secretion coupling; in Cross, Leng, Structure, function and control. Progress in brain research, pp. 281–305 (Elsevier, Amsterdam 1983).
20 Poulain, D.A.; Wakerley, J.B.: Electrophysiology of hypothalamic magnocellular neurones secreting oxytocin and vasopressin. Neuroscience *7:* 773–808 (1982).

21 Poulain, D.A.; Wakerley, J.B.; Dyball, R.E.J.: Electrophysiological differentiation of oxytocin- and vasopressin-secreting neurones. Proc. R. Soc. B *196:* 367–384 (1977).
22 Sakai, K.K.; Marks, B.; George, J.; Koestner, A.: The isolated organ-cultured supraoptic nucleus as a neuropharmacological test system. J. Pharmac. exp. Ther. *190:* 482–491 (1974).
23 Theodosis, D.T.; Legendre, P.; Vincent, J.D.; Cooke, I.: Immunocytochemically identified vasopressin neurons in culture show slow, calcium-dependent electrical responses. Science *221:* 1052–1054 (1983).
24 Verney, E.B.: The antidiuretic hormone and the factors which determine its release. Proc. R. Soc. B *135:* 25–106 (1947).
25 Wakerley, J.B.; Lincoln, D.W.: Phasic discharge of antidromically identified units in the paraventricular nucleus of the hypothalamus. Brain Res. *25:* 192–194 (1971).
26 Wakerley, J.B.; Poulain, D.A.; Brown, D.: Comparison of firing patterns in oxytocin- and vasopressin-releasing neurones during progressive dehydration. Brain Res. *148:* 425–440 (1978).
27 Yamashita, H.: Effect of baro- and chemoreceptor activation on supraoptic nuclei neurons in the hypothalamus. Brain Res. *126:* 551–556 (1977).

Dr. Jean-Didier Vincent, INSERM – U.176, Domaine de Carreire, Rue Camille Saint-Saëns, F-33077 Bordeaux Cedex (France)

Thirst Mechanisms and Antidiuretic Hormone[1]

Stylianos Nicolaïdis

Laboratoire de Neurobiologie des Régulations, CNRS ER 218, Collège de France, Paris (France)

The physiologic stimulation of thirst, like that of secretion of antidiuretic hormone (ADH), is initiated by deficits of two spaces: the intracellular space and the extracellular space. The dipsogenic effect which follows dehydration of either of these two spaces is additive to the dipsogenic effect of dehydration of the other space [1]. This additive effect indicates two parallel systems of detection of deficiencies, with each system having its own receptors which project to specific areas of the central nervous system. Neurons within these areas in the central nervous system receive information on the state of hydration of the spaces. The main component to provide information on hydration is neural, but several humoral factors have also been shown to play a role in drinking by either enhancing or inhibiting thirst. Several questions regarding the role of intracellular and extracellular dehydration on thirst have been investigated: what is the nature of the ultimate stimulus generated by the state of hydration; where is the stimulus applied; how are various stimuli integrated to produce the final expression of thirst; and, how can thirst be extinguished before swallowed water is able to leave the gastrointestinal tract and reach the internal milieu? We will review some of the data available to answer these questions although some of the answers are controversial and some of the data incomplete.

Before examining the role of humoral factors (particularly ADH) on drinking, the mechanisms of thirst generated by intracellular and extracellular dehydration will be summarized.

---

[1] This work was carried out with the contribution of a grant from Volvic Co. and Sandoz S.A.

Physiological Stimuli of Drinking

Intracellular Dehydration

It is believed that thirst does not reflect the increase in osmotic pressure per se but rather the degree of cellular dehydration [2, 3], because of investigations which compared the dipsogenic effect of various salts with an equal osmolar administration of urea which was a much less effective dipsogen [4, 5]. Since urea, unlike sodium, has access to the intracellular space, this agent will cause no cellular dehydration in spite of similar osmolar concentration. Cell-membrane shrinkage is proportional to the 'effective' osmotic pressure difference between the intracellular and the extracellular fluids and no difference is developed by urea which crosses the cell membrane. This concept of effective osmoreceptor was originally put forth by *Verney* [6] and by *Wolf* [7]. The temporary dipsogenic effect of urea is accounted for by the time required for urea to cross the blood-brain barrier thus inducing a transient neuron shrinkage [8].

In addition to hypernatremia, intracellular dehydration may be the consequence of excessive losses of potassium, the ion which is predominately responsible for the intracellular tonicity. In experimental manipulations leading to potassium deficiency as well as in human pathology such as primary hyperaldosteronism abnormally increased drinking is observed [9, 10]. This drinking occurs in spite of overhydration of the extracellular space (which has occurred at the expense of loss of intracellular water) [11]. In thirst due to intracellular dehydration, the thirst is not quenched by drinking pure water because water is unable to reach the intracellular space against the osmolar gradient.

Sodium-Sensitivity

Andersson, who originally demonstrated that there were receptors within the central nervous system which recognized cellular hydration and which led to increased water intake [12], subsequently presented the alternative idea that juxtaventricular sodium-sensitive receptors might be responsible for drinking [13]. Later he wrote: 'The role of CSF sodium as a dominant determinant of receptor activation was intentionally overemphasized in order to draw attention to the CSF as an indirect route by which changes in blood plasma composition might influence the cerebral mechanism regulating ADH secretion and water intake' [14]. This concept was based on experiments which showed that central administration of sodium was the most powerful agent to stimulate drinking and secretion of ADH and that the action of sodium could be inhibited by Na^+/K^+ ATPase inhibitors [14, 15]. However, the results of various ions administered intraventricularly may give contradictory effects on drinking depending upon the species used. For example, increase in tonicity of the CSF caused by administered sucrose is as potent a stimulus of drinking as equiosmolar NaCl in the dog [16], not as effective in goats and sheep [17, 18], and relatively ineffective in the pigeon [19]. The pigeon shows other peculiarities with regard to sodium sensitivity, since unlike the goat [14], but like the rat [20], drinking is stimulated by intraventricularly administered sodium and angiotensin II (see below) in a simple cumulative way without any potentiation of the dipsogenic effect of one by the other [21]. The mechanism by which extracellular sodium stimulates thirst will be difficult to determine since each time sodium increases outside membranes movement of water from intracellular to extracellular space is always accompanied by electrophysiologic changes due to the differences in sodium concentration. The early studies of *Oomura* et al. [22] showed an amazingly short latency of electrophysiologic responses obtained from iontophoretic application of sodium on the surface of osmosensitive hypothalamic units. The near absence of latency supported the idea that thirst was stimulated by an electrochemical membrane-phenomenon rather than due to osmotically induced transport of water from the intracellular to the extracellular space and the consequent membrane shrinkage.

Although it has been suggested that sodium and angiotensin II both act on the Na^+/K^+ ATPase to induce thirst, some electrophysiological studies do not support this mechanism. *Thornton, Ishibashi and Nicolaïdis* [unpublished] applied sodium and angiotensin II simultaneously from 2 iontophoretic micropipettes to 180 separate neurons along the anterior wall of the 3rd ventricle. No potentiation of the effect of the two agents was ever encountered.

Extracellular Dehydration
Dehydration of the extracellular space combines with dehydration of the intracellular space in an additive way to induce the final common response of drinking [1]. The algebraic relationship of this addition is demonstrated by the observations that overhydration of the intracellular space abolishes the thirst induced by dehydration of the extracellular space. It has also been observed that when a volume deficit is due to a

combined loss of water and sodium (as in sweating) but the drinking is pure water, the drinking stops prematurely. This has been designated as 'voluntary dehydration' by *Adolph* et al. [23]. Alternatively, if isotonic saline were drunk thirst lasted until complete volume repletion. In animal studies hypovolemic thirst is accompanied by sodium appetite. When animals with dehydration of the extracellular space are offered a choice, they prefer saline to water as a drinking solution. The physiology of sodium appetite has been recently reviewed by *Epstein* [24] and by *Fitzsimons* [25]. Losses of sodium from urine, feces, sweating, etc., which may be aggravated by pathologic loss of fluid as due to bleeding, diarrhea, vomiting, or nonswallowed salivation would lead to fluid loss which was isotonic or hypertonic and the dehydration would be only extracellular. A system which detected only intracellular dehydration could insure perfect isotonicity of the intracellular fluid while allowing isotonic hypovolemia. But, experimental manipulations including intraperitoneal or extracorporeal dialysis, sodium deprivation or diuresis will lead to extracellular dehydration and thirst due to stimulation via the extracellular volume receptors [26].

Receptors

Although the state of hydration is effectively sensed by receptors located in the intracellular and extracellular spaces, transitional events such as drinking which lead to repletion of the lost fluid are also detected by external oro-gastrointestinal receptors.

Receptors of Intracellular Dehydration and/or of Sodium
Receptors for thirst induced by each dehydration of the intracellular space are independent of the osmoreceptors described by *Verney* [6] for control of release of vasopressin. Receptors for thirst probably respond to their own dehydration and are located close to the vasopressin producing areas in the diencephalon. The localization of the receptors was established by *Andersson* et al. [27, 89] who produced polydipsia in volume-replete goats when minute amounts of salt were injected into various areas of the hypothalamus. The areas were distributed medially in the vicinity of the 3rd ventricle between the decending mammilothalamic tracts and the fornical column. In rats and rabbits the distribution extended more rostrally [28, 29], and in studies using electrophysiologic

response of cells after intracarotid injection of saline the specific receptive cells extend into the preoptic area [30–33]. Other studies using electrophysiologic response and micropipette-iontophoretic techniques have confirmed lateral hypothalamic localization of receptor cells which function as osmometer-like units [34]. In the rat and in the goat, electrical stimulation of these areas of the brain trigger drinking behavior [35, 36]. However this behavioral response to electrical stimulation as well as hypodipsia or adipsia in response to ablation of these areas must be interpreted with caution as definitively localizing the receptor cells because activation or inhibition of any part of the circuitry involved in drinking behavior might produce similar results [37].

The sodium receptor elements are thought to be primarily within the walls of the 3rd ventricle [38] and particularly in the anterior and ventral portion of the 3rd ventricle [17]. In studies of behavior in animals these receptors are potentiated by angiotensin II. Several properties of the sodium-responsive receptors do not support a simple sodium-sensitive mechanism [18, 19, 39], but rather suggest that both sodium and effective osmotic pressure may trigger the receptors of intracellular dehydration. A recent review by *Andersson* [40] dealt extensively with the alternative (or synergistic) possibilities of sodium vs osmotic pressure effects on the receptors. In experimental studies vasopressin has been inhibited and thirst attenuated by slow infusions into the 3rd ventricle of isotonic or even hypertonic nonelectrolyte solutions [41, 42]. However, in these studies such parameters as final sodium concentration in the CSF and intraventricular hydraulic pressure which might have increased because of addition of saccharides to the ventricular space were not controlled. The state of intraventricular turgescence has been suspected to affect periventricular receptors and consequently thirst [43]. Receptors of intracellular dehydration have also been described in more extensive areas of the brain and may exist in more caudal cerebral structures [45]. They have also been described in areas outside the brain as in the portal circulation [43, 44].

Receptors of Extracellular Dehydration

Receptors for extracellular dehydration are frequently identified with receptors for hypovolemia since fluctuations of extracellular space should be reflected by parallel changes in blood volume. Investigations devoted to how extracellular volume changes are detected have been summarized in a review by *Smith* [46]. The Starlings model that the intravascular and extravascular compartments of the extracellular space are in equilibrium would indicate that if volume receptors sense one of the extracellular compartments, receptors in the other extracellular compartment would be redundant and unnecessary. However, not all of the working hypotheses for extracellular volume receptors are oriented to the intra- or peri-vascular location for receptors [46].

Two sets of receptors satisfy a number of criteria which one would expect from extracellular volume receptors: those discovered in the periphery in the atrial walls [47], and those discovered in the brain [48]. However, only the latter seem to play an important role in the initiation of the drinking response. The properties of the vascular receptors have been described by *Henry* et al. [49] for receptors in low-pressure thoracic blood vessels and in the atrium and also have been investigated by the same authors during shock-induced vascular collapse [47]. The properties of these vascular volume receptors and their vagal afferents are well known in relation to ADH secretion but not well investigated in relation to thirst. The only area of the brain sensitive to hypovolemia was found to be located within the paraventricular structures near the anterior portion of the ventral portion of the 3rd ventricle, including the suprachiasmatic area and the tissue around the organum vasculosum of the lamina terminalis [48] (fig. 1). In this area the response to volume as determined by electrophysiology was restricted to the same area in which lesions could be expected to abolish the drinking response to experimental challenges [51, 52]. Additionally, the anteroventral area was the area most sensitive to locally applied angiotensin II [53].

Regulation of the extracellular space might involve more than just volume receptors in the vascular system because there are pathologic situations in which there is dissociation between the hydration of the vascular and extravascular spaces, such as edema with hypovolemia or isolated hypervolemia. For this reason it has been suggested that the vascular and the interstitial spaces may possess independent, although complimentary, receptors [43, 54].

Exteroceptive (Oral and Gastrointestinal) Receptors and Intravenous Drinking

External receptors such as dry mouth have been studied to understand mechanisms controlling drinking. Dry mouth is neither necessary nor sufficient for the onset and offset of drinking, but since the degree of dryness of the mouth correlates well with the state of hydration of the in-

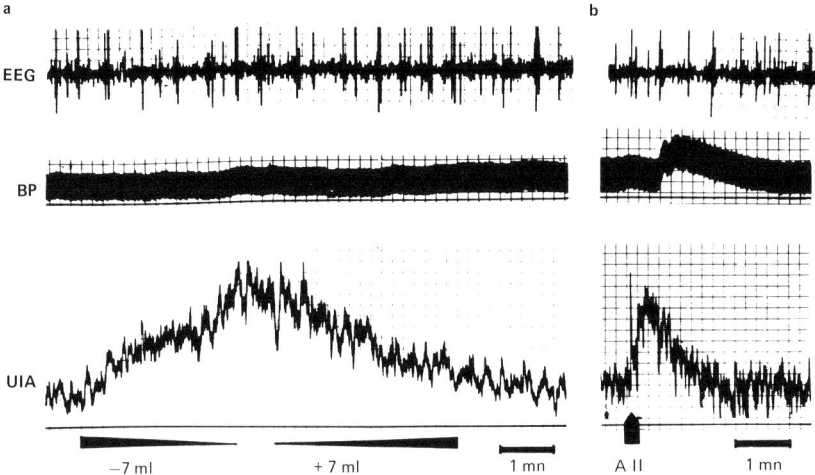

Fig. 1. a Single unit integrated activity (UIA) from a neuron in the suprachiasmatic commissure population caudal to laminae terminalis, during withdrawal from the right atrium of 7 ml of blood followed by its restitution in the cat. The level of blood pressure (BP) cannot account for the volume-dependent increased firing rate. UIA changes are not due to non-specific arousal as shown by the stable EEG. *b* The same unit during intra-carotid injection of angiotensin II (A II). A dramatic short-term response preceded the hypertensive response. The occurrence of hypertension always decreases the background activity of these units. From [106].

ner milieu, this may correlate well with thirst. Sensory information from the oral area may contribute to eliciting and to amplifying the drinking response. External receptors may be a learned response which can be extinguished. For example, when patients suffering from Parkinson's disease receive an anticholinergic agent as treatment their mouth is constantly dry and they complain of excessive drinking. However, with continued treatment normal fluid consumption returns in spite of the persistence of the iatrogenic dryness of the mouth [personal obervation]. Exteroceptive signals may be responsible for a number of drinking responses. They play an important role in the selection of the fluid to be consumed and in adjustment of the amount needed to cover the actual deficit prior to the time the fluid is absorbed. This anticipatory adaptation prior to absorption may be learned but is also innate since at least for some responses preabsorptive adaptation persists even in the anaesthetized rat [32, 48]. The act of drinking water or the gastric instillation of water immediately triggers diuresis while drinking salty solutions results

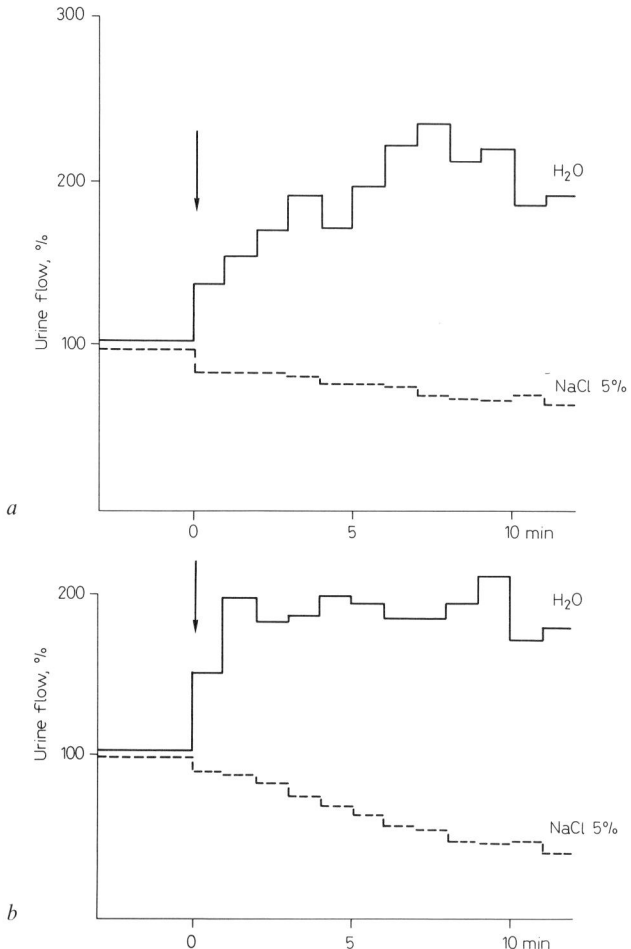

Fig. 2 Effect of administration orally 1–3 ml *(a)* or intragastrically 1 ml *(b)* of tap water (solid line) or 5% NaCl solution (dashed line) on urine flow in hyperhydrated anesthetized rats. Data plotted as mean percent of baseline (100%) urine flow over one minute units. These changes were not blocked by spinal cord transection and have been attributed to ADH reflex-elicited changes. From [48].

in antidiuresis (fig. 2, 3). Orally administered fluid produces a much more appropriate regulation of the deficit than does the same substance administered parenterally [55, 56]. The potodiuretic and potoantidiuretic reflexes use oral and vagal afferents, and probably neural efferent fibers to the kidneys, at least during the early (less than 40 s) changes in

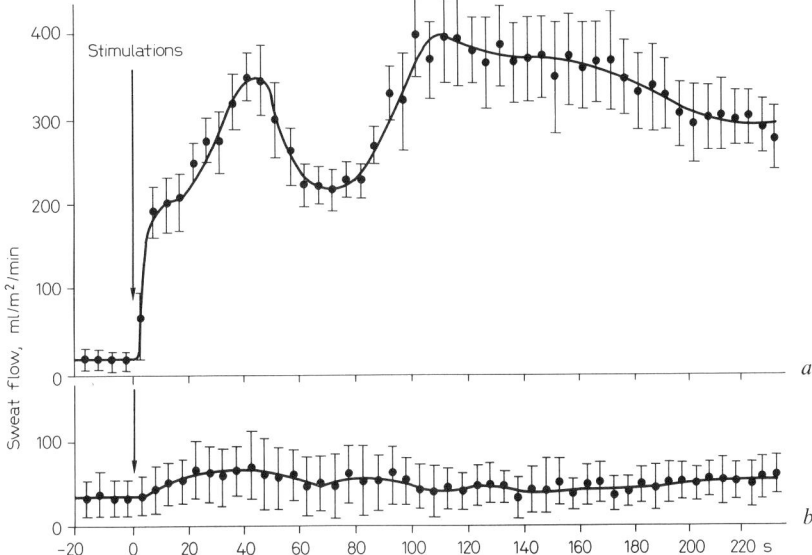

Fig. 3. Effect of drinking water (250 ml at time 0) on sweat flow measured hygrophotographically on the forehead of dehydrated *(a)* and rehydrated *(b)* human subjects. Data plotted as mean ±SE. From [107].

urine flow. Changes in release of ADH occur later as demonstrated recently by *Thrasher* et al. [16]. The role of sensory input in preabsorptive satiation is well documented. Satiety from water ingestion is precisely matched to the water deficit in a number of experimental situations. In rapid drinkers such as humans and dogs, by the time drinking ends almost none of the ingested fluid has reached the internal milieu, so the internal signal to generate thirst is unchanged but the sensation of thirst has been inhibited. The essential information to inhibit the sensation of thirst comes from the oropharynx since satiation is still observed in animals where all absorption is bypassed [58–60]. Other areas of the gastrointestinal tract also signal the central nervous system because short-circuiting the oropharyngeal cavity is still compatible with reasonably accurate adjustment of the ingestion of water to satisfy the deficit [61–63]. However, there is a major difficulty in intake adjustment when oropharyngeal contributions are totally eliminated [64]. Therefore, the neurologic circuitry which leads to early behavioral adaptation of water intake must function through pathways originating mainly in the walls of the oral cavity and to some extent in lower gastric and intestinal levels.

The relative importance of the external (orogastrointestinal) vs systemic receptors can be studied when external receptors are short-circuited by indwelling catheters [65]. In such studies continuous or discontinuous infusions of fluid were administered intravenously and intragastrically. Both infusions reduced water ingestion but the magnitude of the reduction was greater for water given intragastrically than water given intravenously, probably because of the participation of gastrointestinal receptors [66]. In another study external receptors were bypassed completely in the rat by implanting intracardiac catheters connected continuously to a remote injection system. The rat had to press a lever in order to inject an aliquot of fluid into its blood stream. In rats in which this was their sole source of water, chronic dehydration was present but the self-injected volume of fluid was just sufficient to insure indefinite survival. The rat would adjust the rate of lever pressing if the amount delivered per bar press were decreased or during higher ambient temperatures, indicating some regulation. However, there was no increased self-injection in response to experimental challenges such as systemically administered hypertonic saline, hyperoncotic colloid, or angiotensin II.

Integrative Centers

Stereotaxic lesions which affect ingestive behavior have been interpreted as localizing the 'center' of drinking, but such observations only indicate that the lesioned area has something to do with the circuitry implicated in drinking including the motor control of the behavioral execution of water intake. Nonetheless, the areas responsible for drinking as determined by lesions and by techniques of electrical stimulation superimpose rather well on the osmosensitive areas described by *Andersson* [12]. Large lesions of the lateral hypothalamus result in adipsia with aphagia while more selective lesions of the paraformical lateral hypothalamus bring about almost exclusively a deficit of drinking [37, 67]. *Grossman* applied neurochemicals to specific areas of the osmosensitive area and determined that cholinergic agents would cause drinking and adrenergic agents eating [68, 69]. Degeneration of the nigro-striatal system with local application of 6-hydroxy-dopamine was also followed by adipsia and aphagia [70].

The most direct evidence for localization of integrative neurons for drinking are studies using electrophysiologic determinants of firing rates of neurons within the lateral hypothalamic area. In the cat, the neurons

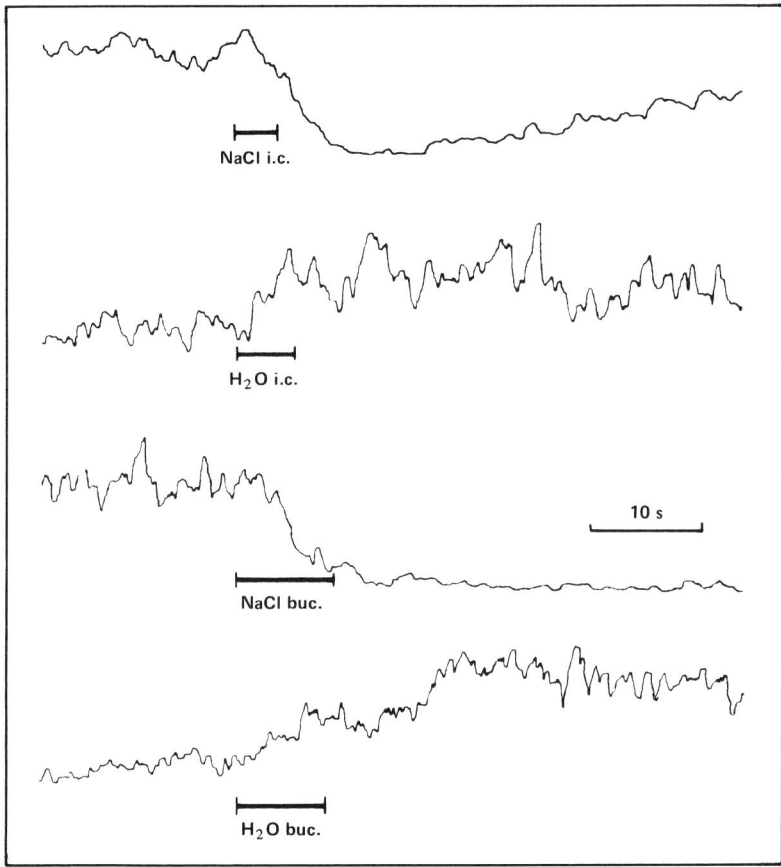

Fig. 4. Integrated unit activity measured at the post-commissural level of the medial forebrain bundle. The same inhibitory effect was obtained after both intracarotid (i.c.) and buccal (buc.) NaCl administration. This response was specific and opposite to the response to the same amount of water administered either into the carotid artery or on the tongue. The data show that external sensory and internal hypo- or hyperosmolar messages converge upon the same integrating neuron.

which increase or decrease their firing response in relation to administered hypertonic saline or distilled water through the carotid artery also respond in the same way when salt or distilled water is applied to the tongue (fig. 4). Similar findings were subsequently reported in the monkey in which the response of such neurons to injection of hypertonic saline was inhibited if the animal was drinking [71].

These findings provide a possible explanation for preabsorptive satiety, i.e. units that discharge in response to stimulation of intracellular dehydration return to their previous quiescent state as soon as gustatory signals inform them of the swallowing of water, thus anticipating that intestinal absorption will occur and that the osmolality will return to normal.

Studies of the electrophysiology of neurons of the cat hypothalamus have defined at least 2 different receptor sites as 'integrative' centers, because neurons within these areas respond to multiple stimulations. The anterior wall of the 3rd ventricle (A3V) extending from the anterior commissure to the suprachiasmatic area and to the organum vasculosum laminae terminalis (OVLT) [48] is one area. This area is terminated by 2 circumventricular organs, the OVLT and the organum subfornicale. These 2 circumventricular organs have large communications with each other by means of both neural and vascular connections [71, 72]. Neurons within the A3V area are often sensitive to both hypertonic NaCl and to hypotension; and, units within the suprachiasmatic area (near the OVLT) respond to hypovolemia, hypertonic NaCl, and angiotensin II [30, 48]. The organum subfornicale has been implicated by *Palkovits* in the regulation of extracellular fluid volume [74], and was shown also to be responsible for the dipsogenic effect of angiotensin II [75] (as was the OVLT [53, 76]). Stimulation of these areas either electrically [77, 78] or chemically [78] also triggers a hypertensive response. Destruction of the organum subfornicale abolishes the drinking response of administered angiotensin II [79]. Well circumscribed lesions in the A3V area not in the OVLT result in prolonged and complete adipsia in the rat [80], in the goat [52, 81], in the sheep [82], and in the dog [83]. The integrated nature of the A3V is illustrated by lesions which not only affect drinking but also secretion of vasopressin, regulation of blood pressure, and nephrogenic hypertension [84].

Neuroendocrinology of Thirst

The Renin-Angiotensin System
Between 1966 and 1969, *Fitzsimons* [50, 106] reported that renal renin-angiotensin systems may have a role in control of thirst because challenges which stimulated secretion of renin also caused thirst and because some drinking responses after such stimulation could be blocked

or attenuated by nephrectomy or by specific peptide antagonists of the renin-angiotensin II system. Brain tissue has been shown to produce an iso-renin [85, 86] and is equipped with the enzymatic apparatus to produce and catabolize angiotensin II. Angiotensin II is a potent stimulator of thirst when injected peripherally or directly into the brain in a large number of vertebrate species including birds and reptiles [87, 88]. The target organs for the action of angiotensin II appear to be within the cerebroventricular walls of the diencephalon. This area plays an important role in the drinking response to administered angiotensin II [14, 89, 90] and the apparent potentiation of drinking induced by sodium [14]. However, the most effective responses to administered angiotensin II are obtained at the level of the two circumventricular organs and the A3V structures of the 3rd ventricle [53, 75, 76].

The mechanism by which angiotensin II induces an increase in drinking has been considered. One explanation, the vascular hypothesis [53], proposes that the vasoconstrictor action of angiotensin II affects the diameter of the vessels in the circumventricular organs. The decrease in vessel size would be similar to that which would happen in a passive deflation with hypovolemia and would subsequently stimulate thirst. Other substances which have little in common with angiotensin II except their vasoconstrictive properties may also cause drinking when injected into the circumventricular organs. Also, vasoplegic drugs will interfere with drinking induced by angiotensin II [53]. One vasoplegic drug, prostaglandin E_2, normally exists in the brain and has been considered by *Kenney and Epstein* as a hormonal factor of water-induced satiety [105]. Within the circumventricular organs angiotensin II may amplify the vasoconstriction induced by extracellular volume depletion, thus explaining the addition of thirst induced by injection of angiotensin II to the thirst produced by extracellular dehydration. Not only thirst, but other responses which tend to correct extracellular dehydration may be induced by administered angiotensin II (possibly through the vasoconstrictor property). These include increase in sympathetic tone, release of aldosterone, release of ACTH and release of vasopressin.

In recent studies by the author it has been demonstrated that approximately 3% of neurons in the wall of the anterior 3rd ventricle will respond to administered vasopressin only in the presence of co-administration of angiotensin II. This finding provides further evidence of neuromodulatory regulation of the electrophysiologic response induced by various peptides upon receptors, thus providing another explanation of how angiotensin II can modulate responses related to vasopressin and water balance [*Nicolaïdis and Jeulin,* unpublished].

Vasopressin and Thirst

Since vasopressin is released in response to water deprivation [5] and extracellular [91] and intracellular [6] dehydration it has been postulated that vasopressin might be a modulator of the accompanying thirst. In other cases such as vasopressin-secreting carcinomas [92] thirst may also be stimulated in spite of hyponatremia which might suggest a role for vasopressin to stimulate thirst (although cellular dehydration and potassium depletion with cellular shrinkage might also be implicated). Similarly, intracerebral injections of either acetylcholine [93] or angiotensin II [94] may induce both release of vasopressin and the sensation of thirst. In most cases vasopressin is released at the same time as thirst is noted and a possible direct effect of vasopressin cannot be determined. However, in patients with diabetes insipidus small doses of pitressin which are not able to eliminate urinary losses of water may diminish thirst [95]. It has been proposed that vasopressin facilitates the transfer of water into the receptor cells which sense dehydration and are responsible for thirst. Thus vasopressin would decrease the intracellular dehydration in the receptor cell and decrease the sensation of thirst. Since vasopressin would have to act directly on the central receptors for thirst to have this effect, it was important to study the effect of directly applied vasopressin on the central nervous system structures. As much as 20 μU of vasopressin applied directly to preoptic area did not elicit drinking [87], but smaller doses [1–10 μU) into the subfornical organ and into the A3V area did induce drinking in some water-repleted rats [53]. In other studies the administration of vasopressin in both intact and nephrectomized animals had no effect on drinking [96]. In all studies where vasopressin has been shown to affect drinking the amount of vasopressin administered has been in the pharmacologic range [97–99].

In some studies where vasopressin was reported not to stimulate thirst facilitation of drinking was observed [11, 58, 100–102]. In the dog modest levels of hormone in the concentration, e.g. 20 μU/ml, were associated with facilitated drinking while higher levels, 400 μU/ml, had an inhibitory effect [11].

Within the anterohypothalamic area where neuroreceptors generate both drinking and secretion of vasopressin it is postulated that vasopressin secreted into the CSF in parallel with its secretion into plasma

[103] increases the water permeability of the brain tissue [95] and stimulates the production of prostaglandins; both the increased tissue water and prostaglandins may tend to limit the effect of vasopressin [104] in a simple negative feed-back.

References

1 Fitzsimons, J.T.; Oatley, K.; Additivity of stimuli for drinking in rats. J. comp. physiol. Psychol. *67:* 273–283 (1968).
2 Mayer, A.: Variations de la tension osmotique du sang chez les animaux privés de liquides. C.r. Séance. Soc. Biol. *52:* 153–155 (1900).
3 Wettendorf, H.: Modifications du sang sous l'influence de la privation d'eau: contribution à l'étude de la soif. Trav. Lab. Physiol., Inst. Solvay *4:* 353–484 (1901).
4 Gamble, J.L.; Putnam, M.C.; McKhann, C.F.: The optimal water requirements in renal functions; in Measurements of water drinking by rats according to increments of urea and of several salts in the food. Am J. Physiol. *88:* 571–580 (1929).
5 Gilman, A.; Goodman, L.: The secretory response of the posterior pituitary to the need for water conservation. J. Physiol., Lond. *90:* 113–124 (1937).
6 Verney, E.B.: The antidiuretic hormone and the factors which determine its release. Proc. R. Soc. Ser. B *135:* 26–106 (1947).
7 Wolf, A.V.: Osmometric analysis of thirst in man and dog. Am. J. Physiol. *161:* 75–86 (1950).
8 Wood, R.J.; Rolls, B.J.; Ramsay, D.J.: Drinking following intracarotid infusions of hyptertonic solutions in dogs. Am J. Physiol. *232:* R88–R92 (1977).
9 Fourman, P.: Depletion of potassium induced in man with an exchange resin. Clin. Sci. *13:* 93–110 (1954).
10 Fourman, P.; Leeson, P.M.: Thirst and polyuria with a note on the effects of potassium deficiency and calcium excess. Lancet *i:* 268–271 (1959).
11 Kozlowski, S.; Szczepanska-Sadowska, E.: Mechanisms of hypovolaemic thirst and interactions between hypovolemia, hyperosmolality and the antidiuretic system; in Peters, Fitzsimons, Peters-Haefeli, Control mechanisms of drinking, pp. 25–35 (Springer, Berlin 1975).
12 Andersson, B.: Polydipsia caused by intrahypothalamic injections of hyptertonic NaCl solutions. Experientia *8:* 157–158 (1952).
13 Andersson, B.: Thirst and brain control of water balance. Am. Scient. *59:* 408–415 (1971).
14 Andersson, B.: Regulation of water intake. Physiol. Rev. *58:* 582–603 (1978).
15 Gutman, J.; An extrarenal effect of hydrochlorothiazide. Experientia *19:* 544–545 (1963).
16 Thrasher, T.N.; Nistral Herrare, J.F.; Keil, L.C.; Brown, C.J.; Ramsay, D.J.: Inhibition of ADH secretion and satiety following water ingestion in water-deprived dogs. Neurosci. Abstr. *5:* 225 (1979).
17 Andersson, B.; Dallman, M.F.; Olsson, K.: Observations on central control of drinking and of the release of antidiuretic hormone (ADH). Life Sci. *8:* 425–432 (1969).

18 McKinley, M.J.; Denton, D.A.; Weisinger, R.S.: Sensors for antidiuresis and thirst-osmoreceptors of CSF sodium detectors? Brain Res. *141:* 89–103 (1978).
19 Fitzsimons, J.T.; Massi, M.; Thornton, S.N.; Permissive effect of cerebrospinal fluid (Na) on drinking in response to cellular dehydration in the pigeon *Columba livia.* J. Physiol., Lond. *315:* 14–15 (1981).
20 Buggy, J.; Hoffmann, W.E.; Phillips, M.I.; Fisher, A.E.; Johnson, A.K.: Osmo-sensitivity of rat third ventricle and interactions with angiotensin. Am J. Physiol. *236:* R75–R82 (1979).
21 Thornton, S.N.: Osmometric drinking in the pigeon *Columba livia.* PhD thesis Cambridge (1980).
22 Oomura, Y.; Ono, T.; Ooyama, H.; Wayner, M.J.: Glucose and osmosensitive neurons of the rat hypothalamus. Nature, Lond. *222:* 282–284 (1969).
23 Adolph, E.F.; Barker, J.P.; Hoy, P.A.: Multiple factors in thirst. Am J. Physiol. *178:* 538–562 (1954).
24 Epstein, A.N.: Mineralocorticoids and cerebral angiotensin may act together to produce sodium appetite. Peptides *3:* 493–494 (1982).
25 Fitzsimons, J.T.: The physiology of thirst and sodium appetite. Monogr. Physiol. Soc., vol. 35 (Cambridge Univ. Press, London 1979).
26 Fitzsimons, J.T.: Drinking in rats depleted of body fluid without increase in osmotic pressure. J. Physiol., Lond. *159:* 297–309 (1961).
27 Andersson, B.; McCann, S.M.: A further study of polydipsia evoked by hypo-thalamic stimulation in the goat. Acta physiol. scand. *33:* 333–346 (1955).
28 Blass, E.M.; Epstein, A.N.: A lateral preoptic osmosensitive zone for thirst in the rat. J. comp. physiol. Psychol. *76:* 378–394 (1971).
29 Peck, J.W.; Novin, D.: Evidence that osmoreceptors mediating drinking in rabbits are in the lateral preoptic area. J. comp. physiol. Psychol. *74:* 134–147 (1971).
30 Nicolaïdis, S.: Réflexe poto-hidrotique et étude de son mécanisme neurophysiolo-gique. J. Physiol., Paris *63:* 361–369 (1971).
31 Nicolaïdis, S.: Réponses des unités osmosensibles hypothalamiques aux stimulations salines et aqueuses de la langue. Cr. hebd. Séanc. Acad. Sci., Paris *267:* 2352–2355 (1968).
32 Nicolaïdis, S.: Early systemic responses to oro-gastric stimulation in the regulation of food and water balance. Functional and electrophysiological data. Ann. N.Y. Acad. Sci. *157:* 1176–1203 (1969).
33 Sessler, M.; Salhi, M.D.: Interaction of hypertonic NaCl and neural stimuli on lateral preoptic neurons. Neurosci. Lett. *26:* 319–324 (1981).
34 Cross, B.A.; Green, J.D.: Activity of single neurones in the hypothalamus: Effect of osmotic and other stimuli. J. Physiol., Lond. *148:* 554–569 (1959).
35 Mogenson, G.J.; Stevenson, J.A.F.: Drinking and selfstimulation with electrical stimulation of the lateral hypothalamus. Physiol. Behav. *1:* 251–254 (1966).
36 Andersson, B.; McCann, S.M.: Drinking, antidiuresis and milk ejection from electrical stimulation within the hypothalamus of the goat. Acta physiol. scand. *35:* 191–201 (1978).
37 Montemurro, D.G.; Stevenson, J.A.F.: Adipsia produced by hypothalamic lesions in the rat. Can. J. Biochem. Physiol. *35:* 31–37 (1957).
38 Olsson, K.; On the importance of CSF Na^+ concentration in central control of fluid balance. Life Sci. *11:* 397–402 (1972).

39 McKinley, M.J.; Blaine, E.H.; Denton, D.A.: Brain osmoreceptors, cerebrospinal fluid electrolyte composition and thirst. Brain Res. *70:* 532–537 (1974).
40 Andersson, B.: Regulation of water intake. Physiol. Rev. *58:* 582–603 (1978).
41 Eriksson, L.: Effect of lowered CSF sodium concentration on the central control of fluid balance. Acta physiol. scand. *91:* 61–68 (1974).
42 Olsson, K.; Fyhrquist, F.; Larsson, B.; Eriksson, L.: Inhibition of vasopressin-release during developing hypernatremia and plasma hyperosmolality: an effect of intracerebroventricular glycerol. Acta physiol. scand. *102:* 399–409 (1978).
43 Haberich, F.J.: Osmoreception in the portal circulation. Fed. Proc. *27:* 1137–1141 (1968).
44 Adashi, A.; Niijima, A.; Jacobs, H.L.: A hepatic osmoreceptor mechanism in the rat. Am J. Physiol. *231:* 1043–1049 (1976).
45 Grill, H.J.; Miselis, R.R.: Lack of ingestive compensation to osmotic stimuli in chronic decerebrate rats. Am J. Physiol. *240:* R81–R86 (1981).
46 Smith, H.W.: Salt and water volume receptors. Am J. Med. *1957:* 623–652.
47 Gauer, O.H.; Henry, J.P.: Neurohormonal control of plasma volume; in Guyton, Cowley, Cardiovascular physiology II. Review of Physiology, vol. 9, pp. 145–190 (Univ. Park Press, Baltimore 1976).
48 Nicolaïdis, S.; Réponses unitaires des aires antérieures et médianes de l'hypothalamus antérieur associées à des variations de pression artérielle et volémie. Cr. hebd. Séanc. Acad. Sci., Paris *270:* 839–842 (1970).
49 Henry, J.P.; Gauer, O.H.; Reeves, J.L.: Evidence of the atrial location of receptors influencing urine flow. Circulation Res. *4:* 85 (1956).
50 Fitzsimons, J.T.: Effect of nephrectomy on the additivity of certain stimuli of drinking in the rat. J. comp. physiol. Psychol. *68:* 308–314 (1969).
51 Johnson, A.K.; Buggy, J.; Housh, M.: Effects of lesions surrounding the anteroventral third ventricle on fluid homeostasis. Neurosci. Abstr. *2:* 301 (1976).
52 Rundgren, M.; Fyhrquist, F.: Transient water diuresis and syndrome of inappropriate antidiuretic hormone secretion (SIADH) induced by forebrain lesions of different location. Acta physiol. scand. *103:* 421–429 (1978).
53 Nicolaïdis, S.; Fitzsimons, J.T.: La dépendance de la prise d'eau induite par l'angiotensine II envers la fonction vasomotrice cérébrale locale chez le rat. Cr. hebd. Séanc. Acad. Sci. *281D:* 1417–1420 (1975).
54 Kaufman, S.; Kaesermann, H.P.; Peters, G.: The mechanism of drinking induced by parenteral hyperoncotic solutions in the pigeon and in the rat. J. Physiol, Lond. *301:* 91–99 (1980).
55 Nicolaïdis, S.: Rôle des réflexes anticipateurs orovégétatifs dans la régulation hydrominérale et énergétique. J. Physiol., Paris *74:* 1–19A (1978).
56 Nicolaïdis, S.: Hypothalamic convergence of external and internal stimulation leading to early ingestive and metabolic responses. Brain Res. Bull. *5:* suppl. 4, 97–104 (1980).
57 Nicolaïdis, S.: Effets dipsogéniques de la stimulation mécanique de l'OSF. J. Physiol., Paris *74:* 7–19A (1978).
58 Bellows, R.T.: Time factors in water drinking in dogs. Am J. Physiol. *125:* 87–97 (1939).
59 Bernard, C.: Leçons de physiologie appliquées à la médecine faites au Collège de France, vol. 2, pp. 50–51 (Baillière, Paris 1856).

60 Towbin, E.J.: Gastric distension as a factor in the satiation of thirst in oesophagostomized dogs. Am. J. Physiol. *159:* 533–541 (1949).
61 Epstein, A.N.; Teitelbaum, P.: Regulation of food intake in the absence of taste, smell and other oropharyngeal sensations. J. comp. physiol. Psychol. *155:* 753–759 (1964).
62 Janowitz, H.D.; Grossman, M.I.: Some factors affecting the food intake of normal dogs and dogs with oesophagotomy and gastric fistula. Am J. Physiol. *159:* 143–148 (1949).
63 Stellar, E.; Hyman, R.; Samet, S.: Gastric factors controlling water and salt solution drinking. J. comp. physiol. Psychol. *47:* 220–226 (1954).
64 Snowdon, C.T.: Motivation, regulation and the control of meal parameters with oral and intra-gastric feeding. J. comp. physiol. Psychol. *69:* 91–100 (1969).
65 Rowland, N.; Nicolaïdis, S.: Metering of fluid intake and determinants of ad libitum drinking in rats. Am J. Physiol. *231:* 1–8 (1976).
66 Nicolaïdis, S.; Rowland, N.: Long-term self-intravenous drinking in the rat. J. comp. physiol. Psychol. *87:* 1–15 (1974).
67 Teitelbaum, P.: Epstein, A.N.: The lateral hypothalamic syndrom: recovery of feeding and drinking after lateral hypothalamic lesions. Psychol. Rev. *69:* 74–90 (1962).
68 Grossman, S.P.: Eating or drinking elicited by direct adrenergic or cholinergic stimulation of the hypothalamus. Science *132:* 301–302 (1960).
69 Fisher, A.E.; Coury, J.N.: Cholinergic tracing of a central neural circuit underlying the thirst drive. Science *138:* 691–693 (1962).
70 Ungerstedt, U.: Adipsia and aphagia after 6-hydroxy-dopamine-induced degeneration of the nigro-striatal dopamine system. Acta physiol. scand. *82:* suppl. 367, pp. 95–122 (1971).
71 Vincent, J.D.; Arnauld, E.; Bioulac, B.: Activity of osmosensitive single cells in the hypothalamus of the behaving monkey during drinking. Brain Res. *44:* 371–384 (1972).
72 Miselis, R.R.: The efferent projections of the subfornical organ of the rat. A circumventricular organ within a neural network subserving water balance. Brain Res. *230:* 1–23 (1981).
73 Duvernoy, H.; Koritke, J.G.: Die Gefässversorgung der Lamina terminalis bei einigen Vögeln. Verh. anat. Ges. *112:* suppl, pp. 391–404 (1963).
74 Palkovits, M.: The role of the subfornical organ in the salt and water balance. Naturwissenschaften *53:* 336 (1966).
75 Simpson, J.B.; Routtenberg, A.: Subfornical organ: site of drinking elicitation by angiotensin II. Science *181:* 1172–1175 (1973).
76 Buggy, J.; Fisher, A.E.; Hoffman, W.E.; Johnson, A.K.; Phillips, M.I.: Ventricular obstruction: effect on drinking induced by intracranial angiotensin. Science *190:* 72–74 (1975).
77 Nicolaïdis, S.; Ishibashi, S.: Hypertension induced by electrical stimulation of the subfornical organ (OSF). Brain Res. Bull. *6:* 135–139 (1981).
78 Mangiapane, M.L.; Simpson, J.B.: Subfornical organ site of pressor and drinking effects of A II. Neurosci. Abstr. *3:* 351 (1977).
79 Simpson, J.B.; Epstein, A.N.; Camardo, J.S.: Localization of receptors for the dipsogenic action of A II in the subfornical organ of the rat. J. comp. physiol. Psychol. *92:* 581–601 (1978).

80 Buggy, J.; Johnson, A.K.: Preoptic-hypothalamic periventricular lesions: thirst deficits and hypernatremia. Am. J. Physiol. *233:* R44–R52 (1977).
81 Andersson, B.; Leksell, L.G.; Lishajko, F.: Perturbations in fluid balance induced by medially placed forebrain lesions. Brain Res. *99:* 261–275 (1975).
82 McKinley, M.J.; Denton, D.A.; Leksell, L.G.; Mouw, D.R.; Scoggins, B.A.; Smith, M.H.; Weisinger, R.S.; Wright, R.D.: Osmoregulatory thirst in sheep is disrupted by ablation of the anterior wall of the optic recess. Brain Res. *236:* 210–215 (1982).
83 Thrasher, T.N.; Keil, L.C.; Ramsay, D.J.: Lesions of the OVLT attenuate osmotically induced drinking and vasopressin secretion in the dog. Endocrinology *110:* 1837–1839 (1982).
84 Johnson, A.K.; Hoffman, W.E.; Buggy, J.: Attenuated pressor responses to intracranially injected stimuli and altered antidiuretic activity following preoptic hypothalamic periventricular ablation. Brain Res. *157:* 161–166 (1978).
85 Ganten D.; Minnich, J.L.; Granger, P.; Hayduk, K.; Brecht, H.M.; Barbeau, A.; Boucher, R.; Genest, J.: Angiotensin-forming enzyme in brain tissue. Science *173:* 64–65 (1971).
86 Fischer-Ferraro, C.; Nahmod, V.E.; Goldstein, D.J.; Finkelman, S.: Angiotensin and renin in rat and dog brain. J. exp. Med. *133:* 353–361 (1971).
87 Epstein, A.; Fitzsimons, J.T.; Rolls, B.J.: Drinking caused by intracranial injection of angiotensin into the rat. J. Physiol., Lond. *200:* 98–100 (1969).
88 Fitzsimons, J.T.; Kaufman, S.: Cellular and extracellular dehydration and angiotensin as stimuli to drinking in the common iguana, *Iguana Iguana.* J. Physiol., Lond. *265:* 443–463 (1977).
89 Andersson, B.: The effect of injections of hypertonic NaCl-solutions into different parts of the hypothalamus of goats. Acta physiol. scand. *28:* 188–201 (1953).
90 Johnson, A.K.; Epstein, A.N.: The cerebral ventricles as the avenue for the dipsogenic action of intracranial angiotensin. Brain Res. *86:* 399–418 (1975).
91 Share, L.: Extracellular fluid volume and vasopressin secretion; in Ganong, Martini, Frontiers of Neuroendocrinology, pp. 183–210 (Oxford Univ. Press, New York 1969).
92 Segar, W.E.; Moore, W.W.: Hyponatremia. Increased antidiuretic hormone and inappropriate thirst in a patient with bronchogenic carcinoma. Minn. Med. *51:* 625–629 (1968).
93 Pickford, M.: The action of acetylcholine in the supraoptic nucleus of the chloralosed dog. J. Physiol., Lond. *106:* 264–270 (1974).
94 Severs, W.B.; Daniels-Severs, A.E.: Effects of angiotensin on the central nervous system. Pharmac. Rev. *25:* 415–449 (1973).
95 Raichle, M.E.; Gruble, R.L.: Regulation of brain water permeability by centrally-released vasopressin. Brain Res. *143:* 191–198 (1978).
96 De Wied, D.: Effect of autonomic locking agents and structurally related substances on the 'salt arousal of drinking'. Physiol. Behav. *1:* 193–197 (1966).
97 Rolls, B.: The effect of intravenous infusion of antidiuretic hormone on water intake in the rat. J. Physiol., Land. *219:* 331–339 (1971).
98 Smith, R.W.; McCann, S.M.: Increased and decreased water intake in the rat with hypothalamic lesions; in Wayner, Thirst. Proc. 1st Int. Symp. on Thirst in the Regulation of Body Water, pp. 381–392 (Pergamon Press, Oxford 1984).

99 Radford, E.P.: Factors modifying water metabolism in rats fed dry diets. Am. J. Physiol. *196:* 1098–1108 (1959).
100 Holmes, J.H.; Gregersen, M.I.: Observations on drinking induced by hypertonic solutions. Am. J. Physiol. *162:* 326–327 (1950).
101 Barker, J.P.; Adolph, E.F.; Keller, A.D.: Thirst tests in dogs and modifications of thirst with experimental lesions of the neurohypophysis. Am. J. Physiol. *173:* 233–245 (1953).
102 Szczepanska-Sadowska, E.; Kozlowski, S.: Equipotency of hypertonic solutions of manitol and sodium chloride in eliciting thirst in the dog. Pflügers Arch. *358:* 259–264 (1975).
103 Szczepanska-Sadowska, E.; Gray, D.; Simon-Oppermann, C.: Vasopressin in blood and third ventricle CSF during dehydration, thirst and hemorrage. Am. J. Physiol. *245:* R549–555 (1983).
104 Zusman, R.M.; Keiser, H.R.; Handler, J.S.: Vasopressin-stimulated prostaglandin E biosynthesis in the toad urinary bladder. J. Clin. Invest. *60:* 1339–1347 (1977).
105 Kenney, N.J.; Epstein, A.N.: The antidipsogenic action of Prostaglandin E (PGE). Neurosci. Abstr. *1:* 469 (1975).
106 Fitzsimons, J.T.: Hypovolaemic drinking and renin. J. Physiol., London *186:* 130P–131P (1966).
107 Nicolaïdis, S.: Réflexe poto-diurétique et rôle dans la régulation de l'équilibre hydrominéral. Archs Sci. physiol. *24:* 381–396 (1970).

S. Nicolaïdis, MD, Laboratoire de Neurobiologie des Régulations, CNRS ER 218, Collège de France, 11, Place Marcelin-Berthelot, F-75231 Paris Cedex 05 (France)

Vasopressin Receptors

Serge Jard

Centre CNRS-INSERM de Pharmacologie-Endocrinologie, Montpellier, France

Introduction

As reviewed in the contribution by *Zimmerman* [this volume], vasopressin is found in several areas of the brain and outside the central nervous system [1–3]. Vasopressin plays a major role in the regulation of plasma osmolality through an action on water and solute reabsorption by the terminal part of the nephron. It is now clear that vasopressin plays a number of other physiological roles. There is evidence that vasopressin: (1) participates in short-term regulation of arterial pressure by its direct vascular actions and interacts with the central nervous system baroreceptor reflex pathways [4]; (2) is one component of the multifactorial regulation of ACTH release by the adenohypophysis [5, 6]; (3) induces contraction of glomerular mesangial cells [7]; (4) increases prostaglandin synthesis by medullary interstitial cells [8]; (5) causes platelet aggregation and release of factor VIII [9]; (6) has a mitogenic effect on several cell types (chondrocytes [10], thymocytes [11], mouse 3T3 fibroblasts [12], and hepatocytes [13]); (7) enhances water and sodium absorption across everted sacs of the rat colon descendens [14]; (8) increases the firing rate of hippocampal neurones [15]; (9) affects several brain functions such as memory consolidation [16].

Facing such a diversity in vasopressin actions, it is pertinent to pose a number of questions regarding vasopressin receptors: (1) Does a single type of vasopressin receptor mediate all known vasopressin effects? (2) If not, what could distinguish vasopressin isoreceptors? (3) How do we avoid inherent side effects when using vasopressin analogues as therapeutical tools in view of the tissue specificity of vasopressin action. The answer to the first question is 'no'. It is well established that cAMP mediates the effects of vasopressin on the renal tubules. The likelihood that several extrarenal effects of vasopressin are not mediated by receptors coupled to adenylate cyclase emerged in 1974 when *Kirk and Hems* [18] reported that the glycogenolytic effect of vasopressin in rat liver was not accompanied by an increase in the intracellular concentration of cAMP. A role for calcium in the mechanism of vasopressin stimulation of liver and smooth muscle was demonstrated [see for instance 19]. On this functional basis *Michell* et al. [20] proposed to distinguish two types of vasopressin receptors: V_2 receptors coupled to adenylate cyclase and V_1 receptors coupled to those poorly understood mechanisms leading to an increase in intracellular calcium concentration. It was established that several structural modifications of natural vasopressin could affect its biological activities in different ways in different target tissue [21]. In particular, it was shown that the ligand selectivities of the vasopressin receptors found in the liver and vasculature were similar but were quite different from the vasopressin receptors in the kidney [22, 23], suggesting that pharmacological criteria could be used to distinguish vasopressin isoreceptors.

The main purpose of the present article is to review experimental data on: the kinetics of vasopressin binding to vasopressin receptors and its relation to physiological response – the transduction mechanisms triggered by vasopressin receptors, and the ligand specificities of these receptors.

Kinetics of Vasopressin Binding to Vasopressin Receptors

Vasopressin receptors from several tissue sources have been characterized: on particulate or plasma membrane fractions from the inner portions of porcine [24], bovine [25, 26], rat [27, 28], gerboa [29], mouse [30] and human [31] kidney; a pig kidney cell line (LLC-PK_1) [32]; isolated rat hepatocytes and rat liver membranes [33]; rat aortic myocytes in primary culture [23]; rat glomerular mesangial cells in primary culture [*Lombard and Jard*, personal communication]; rat adenohypophyseal membranes [6], and rat brain synaptosomal membranes [34]. In all systems, with the exception of LLC-PK_1 cells vasopressin binding was found to obey fairly simple kinetics. The dose dependency for binding at equilibrium can be adequately described by a model involving a reversible binding of the hormone to a single population of sites. As expected form a pseudo first-order reaction the time course of hormone binding as well as the dissociation of vasopressin receptor complexes is an exponential process. On LLC-PK$_1$ cells in monolayer dose-dependent vasopressin binding curves generated curvilinear Scatchard plots. A precise analysis of the kinetics of hormonal binding [35] indicated that neither negative cooperativity nor binding to two or more independent populations of binding sites could adequately account for the observed apparent heterogeneity in the population of vasopressin receptors. The data could be fitted with a model involving a hormone-induced transition in receptor affinity. Convincing experimental evidence indicated that this transition involved the rapid desensitization of adenylate cyclase responsiveness to vasopressin. Indeed receptor transition was not found in EDTA-suspended LLC-PK$_1$ cells which also did not demonstrate desensitization. In other systems as well, there is experimental evidence that vasopressin receptors can exist under different affinity states depending on the presence of modulators of receptor function (magnesium and nucleotides).

Affinity constants (K_{bind}) for vasopressin binding to vasopressin receptors are listed (table I) together with the apparent affinity constants (K_{act}) determined at a primary step (adenylate cyclase activation or increase in inositol lipid breakdown) or at a final step of the biological responses mediated by these receptors. These data deserve several comments:

(1) K_{bind} values differ from one tissue to another in the same species and from one species to another for a given tissue. However, the observed differences do not exceed one order of magnitude and most values are in a nanomolar range. It therefore appears difficult to distinguish several classes of vasopressin receptors on the basis of markedly different affinities for the hormone.

(2) In systems where K_{act} values were determined at an early step of the response, a fairly good correlation was found between K_{act} and K_{bind} values. However, it can be noted that on kidney membranes, adenylate cyclase activation was found to be a saturable function of receptor occupancy (non-linear coupling). Half-maximal activation is obtained for a fractional receptor occupancy less than 0.5. The ratio of K_{bind}/K_{act} values provides a simple estimate of the non-linearity in the coupling of hormone binding to adenylate cyclase activation. This ratio varies somewhat depending on the mammalian species considered: 40, 5, 5, and 1.2

Table I. Affinity constants for the binding and actions of vasopressin on several vasopressin-sensitive tissues

Receptor source	K_{bind} (nM)[a]	K_{act} (nM) primary step	K_{act} (nM) final step
Kidney membranes			
Pig	16 –24 [24]	0.25–0.7	
Ox	1.4– 4 [25]	0.30–0.9	
	0.9 [26]	–	0.02–0.03 [f] [4]
Gerboa	5.3 [29]	1.4 [c]	
Mouse	4.0 [30]	5.0 [c]	
Rat	0.4 [28]	0.4 [c]	
Man	4.4 [31]	3.1 [c]	
LLC-PK$_1$ cells			
In monolayer	3 and 30 [b]	–	–
EDTA-suspended	12 [32]	–	–
Rat hepatocytes	7.9 [37]	8.0 [d] [37]	0.15 [g] [33]
Liver membranes			
Rat	3.2 [33]	–	–
Mouse	21 [30]	–	0.5 [g]
Gerboa	5 – 6 [29]	–	–
Rat aortic myocytes	12 [23]	5.5 [e] [38]	0.1 [h] [4]
Rat mesangial cells	1 [k]	–	–
Rat hippocampal synaptosomal membranes	2.8 [34]	–	–
Rat adenohypophyseal membranes	4 [6]	–	4.3 [i]
Mouse 3T3 fibroblasts	10 [46]	–	3 [j]

[a] All values refer to the binding of arginine vasopressin except for pig and mice kidney, LLC-PK$_1$ cells, mouse liver membranes and mouse 3T3 cells for which they refer to the binding of lysine vasopressin.
[b] Values corresponding to the high-and low-affinity states, respectively.
[c] K_{act} value for adenylate cyclase activation.
[d] K_{act} value for the stimulation of phosphatidylinositide turnover.
[e] Concentration of arginine vasopressin leading to half-maximal phosphatidylinositol labelling in rat aorta.
[f] Plasma AVP levels after 48 h water restriction.
[g] K_{act} for phosphorylase activation.
[h] Plasma AVP level for which half maximal change in mean arterial pressure is observed in baroreceptor-denervated dogs.
[i] K_{act} for AVP-induced ACTH release from isolated adenohypophyseal cells.
[j] K_{act} for stimulation of DNA synthesis.
[k] *Jard and Lombard* [personal communication].

for porcine [17], bovine [25], rat [28] and human [31] adenylate cyclase, respectively. The mechanisms which could account for such a non-linear coupling have been extensively discussed [36]. Shortly, it was demonstrated that the hormonal receptor, the catalytic unit of adenylate cyclase and the GTP-sensitive coupling unit, are distinct molecular entities. The activation process very likely involves the formation of a ternary complex between receptor, coupling and catalytic units under the concerted influences of hormone binding to the receptor and GTP binding to the coupling unit. In such a situation it is easy to show that the K_{bind}/K_{act} ratio can vary from 1 to much higher values depending on the relative concentrations of receptor coupling and catalytic units.

As shown directly in the case of rat hepatocytes [37], and indirectly in the case of vascular smooth muscle cells [23, 38], there is no indication for the existence of a marked amplification in sensitivity between the hormone-receptor interaction step and the primary effect on inositol lipid breakdown.

(3) Comparison of K_{bind} values to the corresponding K_{act} values determined at a distal step of the response reveals marked differences from one vasopressin-sensitive tissue to another. For the kidney a maximum antidiuretic response can occur for plasma vasopressin levels which are much lower than those required for eliciting half-maximal adenylate cyclase activation on acellular kidney fractions in vitro. Indeed, vasopressin plasma levels do not reach values higher than 20–30 pM after 48 h water restriction [4]. Although vasopressin concentrations in the immediate vicinity of vasopressin receptors in the inner portions of the kidney might be higher than systemic concentrations as a consequence of the counter current concentration system, it seems reasonable to conclude that marked amplification in sensitivity occurs between the cAMP formation and final step of the antidiuretic response; i.e. only a small fraction of cAMP formed at complete receptor occupancy is needed to elicit a maximal antidiuretic response. This is validated by the demonstration that vasopressin structural analogues, which behave like partial agonists and have very low intrinsic activity on the adenylate cyclase activation assay, are able to elicit a full antidiuretic response in vivo [28]. On isolated hepatocytes where dose-dependent vasopressin binding and vasopressin-induced phosphorylase activation were determined under identical experimental conditions, a mean K_{bind}/K_{act} ratio of 20 was found for a series of active vasopressin analogues [33]. In this case as well, a marked amplification in sensitivity occurs between the binding

step and the final step. It can be noted that the K_{act} value of 150 pM for vasopressin is in the range of plasma AVP levels found during haemorrhage [4] suggesting that, under some physiological or pathological situations, endogeneous AVP might contribute to the regulation of liver glycogenolysis.

There is still some doubt on the sensitivity of the vascular bed to vasopressin in vivo. Recent studies [4] indicate that when the contributions of the sympathetic nervous system and renin-angiotensin system to the regulation of arterial pressure are eliminated, AVP within the physiological concentrations (1–20 pM) exerts a potent action on vascular resistance. This would indicate the existence in vascular smooth muscle cells of very efficient mechanisms ensuring an amplification in sensitivity to vasopressin.

The case of ACTH-producing cells from adenohypophysis deserves a special comment. For these cells the K_{bind} value for AVP is close to the apparent K_{act} value for AVP to induce corticotropin release. It is noteworthy to mention that pituitary cells are exposed to vasopressin concentrations which are much higher than the systemic vasopressin concentration [34; *Zimmerman*, this volume].

(4) From the above discussion it is clear that a strict correlation between the K_{bind} value and the K_{act} value for the final response cannot be used as a criterion to identify vasopressin receptors. It is also important, however, that even in situations where a large number of spare vasopressin receptors are present one might expect a strict correlation between the K_{bind} value of an antagonist and the so-called K_i value, i.e. the antagonist concentration which reduces the response to a 2 times concentration of agonist to equal the response to a 1 times concentration of agonist in the absence of antagonist. Indeed, if a large receptor reserve exists the agonist will be used in a concentration range far below the K_{bind} value. Therefore, the number of receptors occupied by the agonist will be directly proportional to the total number of free receptors present. The concentration of antagonist which will reduce this number by 50% will be very close to that leading to a 50% receptor occupancy by the antagonist; i.e. very close the K_{bind} value. In the case of isolated hepatocytes a strict correspondence between paired K_{bind} and K_i values for a series of vasopressin antagonists was demonstrated [33]. It was also shown that for the same series of antagonists K_{bind} values for binding to rat aortic myocytes were very close to the corresponding antivasopressor inhibition constants in vivo. The same results were found when

comparing K_{bind} values of two antagonists of the antidiuretic response (d [CH$_2$]$_5$Tyr [Et] VAVP and d [CH$_2$]$_5$Tyr [Et] VDAVP) to the reported inhibition constants of the antidiuretic response in vivo [*Butlen* et al., personal communication]. These observations provide strong evidence that the vasopressin-binding sites detected on kidney membranes, aortic myocytes, and hepatocytes have kinetic properties closely similar to those of the physiological receptors triggering the antidiuretic, vasopressor, and glycogenoloytic responses to vasopressin.

Transduction Mechanisms Triggered by Vasopressin Receptors

As already indicated, two types of vasopressin receptors (V$_1$ and V$_2$) can be distinguished on the basis of a functional criterion. In all species so far studied vasopressin was able to stimulate adenylate cyclase from the medullopapillary portion of the kidney. In all other vasopressin-sensitive tissues, with the possible exception of human platelets (see below), vasopressin did not affect adenylate cyclase activity even under conditions where receptor-mediated activation and/or inhibition events could be demonstrated. The recent report by *Vanderwel* et al. [40] of vasopressin-induced inhibition of adenylate cyclase activity from human platelet particulate fraction is of interest because it might indicate the presence in mammals of a third type of vasopressin receptor.

The molecular mechanisms involved in renal adenylate cyclase activation by vasopressin are probably similar to those which operate in the most extensively studied systems such as the glucagon-sensitive adenylate cyclase from liver or the β-adrenergic receptor-coupled enzyme from turkey erythrocytes [41]. Hormone-sensitive adenylate cyclases are composed of three distinct molecular entities: a receptor unit (R), a catalytic unit (C), and a coupling unit frequently denominated nucleotide regulatory unit (N). According to current views [41], both hormone binding to R and GTP binding to N are required to induce activation of C. The active state of the system decays concomitantly with the hydrolysis of GTP to GDP and P$_i$ at the N regulatory site. Replenishment of N with GTP and the continued presence of hormone at the receptor ensures the ability of the system to reacquire its active cAMP-producing state. This 'on-off' cycle accounts for most of the properties of hormone-dependent adenylate cyclases. In the case of vasopressin-sensitive adenylate cyclase, it was demonstrated that: (1) vasopressin receptors and adenylate cyclase are distinct molecular entities which can be physically separated by various biochemical methods [42]; (2) the guanylnucleotide-binding component involved in adenylate cyclase activation by sodium fluoride and guanylnucleotides has been indirectly identified in renal membranes [42], and (3) GTP and the non-hydrolyzable GTP analogue, 5'-guanylylimidodiphosphate, markedly increase the sensitivity of rat and human [31] kidney adenylate cyclase to stimulation by vasopressin.

There is a general agreement that changes in cell calcium are primarily involved in the extrarenal effects of vasopressin. Available experimental data [43] support the proposal by *Michell* and collaborators that a causal relationship exists between vasopressin-induced inositol lipids breakdown and calcium mobilization. It was recently suggested that two minor inositol lipids phosphatidylinositol-4-phosphate and phosphatidylinositol 4,5-*bis*-phosphate are very rapidly metabolized following vasopressin stimulation of isolated hepatocytes. The time-courses of polyphosphoinositides' breakdown and Ca^{2+}-mediated activation of glycogen phosphorylase are superimposable [37]. It has been suggested that the inositol 1,4,5-triphosphate could act as a second messenger for calcium mobilization.

The kinetics of vasopressin binding to renal and extrarenal vasopressin receptors is affected by magnesium ions and triphosphonucleotides. Binding of [^{3}H]-vasopressin to LLC-PK$_1$ cells [32] and to purified liver membranes [36] exhibits an absolute requirement for the presence of magnesium ions in the incubation medium. The effect of reducing the magnesium concentration is to reduce receptor affinity. On liver membranes it could be shown that the magnesium effect is agonist-specific. Receptor affinity for vasopressin and several analogues active in promoting phosphorylase activation is reduced when magnesium concentration is decreased while, under the same conditions, receptor affinity for antagonists of the glycogenolytic response is unchanged. Magnesium is also required for vasopressin binding to receptors from mesangial cells [*Lombard and Jard,* personal communication] and adenohypophyseal cells [6]. The dose-dependencies determined on liver membranes and LLC-PK$_1$ cells are similar, with an apparent K_m value for magnesium of about 0.5 mM. These data suggest the magnesium ions act at the external border of the membrane.

Kidney, liver and adenohypophyseal vasopressin receptors are sensitive to triphosnucleotides [6, 36]. The nucleotide effect results in an increased dissociation rate of hormone-receptor complexes and a corresponding increase in the equilibrium dissociation constant. The nucleotide effects on kidney and hepatic or adenohypophyseal vasopressin receptors could be clearly distinguished on the basis of their dose dependencies and specificities. ATP and GTP are equally active on liver membranes, with an apparent K_m value of 0.5 mM. It is likely that hydrolysis of ATP or GTP is involved in their effects. Thus, 5′-guanylylimidodiphosphate (Gpp[NH]P) and 5′-adenylylimidodiphosphate are inactive. In contrast, Gpp(NH)P is almost as active as GTP on renal vasopressin receptors with an apparent K_m value in the micromolar range. ATP is far less active than GTP or Gpp(NH)P.

Pharmacological Characterization of Vasopression Receptors

As pointed out in the introduction, vasopressin exerts a variety of biological effects and therefore vasopressin derivatives might be useful pharmacologic and therapeutic tools in a large number of experimental and clinical situations. It is useful to carefully characterize vasopressin isoreceptors to design potent and selective agonists and antagonists for these receptors. An ultimate goal would be the design of vasopressin analogues exhibiting both high selectivity and possibly high activity following oral administration.

Although promising clues to useful vasopressor antagonists [44] were uncovered in the early 1960s, the first effective antagonists of the antidiuretic responses in vivo were designed only 4 years ago [45]. One major difficulty originates from the abundance of spare receptors in most vasopressin-sensitive tissues (see above). From studies on renal vasopressin receptors and vasopressin-sensitive adenylate cyclase, it was shown that a partial agonist of very low intrinsic activity at the adenylate cyclase activation step (about 1% or less of the intrinsic activity of vasopressin) still elicited a maximal antidiuretic response in vivo. Secondly, the use of an in vivo biologic assay as a guide for the design of selective antagonists can be misleading when comparing two target tissues with different proportions of spare receptors. When progressively reducing the intrinsic activity (at the primary step), the transition from full agonism to partial agonism and antagonism will occur first for the target tissue having the lower proportion of spare receptors. As a consequence, a peptide with reduced efficacy at the primary step of the response might be taken as a selective antagonist when considering the biological activities in vivo. Indeed a peptide like d(CH$_2$)$_5$VDAVP was found to be a potent antivasopressor peptide in vivo. But it was not an antidiuretic antagonist in vivo, rather it exhibited a markedly reduced antidiuretic activity. It was therefore considered as a highly selective vasopressor antagonist. When tested in vitro in the renal adenylate cyclase assay, d(CH$_2$)$_5$VDAVP appeared to be a partial agonist of very low intrinsic activity. A further reduction of the intrinsic activity by *O*-alkyltyrosine substitution led to peptides like d(CH$_2$)$_5$Tyr (Me) VDAVP or d(CH$_2$)$_5$Tyr (Et) VDAVP which are anti-antidiuretic and antivasopressor peptides [44].

Salient facts about structure affinity relationships studied at the receptor level follow:

(1) In the rat, the species for which there is the most information, the main structural modifications leading to selective modifications in receptor affinity are the *L/D*-arginine substitution, and introduction of a hydrophobic amino acid in position 4. As shown in table II, these modifications slightly enhance the affinity for renal receptors but markedly reduce the affinity for receptors from liver and aortic myocytes. The effects of the two modifications are partially additive. (4-valine, 8-*D*-arginine)-vasopressin (VDAVP) can be considered as highly selective for renal receptors. Its relative K_{bind} value (AVP being used as a reference) is about 200 for hepatic and vascular receptors as compared to 0.6 for renal receptors. There is, however, indirect experimental evidence that the *L/D*-arginine and glutamine/valine substitutions in positions 8 and 4 also reduce the intrinsic activity at the primary step of vasopressin action on liver and blood vessels [36].

(2) Vasopressin analogues from the above-mentioned series therefore appear to be useful tools for distinguishing vasopressin receptors on the basis of their binding selectivity. In table III are shown the results obtained with dDAVP on six vasopressin receptors from different tissue sources. Closely similar relative affinities were obtained for all extrarenal receptors so far studied (liver, aortic myocytes, mesangial cells, brain and adenohypophysis). Figure 1 indicates that these receptors also have similar affinities for other vasopressin structural analogues including selective antidiuretic or vasopressor peptides. It therefore appears that all V_1 receptors (defined on a functional basis) are closely similar with respect to their binding selectivities.

(3) From presently available data it seems that with porcine kidney receptors excepted V_2 receptors from the different mammalian species

Table II. Relative affinities of some selective antidiuretic peptides for renal and extrarenal vasopressin receptors

Analogue	log-relative affinity (AVP used as a reference)		
	kidney	liver	myocytes
AVP	0.00	0.00	0.00
VAVP	0.11	–0.50	–0.50
dDAVP	0.14	–1.50	–2.04
VDAVP	0.25	–2.20	–2.42

Values in the table are taken from references [23, 28, 33]. VAVP, (4-valine)-arginine vasopressin; dDAVP, deamino-(8-D-arginine)-vasopressin; VDAVP, deamino-(4-valine, 8-D-arginine)-vasopressin.

Table III. Relative affinity of deamino-(8-D-arginine)-vasopressin for renal and extra renal vasopressin receptors

Receptor source	K_{bind} AVP nM	K_{bind} dDAVP nM	Relative affinity for dDAVP
Kidney [28]	0.4	0.25	1.6
Liver [33]	3.2	100	0.032
Myoctytes [23]	12.0	1288	0.009
Mesangial cells[1]	1.0	79	0.013
Brain [34]	2.8	249	0.011
Hypophysis [6]	3.8	342	0.011

[1] *Lombard and Jard* [in preparation].

(rat, ox, man and gerboa) have similar binding selectivities [36]. Data on extrarenal receptors are too scarce to decide if this conclusion also applies to V_1 receptors.

(4) Although several vasopressin analogues with low antidiuretic and high antivasopressor activities in vivo have been designed, studies at the receptor level revealed that these peptides could not be considered as selective blockers of the V_1 (hepatic or aortic myocytes) receptors. Surprisingly, the most efficient antivasopressor peptides were derived from a series of selective antidiuretic agonists with L/D-arginine and glutamine/valine substitutions [44]. The β-β-cyclopentamethylene substituent together with the incorporation of Tyr (Me) or Tyr (Et) in these pep-

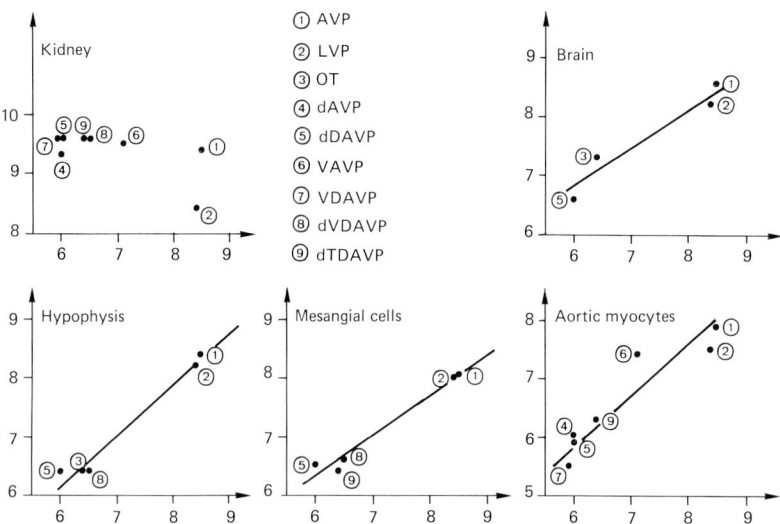

Fig. 1. Similarities and differences in the ligand specificities of vasopressin receptors in the rat. Affinity constants for the binding of vasopressin and analogues to the indicated vasopressin target tissues are expressed in terms of pK_{bind} (pK_{bind}). pK_{bind} values determined on a given system are plotted against the corresponding values determined on liver membranes considered as a reference. Numbers on the graphs refer to the analogues tested. AVP = Arginine vasopressin; LVP = lysine vasopressin; OT = oxytocin; dDAVP = deamino-(8-*D*-arginine)-vasopressin; dVDAVP = deamino-(4-valine, 8-*D*-arginine)-vasopressin; dTDAVP = deamino-(4-thereonine, 8-*D*-arginine)-vasopressin; dAVP = deamino-arginine vasopressin; VAVP = (4-valine)-arginine vasopressin; VDAVP = (4-valine, 8-*D*-arginine)-vasopressin. Data are taken from references [6, 23, 28, 33, 34], and from *Lombard and Jard* [personal communication].

tides restores a high affinity for V_1 receptors. On V_2 renal receptors the same structural modifications lead to a slight decrease in receptor affinity. As a consequence, antagonists of this series are no longer selective for V_2 receptors. Analogues form the series $d(CH_2)_5 X^2 AVP$ where X = *D*-Phe, *D*-Ile, *D*-Leu, *D*-Val [44] appear to have more promise.

Regulation of Vasopressin Receptors

The total number of vasopressin receptors found on membranes from the medullopapillary portions of the kidney is modified in situations where the mean vasopressin blood level is chronically altered. It is lowered in situations where vasopressin blood level is reduced (Battleboro rats, rats with experimentally induced dypsomania or rats receiving repeated gastric water loads). Chronic dehydration and repeated injections of small doses of dDAVP have the opposite effect [27]. These changes in vasopressin receptor number are associated with parallel changes in maximal adenylate cyclase activation by vasopressin. Although a large number of spare vasopressin receptors are present in the kidney, the above-mentioned changes in total receptor number might contribute to a modulation of kidney responsiveness to vasopressin stimulation. Indeed, for any value of free vasopressin concentration the number of occupied receptors is directly proportional to the total number of receptors.

When blood vasopressin levels in the rat are raised to values leading to a high (close to 1) fractional receptor occupancy a rapid down-regulation of kidney vasopressin receptors can be demonstrated [27]. An almost complete disappearance of kidney vasopressin receptors could be observed 40 min after administration of a large dose of vasopressin and complete recovery was observed after 16 h. This down-regulation of vasopressin receptors is very likely of little physiologic relevance since. In none of the physiologic or pathophysiologic situations explored so far in the rat have blood vasopressin levels reached values high enough to ensure a close to 1 fractional receptor occupancy. In the desert rodent *Jaculus orientalis* in which blood vasopressin levels are about 100 times higher than those found in other mammalian species, the down regulation of vasopressin receptors was found to operate in response to changes in the diet water content and subsequent changes in endogeneous blood vasopressin [29].

The down-regulation of hepatic vasopressin receptors could also be demonstrated [29]. However, the half-life of occupied hepatic vasopressin receptors (2 h in the rat) is longer than that of occupied kidney receptors (20 min).

References

1   Buijs, M.; De Vries, G.J.; Van Leuwen, F.W.; Swaab, D.F.: Vasopressin and oxytocin: distribution and putative functions in the brain; in Cross, Leng, Prog. Brain Res., vol. 60, pp. 115–128 (Elsevier, Amsterdam 1983).

2 Wathes, D.C.; Swann, R.W.; Hull, M.G.R.; Drife, J.O.; Porter, D.G.; Pickering, B.T.: Gonadal sources of the posterior pituitary hormones; in Cross, Leng, Prog. Brain Res., vol. 60, pp. 513–519 (Elsevier, Amsterdam 1983).
3 Buijs, R.M.; Van Heerikhinze: Vasopressin and oxytocin release in the brain: a synaptic event. Brain Res. 252: 71–76 (1982).
4 Cowley, A.W.; Barber, B.J.: Vasopressin vascular and reflex effects – a theoretical analysis; in Cross, Leng, in Prog. Brain Res., vol. 60, pp. 415–424 (Elsevier, Amsterdam 1983).
5 Baertschi, A.J.; Beny, J.-L.; Gahwiler, B.H.; Kolodziejczyk, E.: Vasopressin, corticoliberins and the central control of ACTH secretion. Prog. Brain. Res. 60: 505–511 (1983).
6 Gaillard, R.C.; Schoenenberg P.; Favrod-Coune, C.A.; Muller, A.F.; Marie, J.; Bockaert, J.; Jard, S.: Properties of rat anterior pituitary vasopressin receptors: relation to adenylate cyclase and the effect of corticotropin releasing factor (in press).
7 Ausiello, D.A.; Kreisberg, J.I.; Roy, C.; Karnovsky, M.J.: Contraction of cultured rat glomerular cells of apparent mesangial origin after stimulation with angiotensin II and arginine-vasopressin. J. clin. Invest. 65: 754–760 (1980).
8 Zusman, R.M.; Keiser, H.R.: Prostaglandin biosynthesis by rabbit renomedullary interstitial cells in tissue culture. J. clin. Invest. 60: 215–223 (1977).
9 Haslam, R.J.; Rosson, G.M.: Aggregation of human blood platelets by vasopressin. Am. J. Physiol. 233: 958–967 (1972).
10 Miler, R.P.; Husain, F.; Svensson, M.; Lohin S.: Enhancement of ^{3}H-methyl thymidine incorporation and replication of rat chondrocytes grown in tissue culture by plasma tissue extracts and vasopressin. Endocrinology 100: 1365–1375 (1977).
11 Whitfield, J.P.; Macmanus, J.P.; Gillan, I.L: The possible mediation by cyclic AMP of the stimulation of thymocyte proliferation by vasopressin and the inhibition of this mitogenic action by thyrocalcitonin. J. cell. Physiol. 76: 65–76 (1970).
12 Rozengurt, E.; Legg, A.; Curd Pettican, P.: Vasopressin stimulation of mouse 3T3 cell growth. Proc. natn. Acad. Sci. USA 76: 1284–1287 (1979).
13 Russel, W.E.; Bucher, N.L.R.: Vasopressin modulates liver regeneration in the Battleboro rat. Am. J. Physiol. 245: G321–G324 (1983).
14 Bridges, R.J.; Nell, G.; Rommel, W.: Influence of vasopressin and calcium on electrolyte transport across colonic mucosa of the rat. J. Physiol., Lond. 338: 463–475 (1983).
15 Muhlethaler, M.; Dreifuss, J.J.: Excitation of hippocampal neurones by posterior pitiutary peptides: vasopressin and oxytocin compared. Prog. Brain Res. 60: 147–151 (1983).
16 De Wied, D.: Central actions of neurohypophysial hormones; in Cross, Leng, Prog. Brain Res., vol. 60, pp. 155–167 (Elsevier, Amsterdam 1983).
17 Jard, S.; Roy, C.; Barth, T.; Rajerison, R.; Bockaert, J.: Antidiuretic hormone-sensitive kidney adenylate cyclase. Adv. cyclic Nucleotides Res. 5: 31–62 (1975).
18 Kirk, C.J.; Hems, D.A.: Hepatic action of vasopressin: lack of a role for adenosine 3′, 5′-cyclic monophosphate. FEBS Lett. 47: 128–131 (1974).
19 De Wulf, H.; Keppens, S.; Vandenheede, J.R.; Haustraete, F.; Proost, C.; Carton, H.: Cyclic AMP-independent regulation of liver glycogenolysis; in Nunez, Dumont, Hormone and cell regulation (North-Holland, Amsterdam 1980).

20 Michell, R.H.; Kirk, C.J.; Billah, M.M.: Hormonal stimulation of phosphatidylinositol breakdown. with particular reference to the hepatic effects of vasopressin. Biochem. Soc. Trans. 7: 861–865 (1979).

21 Sawyer, W.H.; Crzonka, Z.; Manning, M.: Neurohypophyseal peptides: design of tissue specific agonists and antagonists. Mol. cell. Endocrinol. 22: 117–134 (1981).

22 Keppens, S.; De Wulf, H.: The nature of the hepatic receptors involved in vasopressin-induced glycogenolysis. Biochim. biophys. Acta 588: 63–69 (1979).

23 Penit, J.; Faure, M.; Jard, S.: Vasopressin and angiotensin II receptors in rat aortic smooth muscle cells in culture. Am. J. Physiol. E72–E82 (1983).

24 Bockaert, J.; Roy, C.; Rajerison, R.; Jard, S.: Specific binding of (^{3}H)-lysine-vasopressin to pig kidney plasma membranes. J. biol. Chem. 249: 5922–5931 (1973).

25 Hechter, O.; Terada, S.; Nakahara, T.; Flouret, G.: Neurohypophyseal hormone-sensitive adenylate cyclase. II. Relationship between hormonal occupancy of neurohypo-physeal hormone receptor sites and adenylate cyclase activation. J. biol. Chem. 253: 3219–3229 (1978).

26 Crause, P.; Farhenholz, F.: Affinities of reactive vasopressin analogues for bovine antiduretic receptor. Mol. cell. Endocrinol. 28: 529–541 (1982).

27 Rajerison, R.M.; Butlen, D.; Jard, S.: Effects in vivo treatment with vasopressin and analogues on renal adenylate cyclase responsiveness to vasopressin in vitro. Endocrinology 101: 1–12 (1977).

28 Butlen, D.; Guillon, G.; Rajerison, M.; Jard, S.; Sawyer, W.H.; Manning, M.: Structural requirements for activation of vasopressin-sensitive adenylate cyclase, hormone binding, and antidiuretic action; effects of highly potent analogues and competitive inhibitors. Mol. Pharmacol. 14: 1006–1017 (1978).

29 Butlen, D.; Baddouri, K.; Rajerison, R.M.; Guillon, G.; Cantau, B.; Jard, S,: Plasma A. D. H. levels and liver responsiveness to vasopressin in the jerboa *Jaculus orientalis*. Gen. compar. Endocr. (in press).

30 Assimacopoulos-Jeannet, F.; Cantau, B.; Van de Werve, G.; Jard, S.; Jeanrenaud, B.: Lack of vasopressin receptors in liver, but not in kidney, of ob/ob mice. Biochem. J. 216: 475–480 (1983).

31 Guillon, G.; Butlen, D.; Cantau, B.; Barth, T.; Jard, S.: Kinetic and pharmacological characterization of vasopressin membrane receptors from human kidney medulla: relation to adenylate cyclase. Eur. J. Pharmacol. 85: 291–304 (1982).

32 Roy, C.; Ausiello, D.A.: Characterization of (8-lysine)-vasopressin binding sites on a pig kidney cell line (LLC-PK1). J. biol. Chem. 256: 3415–3422 (1981).

33 Cantau, B.; Keppens, S.; De Wulf, H.; Jard, S.: (^{3}H)-Vasopressin binding to isolated rat hepatocytes and liver membranes: regulation by GTP and relation to glycogenphosphorylase activation. J. Receptor Res. 1: 137–168 (1980).

34 Barberis, C.: ^{3}H-Vasopressin binding to rat hippocampal synaptic plasma membrane. Kinetic and pharmacological characterization. FEBS Lett. 162: 400–405 (1983).

35 Roy, C.; Hall, D.; Karish, M.; Ausiello, D.A.: Relationship of (8-lysine)-vasopressin receptor transition to receptor functional properties in a pig kidney cell line (LLC-PK1). J. biol. Chem. 1256: 3423–3427 (1981).

36 Jard, S.: Vasopressin: mechanisms of receptor activation, in Cross, Leng, Prog. Brain. Res., vol. 60, pp. 383–394 (Elsevier, Amsterdam 1983).

37 Kirk, C.J.; Creba, J.A.; Hawkins, P.T.; Michell, R.H.: Is vasopressin-stimulated inositol lipid breakdown intrinsic to the mechanism of Ca^{2+}-mobilization at V_1 vasopressin receptors? in Cross, Leng, Progr. Brain. Res., vol. 60, pp. 405-411 (Elsevier, Amsterdam 1983).

38 Takhar, A.P.S.; Kirk, C.J.: Stimulation of inorganic phosphate incorporation into phosphatidylinositol in rat thoracic aorta mediated through V_1-vasopressin receptors. Biochemistry, N.Y. *194:* 167–172 (1981).

39 Zimmerman, E.; Carmel, P.; Husain, M.; Ferin, M.; Tannenbaum, M.; Frantz, A.; Robinson, A.: Vasopressin and neurophysin: high concentrations in monkey hypophyseal portal blood. Science *198:* 952 (1973).

40 Vanderwel, M.; Lum, D.S.; Haslam, R.J.: Vasopressin inhibits the adenylate cyclase activity of human platelet particulate fraction through V_1-receptors. FEBS Lett. *164:* 340–344 (1983).

41 Levitsky, A.: Activation and inhibition of adenylate cyclase by hormones: mechanistics' aspects. Trends Pharmacol. Sci. *3:* 203–208 (1982).

42 Guilon, G.; Cantau, B.; Jard, S.: Effects of thiol-protecting reagents on the size of solubilized adenylate cyclase and on its ability to be stimulated by guanylnucleotides and fluoride. Eur. J. Biochem. *117:* 401–406 (1981).

43 Jard, S.: Vasopressin isoreceptors in mammals: relation to cyclic AMP-dependent and cyclic AMP-independent transduction mechanisms. Curr. Top. Membranes Transport *18:* 257–285 (1983).

44 Manning, M.; Sawyer, W.H.: Design of potent and selective in vivo antagonists of the neurohypophysial peptides; in Cross, Leng, Progr. Brain Res., vol. 60, pp. 367–382 (Elsevier, Amsterdam 1983).

45 Sawyer, W.H.; Pang, P.K.T.; Seto, J.; McEnroe, M.; Lammek, B.; Manning, M.: Vasopressin analogues that antagonize antidiuretic responses by rats to the antidiuretic hormone. Science *212:* 49–51 (1981).

46 Collins, M.K.L.; Rozengurt, E.: Vasopressin induces selective desensitization of its mitogenic response in Swiss 3T3 cells. Proc. natn. Acad. Sci. USA *80:* 1924–1928 (1983).

Serge Jard, Centre CNRS-INSERM de Pharmacologie-Endocrinologie,
Rue de la Cardonille, BP 5055, F-34033 Montpellier Cedex (France)

Animal Models of Diabetes insipidus

Heinz Valtin, William G. North, Brian R. Edwards, Miklos Gellai[1]

Department of Physiology, Dartmouth Medical School, Hanover, N.H., USA

Introduction

The term *diabetes insipidus* is derived from the Greek *diabainein*, meaning 'to pass through', and from the Latin *insipidus,* meaning 'not savory'. Freely translated, it refers to the passage of large amounts of hyposmolar (hence, tasteless) fluid. The differential diagnosis of the hypotonic polyurias always includes 'primary polydipsia' because, like 'true diabetes insipidus' and 'nephrogenic diabetes insipidus', it involves the physiology of vasopressin. In fact, the differentiation between 'primary polydipsia' and 'partial hypothalamic diabetes insipidus' presents a major diagnostic problem [*Moses* and *Robertson,* this volume]. Thus, 'primary polydipsia' should be discussed whenever we consider diabetes insipidus; indeed, a case can be made for classifying this entity as 'primary polydipsic (or dipsogenic) diabetes insipidus' [56].

Primary Polydipsic Diabetes insipidus

Mice

In 1961, *Silverstein* et al. [46] reported on a strain of mice in which excessive drinking is the primary event that leads to diabetes insipidus. This strain, called STR/N, arose from the NHO strain of mice which was

---

[1] Work in the authors' laboratories was supported principally by the following grants from the US Public Health Service: AM 08469 (H.V. and M.G.); CA 19613 (W.G.N. and H.V.); CA 00552 and CA 04326 (W.G.N.), and AM 26553 (B.R.E.).

injected with 3-methylcholanthrene in order to induce gastric neoplasia. In 1951, when the STR/N strain was transferred to the National Institutes of Health, USA, there was already some evidence that a few of the mice had a high fluid turnover; after extensive inbreeding, more than 90% of this strain manifested the disorder. High fluid turnover is first detected at approximately 3 months of age and fully developed by about 8 months. It is not known whether the disorder is caused by the 3-methylcholanthrene or whether it arose spontaneously. Figure 1 summarizes data from the study of the STR/N strain by *Silverstein* at al. [46]. There is a rather wide variability, but on average these mice excrete 45% of their body weight as urine compared to 2.4% in control strains. When the diseased mice are eating and drinking at will, the osmolality of their urine is approximately 298 mosm/kg H_2O, while it is 1,750 mosm/kg H_2O in normal strains (fig. 1). Plasma sodium concentrations are not significantly different, averaging 147.2 mEq/l in STR/N mice and 147.0 mEq/l in controls. The kidneys of the diseased mice respond to vasopressin (Pitressin tannate in oil) by a rise in their urine osmolality to approximately 850 mosm/kg H_2O, ruling out nephrogenic diabetes insipidus. Osmolality does not rise to a maximum after a brief exposure to vasopressin, since, by analogy with studies in the Brattleboro rat (see below), tubular fluid is being equilibrated with a relatively low papillary interstitial osmolality [53].

The most crucial tests which established that the STR/N mice have a primary polydipsic disorder involved dehydration and fluid restriction (fig. 1). Not only were these mice able to concentrate their urine to nearly 1,500 mosm/kg H_2O after just 7 h of water deprivation, but they did so while incurring a loss of body weight, which just barely exceeded that in normal mice. This degree of urinary concentration with minimal weight loss can be contrasted with the results obtained in mice with nephrogenic diabetes insipidus under similar conditions (fig. 1). The latter animals (to be described subsequently) manifest a rapid loss of weight after just short periods of water deprivation, and die quickly if they do not have access to fluid [*Valtin,* unpublished observations]. The STR/N mice were also studied during 15 months of fluid restriction (fig. 1). During this period, their intake of drinking water averaged 3.2 ml/day (compared to 5–6 ml/day for normal mice), and they concentrated their urine to the same degree as did normal mice. Although *Silverstein* et al. [46] did not give the body weights of STR/N mice during the prolonged fluid restriction, the mice were apparently healthy and presumably not losing

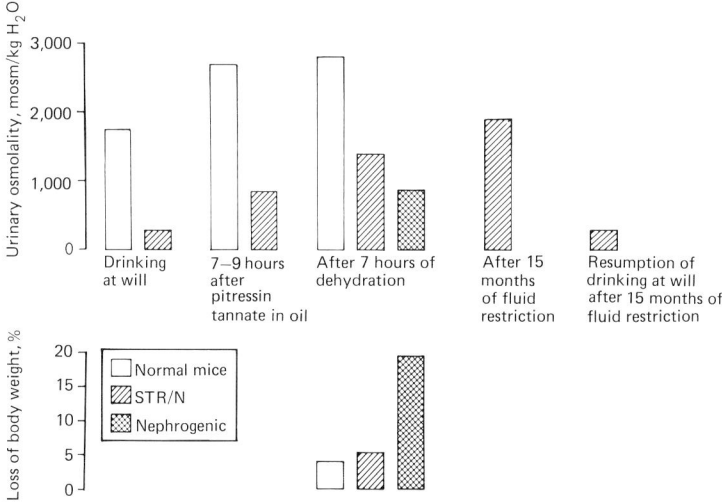

Fig. 1. Measures of fluid turnover in mice of the STR/N strain, with inherited primary polydipsic diabetes insipidus. For comparison, some data are also given for normal mice (a composite of several strains), as well as for mice with hereditary nephrogenic diabetes insipidus (so-called *DI +/+ Severe*, described in conjunction with fig. 5). The data are approximations taken from reference [46]; in some instances, reported values for urinary specific gravity were converted to urinary osmolalities using the calibration chart given in reference [45]. The values for 'nephrogenic' mice are unpublished observations of *Valtin*.

weight. Interestingly, the STR/N mice resumed their excessive drinking just as soon as they were offered water ad libitum, even after more than a year of restricted intake. Within just 3 days of water ad libitum they increased their intake to 19 ml/day, and subsequently to 30 ml/day, with appropriate reduction in urinary osmolality (fig. 1). Measurements of vasopressin have not been reported in STR/N mice, but they have abundant neurosecretory material in their hypothalamo-neurohypophyseal system [46], presumably reflecting production of vasopressin.

Males of the STR/N strain manifest what is perhaps the commonest complication of all forms of adult diabetes insipidus (provided access to adequate fluid is assured), namely hydronephrosis. Beyond the age of 8 months, 70% to 97% of the male mice showed massive hydronephrosis, presumably due to a combination of the high urine flow and urethral plugs (since the plugs were not present in females which have the same degree of polyuria) [45].

Chickens

The discovery of high fluid turnover in the Penn State strain of chickens was reported by *Dunson and Buss* [10] in 1968; subsequently, these authors and their co-workers reported extensive experiments in which they tried to identify the cause of the syndrome [11]. The fact that they were – justifiably – unable to reach a conclusion reflects not only the difficulties inherent in the differential diagnosis of the hypotonic polyurias [34, 50, 61], but also some of the uncertainties about the role of antidiuretic hormone (in this case, arginine vasotocin) in the water economy of birds [6, 48].

The disorder in these chickens is inherited as an autosomal recessive trait. Although there is large variability in the amount of water drunk, especially in females [11], there is almost no overlap with the fluid intake of normal chickens; on the average, the affected birds drink the equivalent of 32% of their body weight each day, as compared to 9% in normal chickens. Plasma sodium concentration is slightly but not significantly higher in the diseased birds (153.6 mEq/l) than in normals (151.0 mEq/l). The cloacal fluid is distinctly hyposmotic to plasma but the precise osmolality during ad libitum intake is not known, since the fluid was tested during the resumption of intake following fluid restriction and was admixed with intestinal contents [47]. There was no rise in the osmolality of cloacal fluid in response to even very high doses of vasopressin. This failure, however, must be interpreted in light of the following: that Pitressin tannate in oil, rather than the native arginine vasotocin, was used; that the hormone was given on a background of very high fluid intake which may have washed out the renal medullary interstitial osmotic gradient; that cloacal fluid, rather than ureteral urine, was tested; and that the role of vasotocin in altering the water permeability of the avian cloaca and kidney is not fully known [6, 48]. As suggested by the authors [11], the apparent lack of response to Pitressin tannate in oil does not argue conclusively for nephrogenic diabetes insipidus in these chickens.

The Penn State strain of chickens can produce antidiuretic hormone (presumably arginine vasotocin), as judged by bioassays for antidiuretic and pressor activities in the neurohypophyses and hypothalami of the diseased birds. Plasma concentrations were not determined.

Further compelling studies which favor primary polydipsia in these chickens involved water deprivation. When no water was given for 15 days (fig. 2), the diseased birds lost body weight at the same rate as did normal birds, and at the end of the deprivation their plasma sodium con-

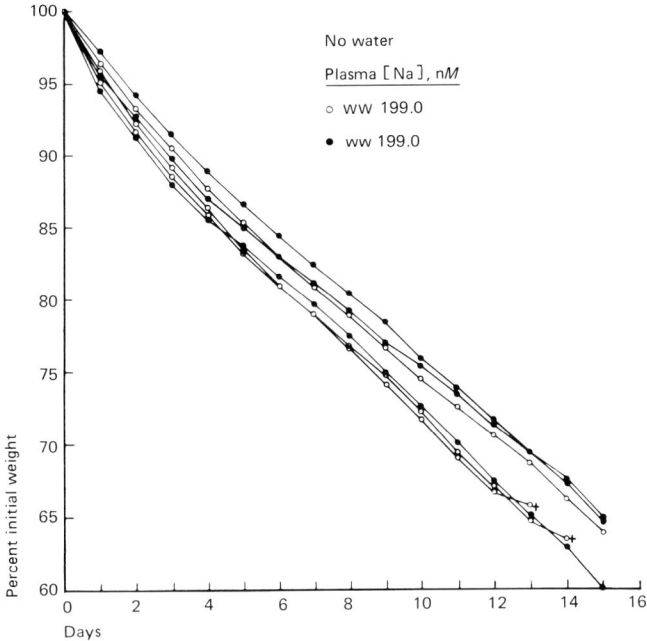

Fig. 2. Effect of withholding drinking water on the loss of body weight in the Penn State strain of chickens with polydipsia/polyuria (●, ww) and in normal chickens (○, WW). Plasma [Na] = plasma sodium concentration on day 15; + denotes 2 normal chickens, which did not survive the full period of water deprivation. From ref. [11]; published with permission.

centration had risen to exactly the same value (199 mEq/l) as the normal chickens. In fact, while 2 of the 3 normal birds died before the period of water deprivation was completed, all 3 of the diseased birds survived. Moreover, affected chickens were able to tolerate a normal fluid intake (200 ml/day) for 5 months without significantly increasing plasma sodium concentration over that of normal birds. They did gain less weight (3% of initial weight) during this period than the normals (14%), but that probably could be accounted for entirely by an initial weight loss (during the first 2 days), which exceeded that of normal birds. As was the case with the polydipsic mice described above, high fluid turnover resumed when free access to water was again permitted.

The data depicted in figure 2 are most consistent with polydipsia as the primary event. If a partial defect in the production of arginine vasotocin were the cause, one would expect the neurohypophysial stores to become depleted during prolonged and total water deprivation, causing the affected birds to lose more weight and to raise their plasma sodium concentration more. The same outcome would be predicted if the disorder were due to an inability of the kidneys or the intestine to respond to vasotocin [47]. Rather, the data in figure 2 show a response of diseased chickens to total withdrawal of drinking fluid which is indistinguishable from the response in normal birds, a finding most compatible – possibly only compatible – with polydipsia as the initiating event.

Lessons

(1) The fact that primary polydipsic diabetes insipidus can occur in experimental animals (certainly in the STR/N strain of mice and possibly in Penn State chickens) implies strongly that the same condition in humans is not necessarily psychogenic. Although most of us realize this fact, the persistence of the terms, compulsive water drinking and psychogenic polydipsia, tend to perpetuate the notion that in humans the disorder has an emotional basis.

(2) Despite our understanding of the physiology of vasopressin, the differential diagnosis of the hypotonic polyurias continues to be difficult in many instances [*Moses* and *Vokes and Robertson,* this volume]. Often, a difficult diagnosis is best resolved by a therapeutic test, such as the response to exogenous vasopressin or to fluid restriction or deprivation [*Moses, Vokes and Robertson,* and *Czernichow* et al., this volume]. An individual or animal with hypothalamic or nephrogenic diabetes insipidus is unlikely to tolerate severe fluid restriction for long periods without losing body weight (fig. 1) and raising serum osmolality excessively. The ability of these animal models to tolerate such restriction for months speaks for a primary polydipsic disorder. The possibility of a fundamental disorder of the thirst center was supported by the immediate resumption of high fluid intake in the STR/N mice and Penn State chickens when prolonged fluid restriction was terminated.

(3) So long as these animals have free access to water they, like adult humans [3], appear to have no physical sequelae except for hydronephrosis. The emphasis here is on *physical* sequelae; the emotional consequences of untreated severe polyuria/polydipsia, of whatever cause, are not to be underestimated, and it is therefore important to consider treatment in all forms of diabetes insipidus. Furthermore, in the pediatric age group, failure to thrive, mental deficiency, and especially retardation of growth are important consequences of untreated diabetes insipidus, as pointed out so clearly in Dr. *Kauli's* contribution to this symposium.

Hypothalamic Diabetes insipidus

Brattleboro Rats

The animal model of diabetes insipidus that has been most extensively studied is the Brattleboro rat, which was discovered in 1961 and which lacks vasopressin (see below) [57]. The inability to synthesize

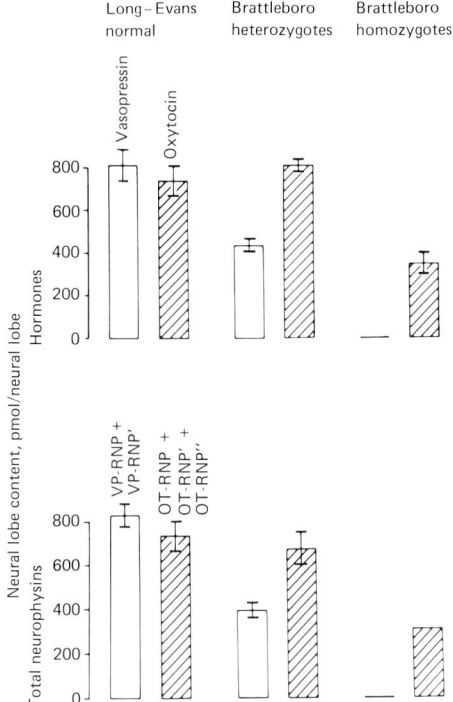

Fig. 3. Neural lobe contents of vasopressin and its specifically associated neurophysins, and of oxytocin and its neurophysins, in three genotypes of rat. The contents were determined by radioimmunoassays [37]. Vasopressin-associated rat neurophysin (VP-RNP) is the primary protein that is initially synthesized as part of the common precursor molecule with vasopressin; VP-RNP is then metabolized to VP-RNP' [38]. A similar situation holds for oxytocin-associated rat neurophysin (OT-RNP) except that two metabolic products (OT-RNP' and OT-RNP'') are formed. In each instance the sum of the neurophysins is shown. From ref. [57].

vasopressin is inherited as an autosomal semirecessive trait at a single gene locus. Although inherited hypothalamic diabetes insipidus is rare, the phenotype expressed by the Brattleboro homozygote is probably representative of most, if not all, forms of this type of diabetes insipidus, and as such, the strain is a most useful animal model of this clinical entity. The broad utilization of this strain in biomedical research has extended far beyond the study of hypothalamic diabetes insipidus. Studies have encompassed disciplines ranging from molecular biology to clinical investigation, and they have included, among many others, topics such as peptide biosynthesis, reproductive biology, central nervous system physiology and brain transplants, renal and systemic hemodynamics, endocrinology beyond the neurohypophyseal peptides, and disorders of behavior. An international symposium to discuss the total biology of the Brattleboro rat was held in September 1981, and as the published volume of that conference [49] records most of the work done on the Brattleboro rat up to 1981, we shall not repeat it here.

Biosynthesis of Vasopressin

The Brattleboro rat has been a key tool in elucidating the biological mechanisms that direct the synthesis of vasopressin and other neurohypophyseal peptides [17, 39]. As discussed in the contribution by *Robinson and Verbalis* [this volume], *Sachs* et al. [43] proposed that vasopressin and neurophysin were both synthesized via the same precursor molecule. If, indeed, vasopressin and its specifically associated neurophysin arise from a common precursor, then one would predict that Brattleboro homozygotes lack not only vasopressin but also the neurophysin. Furthermore, one would predict that the hormone and its neurophysin should occur in some constant molar ratio [38]. The fulfillment of these predictions is shown in figure 3.

The absence of vasopressin-associated neurophysin in Brattleboro homozygotes strongly suggested that neurophysin was part of the precursor molecule. Likewise, the finding that a glycoprotein is normally present within neurosecretory granules but absent from the granules of Brattleboro homozygotes [*North*, unpublished] suggested that a glycopeptide might be part of the precursor for vasopressin. These suggestions have now been documented [28, 40; *Robinson and Verbalis, Richter and Schmale*, this volume].

Enzyme(s) within Neurosecretory Granules

Processing of the precursor within neurosecretory granules to yield vasopressin requires enzymatic activity within the granules. Indirect evidence for such activity was offered by *Burford and Pickering* [39] who suggested that oxytocin-associated rat neurophysin prime (OT-RNP′) was a metabolic product of the primary oxytocin-neurophysin (OT-RNP). As discussed in the contribution by *Robinson and Verbalis* [this volume], direct evidence was brought forth by *North* and his associates after they had purified the rat neurophysins; details of these developments are recounted in [38] and [39]. *North* et al. [38] demonstrated that

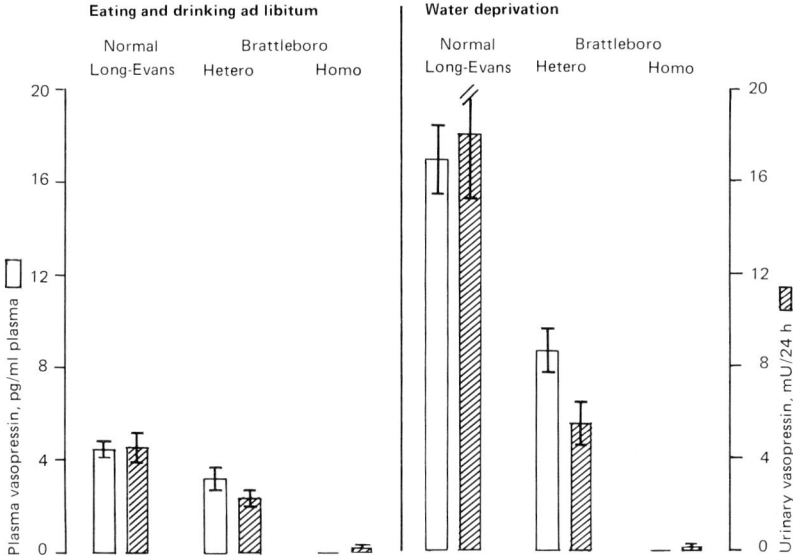

Fig. 4. Results of radioimmunoassays for vasopressin in plasma (open columns) and urine (stippled columns) of Long-Evans (normal) rats, Brattleboro heterozygotes (Hetero), and Brattleboro homozygotes (Homo). Values in the panel to the left were obtained while the animals were drinking at will, those in the right-hand panel, after water deprivation. Plasma values from *Möhring and Möhring* [31]; those on urine from *Miller and Moses* [30]. Interpretation of these data is given in the text. From ref. [55].

the metabolic products VP-RNP′, OT-RNP′, and OT-RNP″ (for definitions, see legend to fig. 3) could be generated in vitro from the primary neurophysins, VP-RNP and OT-RNP, respectively, when the latter were incubated with tiny amounts of neurosecretory granule extract, and provided that the incubation was carried out at the acid pH of these granules, approximately pH 5.

The raison d'être for this enzymatic activity(ies) is presumably not simply the catabolism of the neurophysins but rather the processing of the precursor molecule. Possible routes of such processing have been suggested by *North* et al. [38] as well as by others [*Richter and Schmale*, this volume]. It is likely that the signal peptide, which is part of the precursor molecule [28, 40], is split off before the prohormone is packaged into neurosecretory granules. The first step within the granules may then be the cleavage of vasopressin plus a bridging peptide from the prohormone, to yield a molecule consisting of vasopressin-neurophysin (VP-

NP) and the intact glycopeptide covalently linked by a single, bridging amino acid residue; we have called this molecule 'proneurophysin' (by which term we do not wish to imply that the neurophysin and glycopeptide do not share a common precursor molecule with vasopressin). It now appears that several enzymes must be involved in the overall processing of the precursor [38].

Question of Absolute Lack of Vasopressin

It is difficult to answer this question since measurements only permit us to say that the amount that might be present falls below the detectable limit of the assay. At least two findings are compatible with the possibility that Brattleboro homozygotes may synthesize very small amounts of vasopressin: (a) Hypothalamic mRNA from Brattleboro rats *may* yield a tiny amount of vasopressin-precursor when translated in vitro [40]. (b) Many workers find a small amount of reactive material by radioimmunoassay, but this finding might well be accounted for by some cross-reactivity with material that is not vasopressin. A strong argument that Brattleboro homozygotes do not produce any vasopressin lies in the observation that there is no rise in immunoassayable material upon water deprivation. Figure 4 shows that, whereas dehydration causes sharp increases of vasopressin in both plasma and urine of normal Long-Evans rats and of Brattleboro heterozygotes, no increment can be detected in Brattleboro homozygotes – even though there was some detectable material (possibly due to nonspecific cross-reactivity) in the urine of homozygotes prior to dehydration.

Body Fluid Volumes in Brattleboro Rats

Plasma osmolality and plasma sodium concentration are greater in Brattleboro homozygotes than in Long-Evans normal rats [52]. This fact has been interpreted to reflect chronic, mild, and intermittent volume contraction of Brattleboro homozygotes [57], assuming that in the steady state they are slightly behind in replenishing their obligatory high urine output through drinking. The interpretation seemed to be supported by the fact that plasma sodium concentration and osmolality decreased to normal or near normal when homozygotes were treated with vasopressin [4]. Alternatively, *Friedman and Friedman* [16] showed that extracellular fluid volume, expressed as ml/100 g body weight, was higher in Brattleboro homozygotes than in normal Long-Evans rats. But until recently we rationalized that finding as possibly being misleading, since normal

Table I. Body fluid volumes and fat content in weight-matched, conscious Long-Evans normal rats and Brattleboro homozygotes[1]

	Normal Long-Evans	Brattleboro homozygotes	p
Total body water			
ml/100 g body weight	67.7 ± 0.7	69.2 ± 0.4	NS
ml/100 g fat-free body weight	73.3 ± 0.4	73.4 ± 0.5	NS
Plasma volume			
ml/100 g body weight	3.9 ± 0.2	4.4 ± 0.1	<0.025
Blood volume			
ml/100 g body weight	6.7 ± 0.3	7.3 ± 0.2	<0.07
Total body fat content			
g/100 g body weight	7.6 ± 0.6	5.0 ± 0.4	<0.005
g/100 g dry body weight	23.4 ± 1.3	16.5 ± 1.0	<0.005

[1] Data abstracted from *Robinson* et al. [42].

rats are fatter than Brattleboro rats, and fat has a very low content of water. *Robinson* et al. [42] have recently reinvestigated the question of volume contraction in Brattleboro rats. The results of their study are summarized in table I. Long-Evans normal rats are indeed significantly fatter than Brattleboro homozygotes. Nevertheless, total body water, measured by desiccation, is not significantly different in the two types of animals, even when their body weights have been corrected for content of fat. Plasma volume, measured by the Evans blue dye technique, was significantly greater in Brattleboro homozygotes than in Long-Evans rats, and blood volume showed the same tendency – results that are qualitatively in agreement with those of *Dlouhá* et al. [8].

These data confirm and extend the findings of *Friedman and Friedman* [16] and of *Dlouhá* et al. [8], that animals with hypothalamic diabetes insipidus do not appear to be volume-contracted. Rather, the hypernatremia and hyperosmolality of plasma in Brattleboro homozygotes may reflect a higher than normal body content of solute [16, 42]. The switch in our thinking could importantly alter our explanations for many of the abnormalities that are seen in Brattleboro rats [49].

Lessons

(1) Vasopressin and its specific neurophysin, as well as a glycopeptide, are synthesized via a common precursor molecule. Processing of the precursor into its component parts

probably requires several proteolytic enzymes. Therefore, hypothalamic diabetes insipidus, in experimental animals as well as humans, theoretically could result from a fault in any step from biosynthesis of the precursor, to its cleavage within neurosecretory granules, to its release from the neurohypophysis.

(2) It is not settled whether Brattleboro homozygotes manifest a total lack of vasopressin or whether they may produce a tiny amount of the hormone, or possibly a 'defective vasopressin'.

(3) Homozygotes of the Brattleboro strain are probably not chronically volume-contracted, even though they do have hypernatremia and abnormally high serum osmolality. This finding raises questions about the possible influence of vasopressin on solute balance, and suggests that hypernatremia in human patients with diabetes insipidus cannot tacitly be assumed to reflect volume contraction – although that interpretation probably remains the safest reflex in clinical situations.

Nephrogenic Diabetes insipidus

Mice

Several strains of mice with vasopressin-resistant deficiencies of urinary concentration were described by *Falconer* et al. [14] in 1964. Breeding stocks for these animals were kindly sent to us by Dr. Falconer in 1965, and we have worked with these models since then [35, 36]. *Falconer* et al. [14] described two or more defective genes: one, called *Os*, is a dominant gene which causes oligosyndactyly in addition to the urinary concentrating defect; the other(s) is called *DI*. We have subsequently been able to divide animals with the DI gene(s) into two groups (fig. 5a), one of which, *DI +/+ Severe,* eventually excretes urine that is hypotonic to plasma, and the other, *DI +/+ Non-severe,* which develops only a mild urinary concentrating defect. Consequently, we now deal with four abnormal genotypes, all of which fail to respond to exogenous vasopressin (fig. 5b). The dynamics for the renal disorder in each genotype [25] are reflected in figure 5c, where urinary osmolalities and the simultaneously determined osmolalities of the papillary interstitium have been plotted on the ordinate against the length of the countercurrent system on the abscissa.

Normal mice, for comparison in this study, were the so-called *VII +/+* strain. The length of their countercurrent system (i.e. the distance from the junction of the renal cortex and medulla to the tip of the inner medulla) averages approximately 5.4 mm. The fact that their urinary osmolality of approximately 2,800 mosm/kg H_2O coincides with the interstitial osmolality at the papillary tip indicates osmotic equilibration

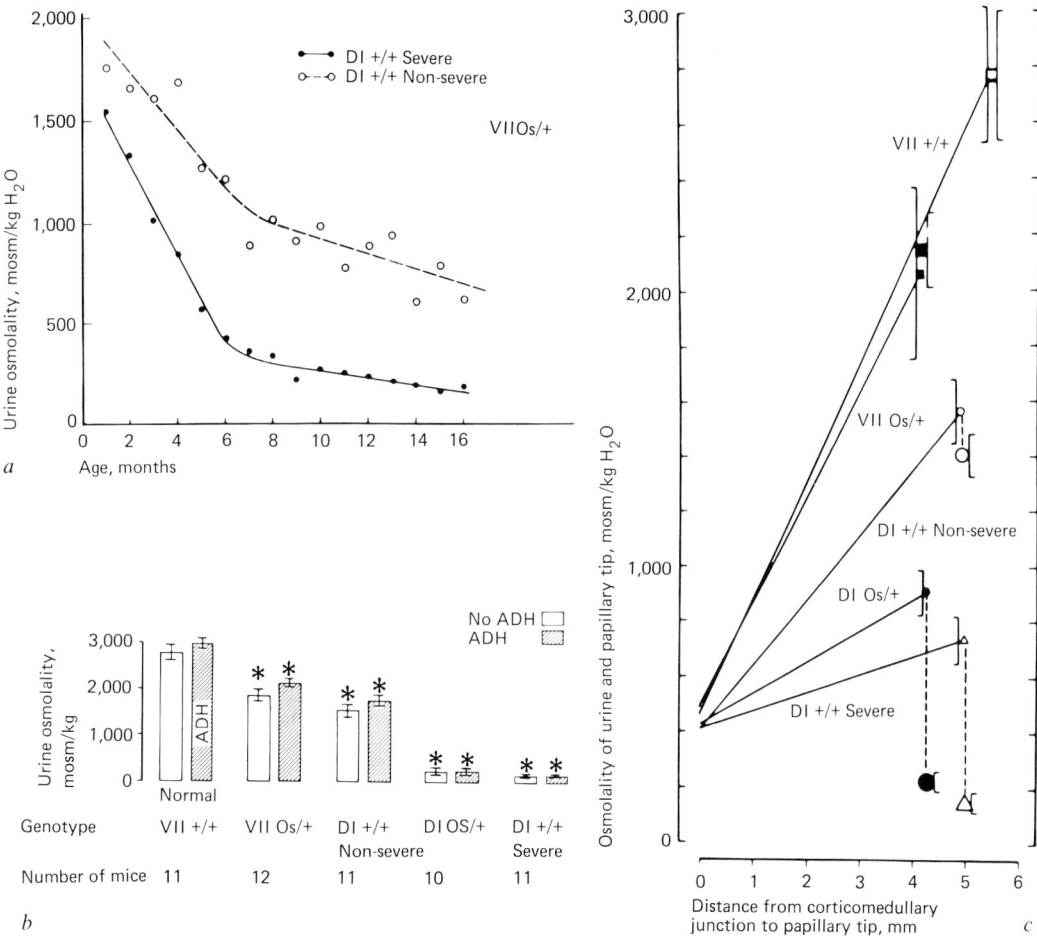

Fig. 5. a Patterns for the decline of urinary osmolalities in the DI strain of mice, which form the basis for classifying these animals into *DI +/+ Non-severe* and *DI +/+ Severe.* From ref. [54]. *b* Mean urinary osmolalities before and after treatment with vasopressin (ADH), in control mice and 4 abnormal genotypes while the animals were eating and drinking at will. The striped columns represent the mean values after daily subcutaneous injections of 0.25 U Pitressin tannate in oil for 3 days. Asterisks denote values that are significantly different from corresponding values in the normal *VII +/+* genotype ($p < 0.05$). From [54]. *c* Urinary osmolalities (large symbols) and interstitial osmolalities at the tip of the inner medulla (small symbols), plotted against the length of the renal countercurrent system in normal *VII +/+* mice and in 4 genotypes with nephrogenic defects of urinary concentration. The differences between the papillary and urinary osmolalities were not statistically significant for *VII +/+, VII Os/+,* and *DI +/+ Non-severe* animals; these differences had p-values of <0.001 for *DI +/+ Severe* and *DI Os/+* mice. From ref. [54].

between collecting duct fluid and the medullary interstitium and thus reflects vasopressin-induced water-permeability of the distal nephron. When the *Os* gene is introduced into the *VII* stock, producing *VII Os/+* animals, there appears not only oligosyndactyly but also a mild deficiency of urinary concentration, shown by an average urinary osmolality of about 2,100 mosm/kg H_2O. This deficiency results from chronic renal failure, for the mice have small kidneys (note their foreshortened countercurrent system) and an elevated blood urea nitrogen concentration [35]; the deficiency is not due to lack of water permeability of the collecting ducts, for the urinary and papillary osmolalities are approximately equal. The *DI* strain of mice does not have small kidneys (the length of their countercurrent system being approximately 5 mm – fig. 5c) and these animals are not in renal failure [35]. The strain is divided into two subgroups: *DI +/+ Severe* animals, which excrete hypotonic urine that is not osmotically equilibrated with the papillary interstitium and, therefore, is due to failure of vasopressin to increase the water permeability of the distal nephron; and *DI +/+ Non-severe* mice, which have a much milder concentrating defect that is presumably due to a partial deficiency of vasopressin-induced water permeability. When the *Os* gene, which by itself causes only a minor deficiency of concentration, is combined with the *DI* strain, the resulting animals, *DI Os/+*, excrete hypotonic urine. This severe diabetes insipidus can occur even if the *Os* gene is introduced into the *DI +/+ Non-severe* stock. The marked concentrating defect probably results from an interplay of several factors: small kidneys, chronic renal failure, and consequent osmotic diuresis per nephron [35], which may prevent osmotic equilibration of collecting duct fluid with the papillary interstitium even though only a moderate deficiency of water permeability may be present (as in *DI +/+ Non-severe* animals).

Deficient Water Permeability

In our more recent work, we have concerned ourselves mainly with *DI +/+ Non-severe* and *DI +/+ Severe* animals, where a deficiency in vasopressin-induced water permeability of the late distal tubules and collecting ducts appears to be the cause of the concentrating defect. We have attacked the problem from two points of view: (a) to try to identify the presumed faulty step in the cellular mode of action of vasopressin; and (b) to apply what is currently probably the most direct indicator of vasopressin-induced water permeability, namely the aggregation of intramembranous particles in the luminal plasma membrane.

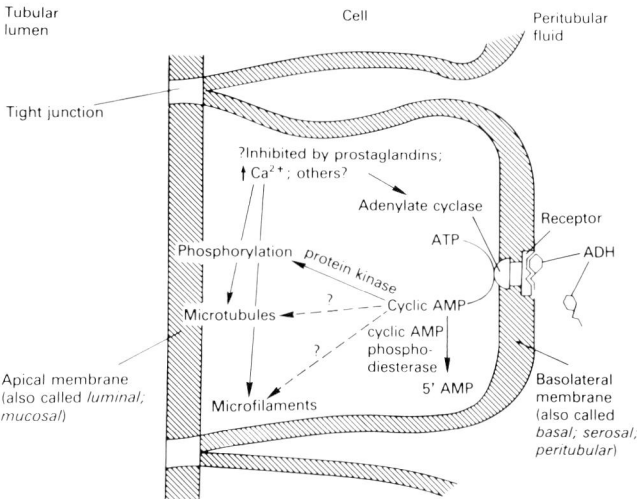

Fig. 6. Chain of events whereby vasopressin (ADH) is thought to increase the water permeability of mammalian late distal tubules and collecting ducts. The schema also applies to other membranes that increase their waterpermeability in response to neurohypophyseal hormones, such as the toad urinary bladder and frog skin. From ref. [58].

Cellular Action of Vasopressin. Figure 6 depicts the chain of events whereby vasopressin is thought to increase the water permeability of sensitive epithelia, including those of the mammalian late distal tubules and collecting ducts. The interaction of vasopressin with specific receptors in the basolateral membrane stimulates the activation of the enzyme, adenylate cyclase, which catalyzes the formation of cyclic 3′, 5′-adenosine monophosphate (cyclic AMP) from adenosine triphosphate (ATP). Cyclic AMP serves as the intracellular mediator for vasopressin, and its concentration within the cell is controlled not only by its rate of formation from ATP but also by its rate of breakdown to adenosine 5′-monophosphate (5′AMP) under the influence of the enzyme, cyclic AMP phosphodiesterase. Once formed, cyclic AMP activates a cascade of events that is still being defined, which culminates in increased water permeability of the apical membrane; the cascade probably includes the phosphorylation of specific membrane proteins under the influence of an enzyme, protein kinase, the intermediation of microtubules and microfilaments, and the incorporation of intramembranous particles into the apical membrane.

The first study, carried out by *Dousa and Valtin* [9], was done on homogenates of the renal medulla in a cell-free system; the second study, performed by *Jackson* et al. [23], utilized the microanalysis for hormonal action developed by *Morel* et al. [32], and thus enabled us to localize defects to specific portions of the nephron. Taken together, the results suggest that the main cause of the diabetes insipidus in *DI +/+ Severe* mice may be an increased activity of cyclic AMP phosphodiesterase preventing intracellular accumulation of cyclic AMP. Some decreased activation of adenylate cyclase by vasopressin may contribute to the dearth of cyclic AMP. Although these results were obtained in segments of collecting ducts from the inner stripe of the outer medulla, the findings can probably be extended to late distal tubules as well as to cortical and inner medullary collecting ducts of *DI +/+ Severe* mice. In addition, we found a striking decrease of vasopressin-stimulated activation of adenylate cyclase in medullary thick ascending limbs of Henle in *DI +/+ Severe* animals. Inasmuch as vasopressin augments reabsorption of solute from these limbs in mice in vivo [20, 44], it is possible that this additional abnormality further contributes to the concentrating defect by diminishing buildup of the corticopapillary interstitial osmotic gradient. In neither the collecting ducts nor the limbs of Henle does the deficiency appear to be caused by lack of substrate, as of ATP (fig. 6) or of nicotinamide adenine dinucleotide (NAD) [27].

There is also the finding of *Strewler* et al. [51], of impaired phosphorylation of two specific proteins in the particulate fraction from the renal medulla of *DI +/+* mice, both *Severe* and *Non-severe*. According to the cascade shown in figure 6, this result might be expected from the lack of vasopressin-stimulated cyclic AMP in collecting ducts, and it could be related to the lack of intramembranous particles (discussed next). On the other hand, since thick ascending limbs far outnumber collecting ducts in the renal medulla, the finding of defective protein phosphorylation could be unrelated to the diminished water permeability.

Intramembranous Particles. The discovery that increased transepithelial flow of water induced by neurohypophyseal hormones is associated with aggregation of particles within apical membranes [5, 24] offered the possibility of measuring vasopressin-induced changes of water permeability directly in vivo. This possibility turned into reality when we demonstrated a dose-response curve between vasopressin and

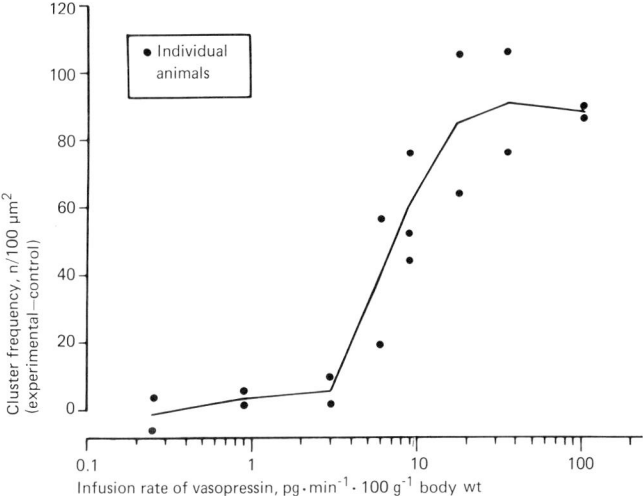

Fig. 7. Relationship between rate of vasopressin infusion to Brattleboro homozygotes and aggregation of intramembranous particles within the luminal membrane of papillary collecting ducts. It was calculated that the infusion rates led to plasma concentrations for vasopressin, which spanned the physiological range. From ref. [21].

intramembranous particle clusters in luminal membranes of mammalian collecting ducts [21]. The relationship shown in figure 7, obtained on Brattleboro homozygotes, was associated with the excretion of hypotonic urine at vasopressin infusions of 2 pg/min·100 g body weight or lower, with maximally concentrated urine at infusion rates of 11 pg/min·100 g body weight, and with intermediate urine osmolalities at infusion rates that fell between 2 and 11 pg/min. We have observed similar changes in both cluster frequency and urine osmolality in normal Long-Evans rats during water loading and dehydration, whereas the frequency remained low in dehydrated Brattleboro rats despite a marked rise in urine osmolality [13]. (The ability to concentrate urine in the apparent absence of vasopressin is, of course, well known [1, 12, 18]).

From the above facts, we predicted that, when examined by freeze-fracture electron microscopy, the apical membranes from collecting ducts of *DI +/+ Severe* mice might show no clusters or very few clusters of intramembranous particles, *VII +/+* (normal) mice might show the greatest number of clusters, and *DI +/+ Non-severe* animals would reveal an intermediate number of clusters. This is precisely what was found [2]: none of the cells examined in *DI +/+ Severe* mice showed aggregation of intramembranous particles; 47% of the cells from *VII +/+* animals showed aggregation, and 28% of the cells from *DI +/+ Nonsevere* mice had aggregates. This finding probably constitutes the most direct proof yet that deficient water permeability of the distal nephron is the ultimate cause of the diabetes insipidus in *DI +/+ Severe* mice and, by extension, in some other forms of nephrogenic diabetes insipidus.

Antagonists to Vasopressin

A relatively new experimental model of diabetes insipidus utilizes various synthetic antagonistic analogues of vasopressin [29]. These compounds have now been developed to such a degree of specificity that they can inhibit the pressor action of vasopressin without influencing its antidiuretic action, or vice versa [19]. When the latter situation is applied to unanesthetized Long-Evans normal rats, urinary flow rises from 9.6 ± 1.7 μl/min · 100 g body weight to 43.5 ± 6.4 μl/min, and urinary osmolality falls from 877 ± 97 to 188 ± 23 mosm/kg H_2O – all while body weight is held constant. The antagonistic effects appear to be localized to those areas of the nephron where vasopressin activates adenylate cyclase [26], and it appears likely that they act as competitive inhibitors. The major importance of this latest model may be that it enables us through a simple maneuver, to produce diabetes insipidus without surgical intervention and without expanding body fluid volumes.

Lessons

(1) Nephrogenic defects of urinary concentration may have either of two fundamental causes, or a combination of the two: (a) a deficiency of vasopressin-induced water permeability of the distal nephron, and (b) reduced buildup of the corticopapillary interstitial osmotic gradient.

(2) The molecular basis for either cause could lie at any step along the cascade from combination of vasopressin with its receptor to stimulation of solute transport out of ascending thick limbs of Henle (in the case of *b* above) or to an increase in water permeability (in the case of *a* above). It is likely that, in the future, we will identify various molecular lesions giving rise to the same phenotype of nephrogenic diabetes insipidus in humans; we will then be in a better position to apply specific therapy.

(3) An experimental model of diabetes insipidus, which utilizes synthetic antagonistic analogues of vasopressin, may constitute a valuable tool in that it enables us to induce florid diabetes insipidus without expanding the animal's body fluid volumes.

References

1 Berliner, R.W.; Davidson, D.G.: Production of hypertonic urine in the absence of pituitary antidiuretic hormone. J. clin. Invest. *36:* 1416–1427 (1957).
2 Brown, D.; Valtin, H.; Morris, J.; Orci, L.: Lack of intramembrane particle aggregates in collecting ducts of mice with nephrogenic diabetes insipidus. Abstr. 9th Int. Congr. Nephrol., p. 407A (1984).
3 Burka, E.R.: Renal function in diabetes insipidus. Archs. intern. Med. *109:* 717–723 (1962).
4 Cheng, S.W.T.; North, W.G.; Gellai, M.: Replacement therapy with arginine vasopressin in homozygous Brattleboro rats. Ann. N.Y. Acad. Sci. *394:* 473–480 (1982).
5 Chevalier, J.; Bourguet, J.; Hugon, J.S.: Membrane associated particles: distribution in frog urinary bladder epithelium at rest and after oxytocin treatment. Cell Tiss. Res. *152:* 129–140 (1974).
6 Dantzler, W.H.: Some renal glomerular and tubular mechanisms involved in osmotic and volume regulation in reptiles and birds; in Jorgensen, Skadhauge, Osmotic and volume regulation, pp. 187–208 (Academic Press, New York 1978).
7 Dashe, A.M.; Cramm, R.E.; Crist, C.A.; Habener, J.F.; Solomon, D.H.: A water deprivation test for the differential diagnosis of polyuria. J. Am. med. Ass. *185:* 699–703 (1963).
8 Dlouhá, H.; Křeček, J.; Zicha, J.: Postnatal development and diabetes insipidus in Brattleboro rats. Ann. N.Y. Acad. Sci. *394:* 10–20 (1982).
9 Dousa, T.P.; Valtin, H.: Cellular action of antidiuretic hormone in mice with inherited vasopressin-resistant urinary concentrating defects. J. clin. Invest. *54:* 753–762 (1974).
10 Dunson, W.A.; Buss, E.G.: Abnormal water balance in a mutant strain of chickens. Science *161:* 167–169 (1968).
11 Dunson, W.A.; Buss, E.G.; Sawyer, W.H.; Sokol, H.W.: Hereditary polydipsia and polyuria in chickens. Am. J. Physiol. *222:* 1167–1176 (1972).
12 Edwards, B.R.; Gellai, M.; Valtin, H.: Concentration of urine in the absence of ADH with minimal or no decrease in GFR. Am. J. Physiol. *239:* F84–F91 (1980).
13 Edwards, B.R.; Harmanci, M.C.: Intramembranous particle clusters in collecting duct cells of rats. Influence of water balance. Renal Physiol. *6:* 275–280 (1983).
14 Falconer, D.S.; Latyszewski, M.; Isaacson, J.H.: Diabetes insipidus associated with oligosyndactyly in the mouse. Genet. Res. *5:* 473–488 (1964).
15 Forssman, H.: On hereditary diabetes insipidus. Acta med. scand. suppl. 159, pp. 9–196 (1945).
16 Friedman, S.M.; Friedman, C.L.: Salt and water distribution in hereditary and in induced hypothalamic diabetes insipidus in the rat. Cand. J. Physiol. Pharmacol. *43:* 699–705 (1965).
17 Gainer, H.: Precursors of vasopressin and oxytocin; in Cross, Leng, The neurohypophysis: structure, function and control, pp. 205–215 (Elsevier, Amsterdam 1983).
18 Gellai, M.; Edwards, B.R.; Valtin, H.: Urinary concentrating ability during dehydration in the absence of vasopressin. Am. J. Physiol. *237:* F100–F104 (1979).
19 Gellai, M.; Bankir, L.; Grünfeld, J.-P. (intr. by H. Valtin): Vasopressin and renal hemodynamics: vascular and tubular effects. Abstr. 16th Annu. Meet. Am. Soc. Nephrology, 1983, p. 150A.

20 Hall, D.A.; Varney, D.M.: Effect of vasopressin on electrical potential difference and chloride transport in mouse medullary thick ascending limb of Henle's loop. J. clin. Invest. 66: 792–802 (1980).
21 Harmanci, M.C.; Stern, P.; Kachadorian, W.A.; Valtin, H.; DiScala, V.A.: Vasopressin and collecting duct intramembranous particle clusters: a dose-response relationship. Am. J. Physiol. 239: F560–F564 (1980).
22 Hays, R.M.; Levine, S.D.: Pathophysiology of water metabolism; in Brenner, Rector, The kidney; 2nd ed., p. 802 (Saunders, Philadelphia 1982).
23 Jackson, B.A.; Edwards, R.M.; Valtin, H.; Dousa, T.P.: Cellular action of vasopressin in medullary tubules of mice with hereditary nephrogenic diabetes insipidus. J. clin. Invest. 66: 110–122.
24 Kachadorian, W.A.; Wade, J.B.; DiScala, V.A.: Vasopressin: induced structural change in toad bladder luminal membrane. Science 190: 67–69 (1975).
25 Kettyle, W.M.; Valtin, H.: Chemical and dimensional characterization of the renal countercurrent system in mice. Kidney int. 1: 135–144 (1972).
26 Kim, J.; Dillingham, M.; Summer, S.; Schrier, R.: Effects of vasopressin (VP) antagonist on the cellular action of VP in isolated tubules. Abstracts, 16th Annu. Meet. Am. Soc. Nephrology, 1983, p. 167A.
27 Kusano, E.; Yusufi, N.K.; Werness, J.; Dousa, T.P.: Dynamics of nucleotides in distal tubular segments of mice with hereditary nephrogenic diabetes insipidus (NDI). Abstracts, 16th Annu. Meet. Am. Soc. Nephrol., 1983, p. 169A.
28 Land, H.; Schütz, G.; Schmale, H.; Richter, D.: Nucleotide sequence of cloned cDNA encoding bovine arginine vasopressin-neurophysin II precursor. Nature, Lond. 295: 299–303 (1982).
29 Manning, M.; Klis, W.A.; Olma, A.; Seto, J.; Sawyer, W.H.: Design of more potent and selective antagonists of the antidiuretic responses to arginine-vasopressin devoid of antidiuretic agonism. J. med. Chem. 25: 414–419 (1982).
30 Miller, M.; Moses, A.M.: Radioimmunoassay of urinary antidiuretic hormone with application to study of the Brattleboro rat. Endocrinology 88: 1389–1396 (1971).
31 Möhring, B.; Möhring, J.: Plasma ADH in normal Long-Evans rats and in Long-Evans rats heterozygous and homozygous for hypothalamic diabetes insipidus. Life Sci. 17: 1307–1314 (1975).
32 Morel, M.; Charbardès, D.; Imbert-Teboul, M.: Methodology for enzymatic studies of isolated tubular segments: adenylate cyclase. Methods Pharmacol. 4B: 297–323 (1976).
33 Moses, A.M.; Streeten, D.H.P.: Differentiation of polyuric states by measurement of responses to changes in plasma osmolality induced by hypertonic saline infusions. Am. J. Med. 42: 368–377 (1967).
34 Moses, A.M.; Miller, M.; Streeten, D.H.P.: Differential diagnosis of polyuria. New Engl. J. Med. 307: 125 (1982).
35 Naik, D.V.; Valtin, H.: Hereditary vasopressin-resistant urinary concentrating defects in mice. Am. J. Physiol. 217: 1183–1190 (1969).
36 Naik, D.V.; Sokol, H.W.: The hypothalamohypophyseal neurosecretory system in mice with vasopressin-resistant urinary concentrating defects. Gen. compar. Endocr. 15: 59–69 (1970).
37 North, W.G.; LaRochelle, F.T.; Jr.; Hardy, G.R.: Radioimmunoassays for individual rat neurophysins. J. Endocr. 96: 373–386 (1983).

38 North, W.G.; Valtin, H.; Cheng, S.; Hardy, G.R.: The neurophysins: production and turnover; in Cross, Leng, The neurohypophysis: structure, function and control, pp. 217–225 (Elsevier, Amsterdam 1983).
39 Pickering, B.T.; North, W.G.: Biochemical and functional aspects of magnocellular neurons and hypothalamic diabetes insipidus. Ann. N.Y. Acad. Sci. *394:* 72–81 (1982).
40 Richter, D.: Vasopressin and oxytocin are expressed as polyproteins. Trends Biochem. Sci. *8:* 278–281 (1983).
41 Robertson, G.L.: Role of radioimmunoassay of vasopressin in diagnosis of diabetes insipidus (this volume).
42 Robinson, D.H.; Jr.; Conrad, K.P.; Edwards, B.R.: Comparison of body fluid compartment sizes in Brattleboro homozygous and Long-Evans rats. Am. J. Physiol. *247:* F234–F239 (1984).
43 Sachs, H.; Fawcett, P.; Takabatake, Y.; Portanova, R.: Biosynthesis and release of vasopressin and neurophysin. Recent Prog. Horm. Res. *25:* 447–491 (1969).
44 Sasaki, S.; Imai, M.: Effects of vasopressin on water and NaCl transport across the in vitro perfused medullary thick ascending limb of Henle's loop in mouse, rat, and rabbit kidneys. Pflügers Arch. *383:* 215–221 (1980).
45 Silverstein, E.: Urine specific gravity and osmolality in inbred strains of mice. J. appl. Physiol. *16:* 194–196 (1961).
46 Silverstein, E.; Sokoloff, L.; Mickelsen, O.; Jay, G.E., Jr.: Primary polydipsia and hydronephrosis in an inbred strain of mice. Am. J. Path. *38:* 143–159 (1961).
47 Skadhauge, E.; Schmidt-Nielsen, B.: Renal function in domestic fowl. Am. J. Physiol. *212:* 793–798 (1967).
48 Skadhauge, E.; Maloiy, G.M.O.: The intestine and osmoregulation; in Jorgensen, Skadhauge, Osmotic and volume regulation, pp. 325–343 (Academic Press, New York 1978).
49 Sokol, H.W.; Valtin, H. (eds): The Brattleboro rat. Ann. N.Y. Acad. Sci. *394:* 1–828 (1982).
50 Stern, P.; Valtin, H.: Verney was right, but... New Engl. J. Med. *305:* 1581–1582 (1981).
51 Strewler, G.J.; Fallon, B.G.; Orloff, J.: Defective protein phosphorylation in renal medulla of vasopressin-resistant mice. Biochem. biophys. Res. Commun. *103:* 713–720 (1981).
52 Valtin, H.; Schroeder, H.A.: Familial hypothalamic diabetes insipidus in rats (Brattleboro strain). Am. J. Physiol. *206:* 425–430 (1964).
53 Valtin, H.: Sequestration of urea and nonurea solutes in renal tissues of rats with hereditary hypothalamic diabetes insipidus: effect of vasopressin and of dehydration on the countercurrent mechanism. J. Clin. Invest. *45:* 337–345 (1966).
54 Valtin, H.: Inherited causes of nephrogenic diabetes insipidus in mice; in Wesson, Fanelli, Recent advances in renal physiology and pharmacology, pp. 187–196 (University Park Press, Baltimore 1974).
55 Valtin, H.: Genetic models for hypothalamic and nephrogenic diabetes insipidus; in Andreoli, Grantham, Rector, Disturbances in body fluid osmolality, pp. 197–215 (Am. Physiological Soc., Bethesda 1977).
56 Valtin, H.: Renal dysfunction: mechanisms involved in fluid and solute imbalance, pp. 31–38 (Little, Brown, Boston 1979).

57 Valtin, H.: The discovery of the Brattleboro rat, recommended nomenclature, and the question of proper controls. Ann. N.Y. Acad. Sci. *394:* 1–9 (1982).
58 Valtin, H.: Renal function: mechanisms preserving fluid and solute balance in health; 2nd ed. pp. 176, 185 (Little, Brown, Boston 1983).
59 Verney, E.B.: The antidiuretic hormone and the factors which determine its release. Proc. R. Soc. *135:* 25–106 (1947).
60 Williams, R.H.; Henry, C.: Nephrogenic diabetes insipidus: transmitted by females and appearing during infancy in males. Ann. intern. Med. *27:* 84–95 (1947).
61 Zerbe, R.L.; Robertson, G.L.: A comparison of plasma vasopressin measurements with a standard indirect test in the differential diagnosis of polyuria. New Engl. J. Med. *305:* 1539–1546 (1981).

Heinz Valtin, Department of Physiology, Dartmouth Medical School,
Hanover, NH 03756 (USA)

Physiology of Secretion of Vasopressin

Tamara Vokes, Gary L. Robertson

University of Chicago, Chicago, Ill., USA

Introduction

Regulation of vasopressin (VP) secretion has now been studied for nearly half a century. During the first few decades, it was discovered that the osmotic pressure of body water and changes in blood pressure or volume were the main determinants of VP release. At the same time, a role for a number of less important factors has been proposed. With the development of a sensitive and specific radioimmunoassay [1], previous concepts of VP physiology, based on bioassays for antidiuretic hormone and measurements of urine flow, were verified, revised and extended. The last decade has advanced the understanding of the mechanisms that are involved in the regulation of VP secretion by discovering or defining a number of physiological and pathological factors that affect it. In the following sections, the most important regulators will be presented in some detail while other influences will be discussed briefly, with an attempt to define their relative significance.

Osmoregulation

The osmotic pressure of body fluids is normally the major determinant of VP secretion [2]. The functional properties of this regulatory system recently have been elucidated (fig. 1) [3, 4]. Below a certain level of plasma osmolality, conveniently termed osmotic threshold, VP is low or undetectable. With an increase in osmolality above the threshold value, the level of VP rises in a linear fashion. The slope of the regression line of

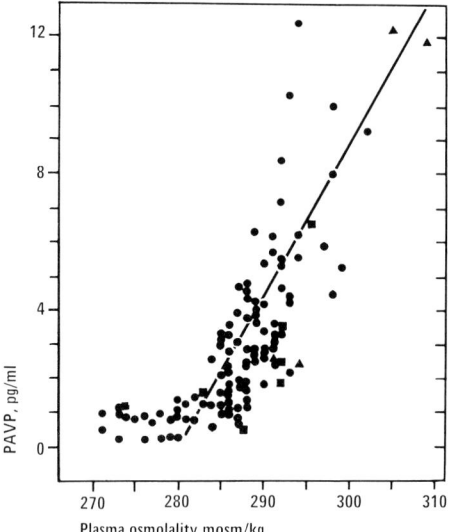

Fig. 1. The relationship of PAVP to plasma osmolality in healthy adults.

VP on plasma osmolality defines the sensitivity of the osmoregulatory system while the x-axis intercept provides a measure of its threshold or set point. It is still unclear if the osmoregulatory system exhibits true threshold behavior or functions in a continuous manner [3, 5]. However, the concept of an osmoreceptor threshold remains useful for clinical and physiological purposes because it describes an important physiologic variable – the level of plasma osmolality at which a maximum water diuresis develops. In so doing, it provides a measure of the level of tonicity at which body water is maintained.

The osmoregulatory system functions with a remarkable sensitivity: an increase in plasma osmolality of only 1% will cause a rise in VP of a magnitude sufficient to effectively change the rate of water excretion. The system also functions with great precision. This property is not evident from figure 1 where the scattering of points is due largely to individual differences in the set and/or sensitivity of the osmoregulatory system. However, it is apparent when multiple determinations of plasma osmolality and VP are made during the infusion of hypertonic saline and the data from each individual are subjected to regression analysis. With this approach, a high degree of correlation is consistently observed (correlation coefficient ranging from 0.88 to 0.99) even though there are

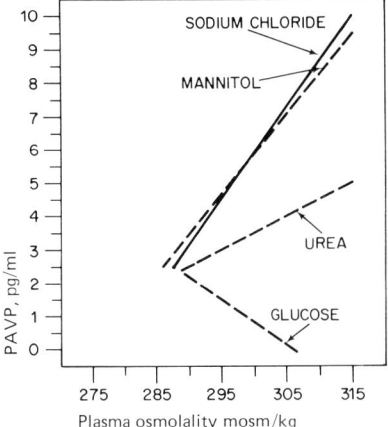

Fig. 2. The relationship of PAVP to plasma osmolality in healthy adults during the infusion of hypertonic solutions of various solutes.

large individual differences in both the slope (0.12–1.66 pg/ml/osm/kg) and threshold (276–291 mosm/kg) [2, 6–8].

The basis of these individual differences is not well understood. However, while the threshold of osmoregulation may vary on repeated testing in the same person, the sensitivity seems to remain relatively constant over a period as long as 5 years [7]. In addition, studies in twins have shown evidence of a significant genetic variance in both the threshold and sensitivity [7]. This suggests that the characteristics of osmoregulation are, at least in part, genetically determined.

The specificity of osmoregulation has been investigated in recent years [9]. Sodium and its anions, which normally contribute more than 95% of the total osmotic pressure of plasma are the most potent solutes known to stimulate VP secretion [1, 10]. Certain sugars such as sucrose and mannitol are equally effective (fig. 2), suggesting that the osmoreceptor responds generally to changes in plasma osmolality rather than, specifically, to changes in sodium concentration. However, when hypertonic urea is infused, plasma VP (PAVP) increases little relative to the rise in measured plasma osmolality, and the infusion of hypertonic glucose results in a decrease in circulating VP (fig. 2). Hence, the osmoreceptor appears to be stimulated by a rise in the plasma concentration of some solutes but not of others. This specificity is of considerable practical importance since it means that neither plasma osmolality nor sodium concentrations can be relied upon to provide a universally valid reference point for assessing VP function. Thus, the monogram shown in figure 1 can be utilized only when blood urea and glucose are within the normal range. Measurements of plasma sodium may not provide a totally reliable correlation either because other naturally occurring solutes, such as glycerol or ketone bodies, may affect VP secretion.

The mechanism by which the osmoreceptor so effectively discriminates between different solutes is not quite clear. Since sodium and mannitol are known to be obligatory extracellular solutes while urea and glucose can penetrate cells, it is likely that the osmoreceptor is stimulated by an osmotically induced decrease in the cellular water content. According to this hypothesis, the capacity of a given solute to stimulate VP would be inversely related to the rate at which it passes from plasma into the osmoreceptor [10]. This penetration rate is probably determined by the permeability characteristics of the osmoreceptor cell membrane since it now seems clear that these neurons are outside the blood-brain barrier. Thus, *Zerbe and Robertson* [9] and *McKinley* et al. [11] have shown that urea had little or no effect on PAVP secretion even though it penetrates the blood-brain barrier poorly. Others have suggested that there may be a separate sodium or osmoreceptor in contact with cerebrospinal fluid (CSF) and beyond the blood-brain barrier [12, 13]. However, these receptors probably have little or no importance for the systemic osmoregulation since: (a) intravenous infusion of hypertonic saline causes release of VP before an increase in CSF sodium occurs [13], and (b) systemic administration of urea does not lead to a significant increase in VP even though it increases CSF sodium [11].

The exact location of the osmoreceptor is still not known with certainty despite continued use of refined anatomical and functional probes. The pioneering work of *Verney* has suggested that osmoreceptor neurons are located somewhere in the anterior hypothalamus, in an area supplied by the anterior communicating artery [3]. This view is consistent with subsequent findings in animals [14] and in a patient who developed hypernatremia with absent osmoregulation of thirst and VP following ligation of the anterior communicating artery [15]. Interestingly, this patient could secrete normal amounts of VP in response to hemodynamic and emetic stimuli. This observation suggested that the osmoreceptor and neurosecretory neurons were anatomically separate, a hypothesis consistant with numerous ablative, electrophysiologic and iontophoretic studies [3, 4].

The exact structure that is responsible for osmoreception is still under dispute. In animals, hypothalamic knife cuts of the periventricular tissue of the anteroventral third ventricle (AV3V), nucleus medianus and particularly the organum vasculosum laminae terminalis have all been reported to result in abnormalities in VP secretion and drinking behavior [16–19]. However, it remains unclear if these lesions destroy the osmoreceptor per se, sever the connection between the osmoreceptor and neurosecretory cells or affect some other structure necessary for the function of the osmoregulatory system.

Baroregulation

The secretion of VP can also be affected by alterations in blood volume and/or pressure [2–4, 22–24]. These hemodynamic changes are thought to be detected by the stretch receptors in the left atrium, aorta and carotid sinus [20–24]. The signals are then transmitted via vagal and glossopharyngeal nerves probably to the nucleus tractus solitarius of the medulla oblongata (fig. 3). From there, post-synaptic pathways, that are partly noradrenergic, project to the magnocellular elements in the supraoptic and/or paraventricular nuclei [3, 25]. Input from the baroreceptor appears to be primarily inhibitory under basal conditions since vagotomy results in a rise in VP secretion. Thus, hypotension and hypovolemia probably stimulate VP secretion by releasing the tonic inhibition. Conversely, an increase in blood pressure or volume would suppress VP through an activation of the inhibitory pathways.

The functional properties of the baroregulatory system differ from those of the osmoregulatory [26] (fig. 4). While the relationship between plasma osmolality and VP appears to be linear, the relationship between blood volume or pressure and PAVP is best described by an exponential function (fig. 4). Therefore, under ordinary physiological circumstances, changes in VP are not observed until blood pressure changes by at least 5–10%. Beyond this level, PAVP rises exponentially in proportion to hypotension. A large decrease in blood pressure (30% or more) leads to very high levels of VP which, by stimulating arteriolar contraction, may indeed help to stabilize cardiovascular function. The effects of hypovolemia are more difficult to characterize precisely. However, it appears that the stimulus-response pattern is nearly identical to that observed with hypotension (fig. 5). For example, removal of up to 7% of total

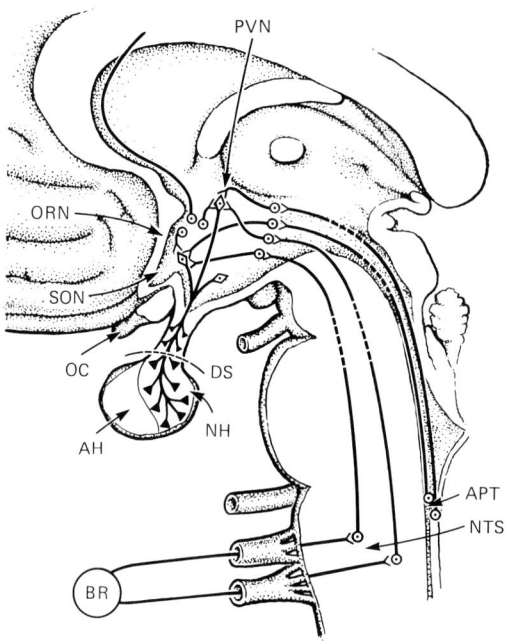

Fig. 3. Schematic representation of the neurohypophysis and its principal regulatory afferents. AH = Adenohypophysis; NH = neurohypophysis; OC = optic chiasm; SON = supraoptic nucleus; PVN = paraventricular nucleus; ORN = osmoreceptor neurons; BR = baroreceptors; NTS = nucleus tractus solitarius; APT = area postrema emetic center; DS = diaphragma sellae. From ref. [4].

blood volume has no effect on PAVP in healthy adults [23], while upright posture, which reduces central or effective blood volume by 10–15%, produces a slight increase. However, the combination of orthostasis and phlebotomy, which probably reduces circulatory blood volume by more than 20% and causes a significant fall in arterial pressure, produces marked increases in PAVP [23, 24]. Similarly, upright posture in patients with impaired compensatory pressure response to orthostasis will produce a dramatic rise in VP unless the orthostatic hypotension is associated with a defect in the afferent or central connection of the baroregulatory reflex (fig. 6) [27]. Hypervolemia and/or hypertension appear to inhibit VP release via opposite effects on the same baroreceptor mechanisms.

The relative insensitivity of VP to modest changes in blood volume or pressure (when compared to its sensitivity to small osmotic changes) has certain practical implications. It means that baroregulatory mech-

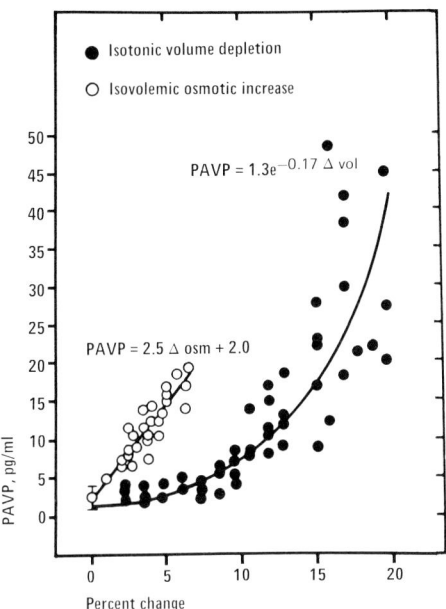

Fig. 4. The effect of hypovolemia or hypotension on relationship of PAVP to plasma osmolality.

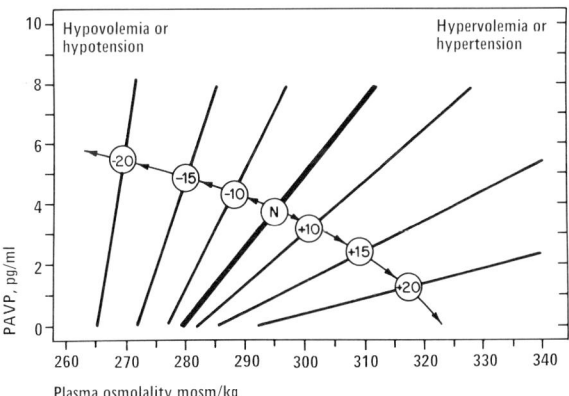

Fig. 5. The effect of hypovolemia or hypotension on the relationship of PAVP to plasma osmolality in conscious rats. Blood volume was reduced 15% by intraperitoneal injection of polyethylene glycol. Mean arterial pressure was reduced approximately 15% by subcutaneous injection of isoproternal hydrochloride and osmolality varied with intraperitoneal injection of hypotonic, isotonic or hypertonic saline. From ref. [2].

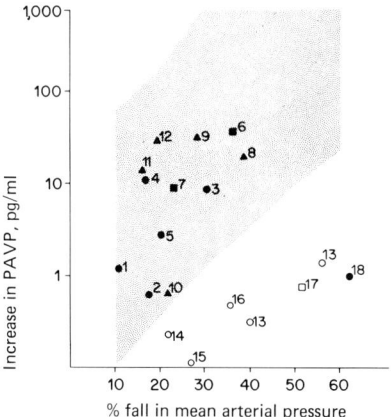

Fig. 6. PAVP increase in patients with orthostatic hypotension. The shaded area represents the 99% confidence interval for hormonal individuals. The etiologic diagnoses are indicated by the following symbols: ○ = Shy-Drager syndrome; □ = baroreceptor deafferentation; ● = periphernol antonomic neuropathy; ■ = dehydration; ▲ = unknown. From ref. [27].

anisms play little or no role in the physiological control of water balance in man. Under ordinary circumstances, total body water rarely changes by more than 1–2%, an amount far too small to affect VP secretion by hemodynamic pathways. This concept is consistent with findings that patients in whom osmoreceptor function has been destroyed exhibit a markedly deficient VP response to dehydration even though their baroregulatory afferents and neurohypophysis remain intact [15, 24]. It also explains an older observation that sodium depletion produced by sweating does not result in a significant reduction in the tonicity of body water until extracellular fluid volume falls by more than 10%, the point at which volume stimuli may be expected to exert an increasingly important influence on VP secretion and renal water excretion [4].

Transient changes in blood pressure and/or effective blood volume may explain an observation made by some [28–30] but not other investigators [31] that VP is secreted episodically, in spurts which do not correspond to changes in plasma osmolality. [4].

Interaction of osmotic and hemodynamic influences has been reviewed extensively in the recent literature [2–4, 23, 24, 32]. Hemodynamic influences appear to alter the set of the osmostat (fig. 5) indicating that the two control systems converge and act on the same population of neurosecretory cells [4]. This concept is further supported by electrophysiologic studies in rats which show that most individual neurohypophyseal neurons alter firing rates in response to both osmotic and hemodynamic stimuli [33]. According to this concept, a decrease in plasma volume and/or pressure will permit normal osmoregulation of VP but the threshold for release of VP will be lower by an amount proportional to the degree of hypovolemia or hypotension. While all studies show a lowering of the osmotic threshold in response to hypovolemia, some also show an increase in the slope or sensitivity of the VP response to osmotic stimuli [34]. This means that under conditions of hypovolemia or hypotension, the level of VP for any given value of plasma osmolality will be higher than under normovolemic (normotensive) conditions. Thus, if plasma osmolality is sufficiently lowered, it should suppress VP. In confirmation of this concept, in recent studies in normal volunteers made hypovolemic by administration of furosemide and upright posture, water loading maximally suppressed VP but at a lower level of plasma osmolality than in a euvolemic state. Moreover, the regression line of PAVP on plasma osmolality obtained during the hypertonic saline infusion was shifted to the left, indicating lowering of the threshold of osmoregulation [*Robertson*, unpublished observations].

The resetting of the osmostat due to the changes in plasma volume is also manifested in certain pathophysiological conditions. For instance, in a patient with hypovolemia and hyponatremia due to an isolated aldosterone deficiency, both VP and thirst were normally responsive to osmotic stimuli but the threshold for each response was lowered [24]. The resetting of the osmostat was completely corrected by volume expansion with isotonic saline and fludrocortisone. An example of the opposite effect of hypervolemia or hypertension is seen in patients with primary aldosteronism and low renin hypertension. It is likely that the upward resetting of the osmostat is caused by hypervolemia (since it is corrected by diuretic therapy) but the possible effect of hypokalemia cannot be excluded.

Emetic Stimulation

Serendipitous observations made during administration of water load or intravenous ethanol to normal volunteers have led to the discovery that nausea is one of the most potent stimuli for release of VP [2].

A systematic study revealed that administration of apomorphine to human volunteers resulted in a marked increase in PAVP in 10 subjects who experienced nausea while no increase was observed in 3 that remained asymptomatic [35]. Pretreatment with antiemetic agents, fluphenazine and haloperidol, prevented both nausea and the rise in VP. It is of interest that rats, which lack an emetic reflex, show little or no increase in VP even when given very high doses of apomorphine. VP release appears to be independent of any other known stimuli and seems to be mediated by the emetic center since it occurs when nausea is induced by vestibular mechanisms as well as via the chemoreceptor trigger zone [3, 35]. The stimulus for VP release appears to be nausea per se since an increase in VP is observed even in subjects who do not vomit.

Glucopenic Stimulation

Hypoglycemia has long been known to cause the release of many hormones. Recently, it has been shown that symptomatic, insulin induced hypoglycemia also stimulated secretion of VP. This effect occurs in both rats [36] and man [37, 38] and appears to be proportionate to the degree of hypoglycemia achieved. The response of VP to hypoglycemia is not mediated by osmotic and hemodynamic stimuli since it was observed in a patient with an osmoreceptor ablation and, if anything, is associated with an increase in blood pressure [38]. However, hypoglycemia appears to act synergistically with osmotic stimuli since its effect on VP secretion is totally abolished by water loading [36]. The mechanism by which hypoglycemia causes a release of VP is unclear but it appears to depend on intracellular glucopenia since a similar effect can be produced by administration of 2-deoxy-glucose to both the rat [39] and man [40].

Other Influences

Neurogenic Control of Vasopressin
Central nervous system mediators of VP release have recently been reviewed in detail [19]. This subject must be approached cautiously because studies of neurotransmitters and other centrally active substances are hampered by many difficulties. Firstly, many of the substances employed may have an indirect effect on VP, by changing the blood pres-

sure, stimulating the emetic center or altering the concentration of another mediator of VP secretion. Secondly, our knowledge of the function of the central nervous system is still limited and thus many of the treatments that are given with an intention of simulating a physiological event may in effect produce pharmacological effect or even damage the structures that are being examined. Finally, the routes of administration that are employed may differ from those that are, under normal circumstances, accessible to the substance in question. In spite of the above-mentioned reservations, our understanding of neural influences on VP secretion has advanced significantly in recent years and promises to be even more interesting in the future.

Catecholamines have been known for many years to influence VP secretion. Norepinephrine, an α-agonist, suppresses VP probably indirectly by increasing the mean arterial pressure [4]. The β-agonist isoproterenol has the opposite effect on both VP and blood pressure [4]. However, catecholamines may also have central effects since norepinephrine containing nerve terminals have been found in supraoptic and paraventricular regions of the hypothalamus [19]. At least some of these noradrenergic connections may arise in the hypothalamus and medulla and mediate hemodynamic influences [25]. Intracerebroventricular administration of norepinephrine in dogs caused a decrease of plasma VP as well as blood pressure, suggesting that norepinephrine has a direct inhibitory effect [41]. This hypothesis is supported by the observation that clonidine, a centrally active α_2-agonist, also decreases plasma VP [42]. However, other studies suggest that central noradrenergic pathways are stimulatory rather than inhibitory. Thus, when brain catecholamines in rats were depleted 67% by intraventricular injection of 6-hydroxydopamine, the response of VP to both osmotic and nonosmotic stimuli was diminished [43]. On the other hand, when hypothalamic norepinephrine was reduced 50% by a chronic subcutaneous administration of α-methyl-*p*-tyrosine, a much less toxic drug without hemodynamic or osmotic effects, there was no effect on the response to osmotic stimuli [44]. The difference between the two studies is unexplained. It may reflect differences in the extent of catecholamine content reduction. Or it could be due to the fact that 6-hydroxydopamine causes a nonspecific damage to periventricular area [45]. Thus, the information available suggests that catecholamines are relevant in central mediation of VP release but the exact role and the mechanisms involved are not yet clear.

Acetylcholine stimulation of VP release has been suggested by several in vivo and in vitro studies [19]. However, it is presently unclear which cholinergic receptor may be responsible for this effect, which regulatory afferent it may subserve and what mechanism is involved.

Opiate effects on VP have been a matter of debate for many years. Initial studies proposed that opiates stimulated VP secretion, although it was not clear if the effect was direct or through activation of hemodynamic and/or emetic influences [3]. Some of the reports, however, indicated that opioid compounds decreased VP levels [3]. In the more recent literature, the controversy continues with some investigators reporting a stimulatory [46–50] and others an inhibitory effect of opioids on VP [51–57]. Although the reasons for the discrepancy are not completely clear, there are several possible explanations: (1) If a compound employed has both agonist and antagonist properties, it is, to some extent, a matter of choice whether the observed effect will be attributed to the agonist or the antagonist. This is well illustrated in a study of *Miller* [46] who found that butorphanol, an opiate with mixed properties, decreased the amount of VP excreted in the urine and concluded that this effect was mediated by the antagonist property. However, a later study [56] has shown that the effect of butorphanol was actually similar to that of morphine and that both were blocked by naloxone, indicating that butorphanol induced suppression of VP was mediated by its agonist property. (2) It is very likely that a certain class(es) of opiate receptors has a considerably greater influence on VP than the others. Thus, some of the discrepancies in the published data may be due to the fact that the different opiates used acted on different receptors. (3) In many of the studies, opiate induced hemodynamic, emetic and other potentially important stimuli of VP release are not excluded. In addition, if an effect on VP is judged from the effect on urine flow, erroneous conclusions can be made since opiates influence urine production by effects that are partially VP-independent [58, 59]. Generally speaking, the studies that report an inhibitory effect of opiates on VP have been more recent, relied on radioimmunoassay rather than bioassay or measurements of urine flow and have been more careful in excluding hemodynamic and emetic effects. In addition, a preliminary study has revealed that the opioid antagonist naloxone partially reverses alcohol-induced suppression of VP, further supporting the contention that opiates have a primarily inhibitory effect on VP [60]. This hypothesis would also postulate that the effect on VP is similar to the generally inhibitory influence of opiates on other hypophyseal and hypothalamic hormones and neural function in general, with the stimulatory effects being observed primarily through the inhibition of an inhibitory pathway or hormone [61]. Thus the weight of evidence suggests that endogenous opiates have an inhibitory role in the regulation of VP secretion. The nature of this inhibitory effect appears to be an upward resetting of the osmostat [56]. It is likely that this role is physiologically important since encephalinergic neurons and receptors are present in relatively high concentrations in the intermediate and posterior lobes of the pituitary [19].

Angiotensin II has been shown to stimulate VP release in some studies in animals [4, 19, 62] and man [4, 22], while others have failed to detect such an effect [19]. The explanation for the discrepancy may be in the fact that angiotensin seems to stimulate VP only when combined with an osmotic stimulus [19]. Thus, angiotensin may act by sensitizing the osmoreceptor to osmotic stimuli. This hypothesis is supported by several other findings: (1) Exposure to angiotensin of a complete hypothalamo-neurohypophyseal system, but not of the isolated neural lobe, resulted in a release of VP, suggesting that a hypothalamic structure mediates the release [63]. (2) Lesions of the A3V3 region of the hypothalamus, which is thought to either contain the osmoreceptor or play a vital role in the transmission of the information from the osmoreceptor to the neurosecretory cells, reduces the VP response to angiotensin [64]. (3) Autoradiographic studies have demonstrated binding of angiotensin to circumventricular organs and in particular to the organum vasculosum laminae terminalis [65] which is presumed to be the site of the osmoreceptor and possibly the thirst center [19].

Prostaglandines are well known to counteract the hydro-osmotic response to VP at the level of the kidney. However, recently they have been found to have a direct stimulatory effect on VP release when given peripherally, intracerebroventricularly or in vitro [19]. The administration of inhibitors of prostaglandin synthesis led to an attenuation of VP response to hypertonicity in some studies [19, 66, 67] but not in others [68].

Other neurotransmitters have also been found to effect VP [19]. γ-Aminobutyric acid (GABA) and glycine, which are thought to be inhibitory neurotransmitters, attenuate the VP response to hypovolemia [19]. Substance P, a peptide widely distributed in the brain, stimulates VP release [19]. VP itself was reported to have an inhibitory effect on plasma VP levels when infused intracerebroventricularly [69]. On the other hand, systematic administration of VP produced a delayed rise in endogenous VP secretion, suggesting that under certain circumstances it may have a positive feedback effect [70].

Effect of Menstrual Cycle and Pregnancy

The effect of normal menstrual cycle on VP has been investigated by several laboratories. *Forsling* et al. [71] found that basal levels of VP were significantly higher at midcycle while *Punnonen* et al. [72] observed a similar trend that did not reach statistical significance. We have examined a response of VP to osmotic stimulation by hypertonic saline infusion and osmotic suppression by an oral water load, and found that both the threshold for VP release and thirst threshold as well as basal plasma osmolality were significantly lower in the luteal than in the follicular phase [73]. An even larger decrease in threshold for thirst and VP release was observed both in the rat [74] and human *pregnancy* [75]. The reason and mechanism for the downward resetting of the osmostat in pregnancy and the luteal phase of the menstrual cycle are unclear. In cannot be attributed to volume depletion since it has been shown that pregnancy in both humans and rats lead to an increase in plasma volume. Furthermore, recent studies by *Barron* et al. [76] have shown that pregnant rats, despite a near doubling of the intravascular space, secrete VP in response to volume depletion in a manner similar to nonpregnant controls, suggesting that the relationship between total blood volume and VP secretion is altered during gestation in such a way that the expanded blood volume is recognized as normal [*Amico,* this volume].

Effects of Hormones and Endocrine Disorders

Skowsky et al. [77] reported that *androgens* decreased while *estrogens* increased PAVP in rat. The latter effect, however, was not reproduced by others [74, 78]. *Forsling* et al. [79] observed that administration of estrogen to postmenopausal women increased basal VP, the effect being partially reversed by progesterone.

Diabetes mellitus, occurring naturally in humans or induced by streptozotocin in rats, has been associated with changes in the secretion of VP. *Walsh* et al. [80] have reported that VP levels were markedly increased, for the level of corrected plasma osmolality, in patients with diabetic ketoacidosis and that the levels decreased with treatment. In another study, a similar increase in VP was observed in both ketotic and nonketotic patients with uncontrolled diabetes [81]. In streptozotocin-treated rats, levels of VP significantly higher than in controls were observed already 24–48 h after the administration of the drug [*Vokes and Robertson,* unpublished observations]. In another group of similarly treated rats, the response of VP to osmotic stimuli induced by administration of hypertonic saline was considerably greater in diabetic rats than in the healthy controls [82]. Likewise, hypertonic saline infusion led to an increase in VP both in normal controls and in hyperglycemic, insulinopenic patients with type I diabetes but the calculated threshold for VP release was lower in the diabetics [83]. Finally, in a group of diabetic patients, the sensitivity of osmoregulation was increased under insulinopenic, hyperglycemic conditions when compared to euglycemic, insulin-replete ones [84].

Effect of Pharmacologic Agents

Several drugs have been found to directly or indirectly influence secretion of VP. In addition to those that were discussed under neural control of VP, several other drugs are known to have an effect and their use should be excluded when unexplained changes in VP response are encountered.

Anesthetics are known to elevate levels of VP both in plasma and in CSF. This effect is thought to be caused by suppression of autonomic influences that exert tonic inhibition on VP release.

Lithium, which is known to cause nephrogenic diabetes insipidus, has recently been found to increase the sensitivity of VP response to osmotic stimulation [85].

Carbamazepine, which can at times produce hyponatremia presumably by enhancing the renal sensitivity to VP, was shown to reduce the sensitivity of VP response to osmotic stimulation [86]. This effect occurs independent of changes in blood volume or pressure.

Nicotine stimulation of VP secretion has been known for many years. Originally, it was attributed to a direct effect of circulating nicotine on the hypothalamus via cholinergic receptors believed to be involved in the control of VP release [3]. However, the effect may actually be more indirect since the doses required to release VP almost always result in nausea or hypotension [3]. More recent studies indicate that the VP response elicited by smoking is not due to emetic or hypotensive stimuli [3] and may not even be mediated by circulating nicotine but rather by an airway-specific mechanism [87].

Miscellaneous

Nociceptive influences and stress in general have long been believed to stimulate VP secretion [88]. However, in controlled experiments in rats, light ether anesthesia, water immersion and pain failed to increase VP even though all of them produced very large rises in plasma corticosterone [89]. In another study, manual compression of rats and electric shock resulted in VP release while a number of other 'stressful' stimuli did not [90]. The effect of the electric shock could have been by direct electric stimulation of the neurohypophysis, while the manual compression probably caused a decrease in circulating blood volume, operating as a hemodynamic stimulus for VP release. In a recent study in healthy human volunteers, immersion of an extremity in ice water produced pain and an increase in cortisol, pulse and blood pressure but had no effect on PAVP. It is likely that in the majority of older studies, VP release that was attributed to stress was actually caused by hypotension or nausea and that pain and stress have no direct effect on VP secretion [3].

Aging in healthy humans appears to affect the regulation of VP secretion [91–93]. Administration of hypertonic saline to a group of young (22–48 years) and a group of old people (52–66 years) revealed that there was an increase in the sensitivity of the osmoregulatory response in elderly [91]. This may be an attempt to compensate for reduced renal sensitivity to VP in elderly. In contrast to hypersensitivity to osmotic stimuli, VP response to orthostasis was subnormal in many elderly subjects in spite of normal cardiovascular responses [93]. This suggests that in the elderly, there may be an abnormality somewhere in the post-synaptic pathway between medulla and neurohypophysis which would diminish VP responsiveness to changes in blood volume or pressure. Such an abnormality could also explain the exaggerated VP response to hypertonic saline infusion since it would eliminate the mild inhibitory effect of the volume expansion normally associated with this procedure.

Hypoxia due to high altitude [94] or lung disease [95] is believed to produce significant changes in salt and/or water excretion. In animals, experimental hypoxia was reported to produce an increase in VP [96, 97]. However, in controlled studies such an effect could be observed in humans only if hypoxia was accompanied by hypotension or nausea [98–100].

Hypercalcemia is known to diminish renal responsiveness to VP

[101]. Recently, it has been shown that hypercalcemia increased the sensitivity of the VP response to osmotic stimulation [102]. The administration of calcium channel blocking agents, nifedipine and verapamil reduced the response of VP to both osmotic and hemodynamic stimuli in conscious rats [19]. Similarly, in vitro studies on explanted neural lobes or neurohypophyses demonstrated that calcium stimulated while slow channel blockers inhibited release of VP [19]. All of these studies indicate that calcium ion is necessary for the stimulation of the hypothalamus and/or secretion from the neurohypophysis. This conclusion is in agreement with the knowledge that calcium is involved in membrane activation, synaptic transmission and hormone release.

Hypokalemia, which also causes renal resistance to VP action, was found to decrease the sensitivity of VP response to osmotic stimuli in rats given low potassium diet [103].

Finally, it has been observed that some pharmacological agents or pathological conditions that decrease renal sensitivity to VP at the same time stimulate its release from the neurohypophysis. This effect is seen with lithium, hypercalcemia, old age and some cases of nephrogenic diabetes insipidus. Conversely, therapy with carbamazepine increases the renal sensitivity to VP but decreases its secretion. The factor that influences pituitary release is obviously not plasma osmolality since the release of VP is greater than expected for the given level of plasma osmolality. Since this *putative mediator* 'informs' the pituitary of the renal responsiveness to VP, it probably originates in the kidney. One possible candidate would be prostaglandin E_2 which opposes the renal effect of VP but stimulates its release from the neurohypophysis. However, urinary prostaglandin excretion was not affected when VP action was inhibited by lithium, catecholamines, probenecid, demeclocycline or calcium [104], two of which are known to alter the VP response to osmotic stimuli. Although urinary measurements of prostaglandins may not reflect their changes in the circulation, these findings make it less likely that prostaglandins are the putative direct feedback signal between the kidney and VP secretion.

Vasopressin in the Cerebrospinal Fluid

The growing interest in the interaction between various hormones and brain and the development of improved techniques for sampling and assaying CSF has prompted new investigative interest in the dynamics and physiologic significance of CSF VP [105].

Little, if any, CSF VP orginates from the blood since only about 1% of exogenously administered arginine VP (AVP) can be recovered in the CSF. Thus, most CSF VP must be derived by direct secretion. In support of this concept is the finding that vasopressinergic hypothalamic neurons send projections to median eminence, lateral amygdala and other parts of brain [3, 105]. It is likely that at least part of the CSF VP is generated by neurosecretory cells different from those that secrete VP into the systemic circulation since patients with severe diabetes insipidus and no detectable VP in plasma have significant amounts of VP in their CSF [105]. In most other cases, including normal subjects and patients with syndrome of inappropriate secretion of antidiuretic hormone (SIADH) [105] or affective illness [106], the levels of AVP in CSF are slightly but significantly lower than those found in plasma. However, a plasma-CSF gradient in the opposite direction has been reported by another laboratory [107]. The reason for this discrepancy is not yet clear. Because of possible role in cognitive function [105], CSF VP has also been studied in patients with various forms of psychiatric illnesses [106]. In affective illness, the levels of VP in CSF were significantly lower in nonpsychotic patients with unipolar or bipolar depression than in controls and also lower in patients with bipolar depression than in patients with bipolar disorders who were in the manic phase [106]. In patients with anorexia nervosa there was an absolute increase in the levels of VP in the CSF or reversal of the normal (<1.0) CSF plasma ratio of VP [108]. The cause and significance of these observations are presently unclear.

Regulation of VP secretion into CSF is poorly understood but appears to respond, at least in part, to the same factors that influence secretion into plasma [105, 109]. Thus, there is a significant positive correlation between plasma and CSF levels of VP. The only exception appears to be diabetes insipidus due to destruction of pars nervosa [105]. *Szczepauska-Sadowska* et al. [110] reported that in conscious dogs, hemorrhage, dehydration and intravenous infusion of hypertonic saline caused an increase in VP, both in plasma and CSF, collected from the anterior part of the third cerebral ventricle. Dehydration and intravenous infusion of hypertonic saline also produced a rise in both plasma and CSF VP but the latter was not statistically significant [110]. In a study of *Wang* et al. [13], in anesthetized dogs, VP concentration in CSF collected from cisterna magna increased promptly when hypertonic saline was administered intracerebroventricularly but not when administered intravenously (PAVP increased in both cases), suggesting that additional cerebroventricular osmoreceptor may regulate a release of VP into CSF in response to an increase in CSF osmolality. If so, the level of CSF sodium required to stimulate the hormone must be very high since CSF osmolality rises in parallel with plasma during hypertonic saline infusion. However, certain factors have been reported to influence VP levels in CSF but not in plasma. Thus, in rats [110] and cats [111–113], diurnal variations with peak levels during the daylight hours are observed in CSF but not in plasma. Change of the timing of the light-dark cycles resulted initially in a disturbance and later readjustment of the pattern to the new timetable.

The physiologic role of VP in the CSF is likewise unclear although a number of attractive hypotheses have been advanced [105]. It has been proposed that VP serves as a neurotransmitter, functions as a releasing factor for anterior pituitary hormones (particularly adrenocorticotropic hormone, ACTH) or mediates the cortical perception of thirst. The latter two possibilities are unlikely to be of any importance, however, since Brattleboro rats, which completely lack VP, have a relatively normal pituitary-adrenal response to stress and normal thirst. An earlier hypothesis that CSF VP may serve as a direct feedback regulator of the supraoptic nucleus has recently been supported by previously mentioned finding [69] that administration of VP into lateral ventricles decreased PAVP. *Rodriquez and Heller* [cited 105] have proposed that VP may act on the choroid plexus to modify the rate of formation or chemical composition of CSF [105]. *Raichle and Grubb* [114] have suggested that VP may regulate brain water content since AVP administered into the lateral ventricle of a monkey induced a transient increase in cerebral capillary permeability without affecting cerebral blood flow. Although this is an interesting hypothesis, the results presented should be viewed with caution since the variables examined are difficult to quantitate. Several authors have postulated a role for CSF VP in the central sympathetic regulation of the vasomotor function [for review, see 115]. *De Wied* and co-workers [116–118] have proposed that AVP was 'a memory consolidation hormone' since when it was administered into the cerebral ventricles of rats, it induced resistance to extinction of a conditioned avoidance response. Finally, the finding of altered concentration of VP in the CSF in patients with affective illness [106, 108] and improvement of

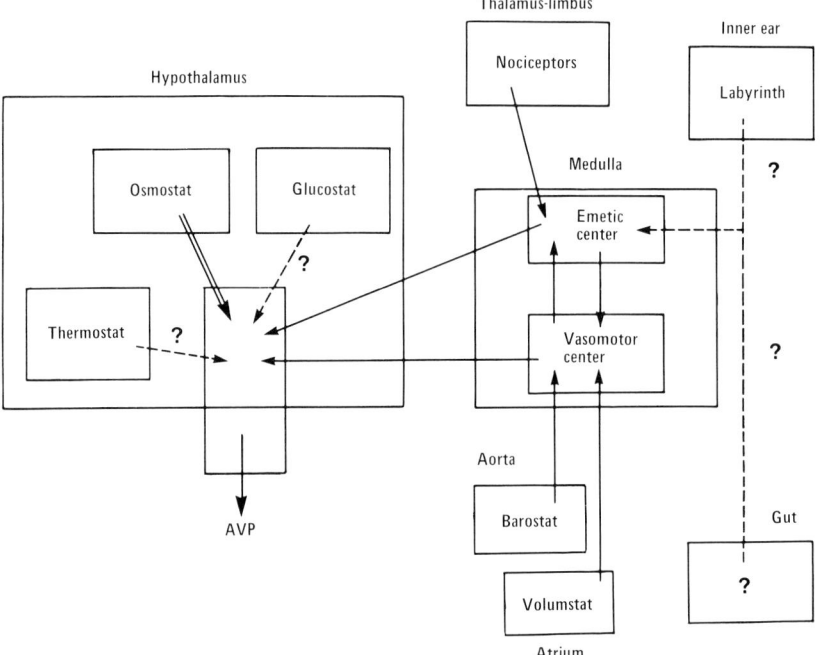

Fig. 7. Schematic representation of the variables and receptors that influence VP secretion. From ref. [24].

depressive symptomatology during therapy with the VP analog 1-desamino-8-*D*-arginine-vasopression (DDAVP) [106] suggest that VP may be involved in neurobiochemical processes of normal behavior.

Conclusion

A schematic representation of the variables that influence VP secretion is shown in figure 7. Under normal circumstances, secretion of VP is regulated primarily by small changes in plasma osmolality. Decrease in blood pressure and/or volume is a potent stimulus but only after the reduction has reached a certain magnitude (10–15%). Hemodynamic stimuli do not disrupt the function of the osmoregulatory system but reset it to a new level. A variety of mediators, drugs, hormones and other influences can affect VP release and also appear to do so by resetting

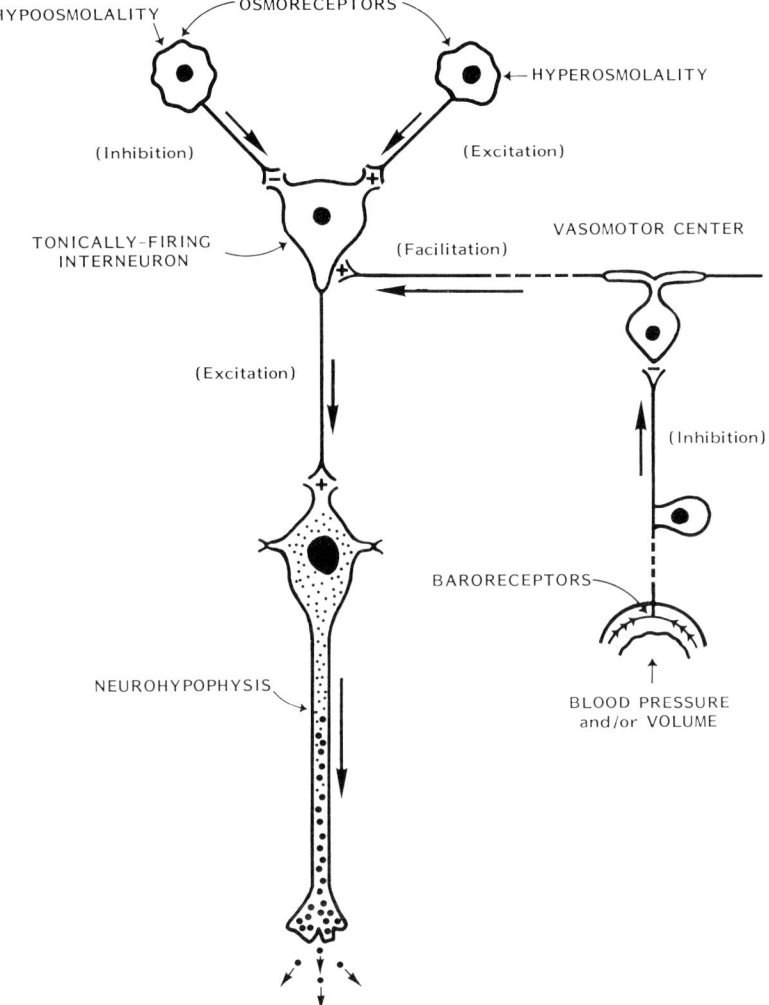

Fig. 8. Hypothetical model of the organization of the neurohypophysis and its principal regulatory pathways. From ref. [24].

rather than interrupting the osmoregulatory mechanism. The most potent nonosmotic stimulus is nausea while many others produce relatively small alterations of the osmoregulation and, although helpful for the understanding of the physiology of VP secretion, probably have minor clinical significance.

The recently acquired knowledge of secretion of VP in normal and pathological conditions permits several conclusions and hypotheses, concerning the functional organization of the neurohypophysis and its regulatory pathways (fig. 8) [24]. Firstly, it is now clear that the osmoreceptors for thirst and VP secretion are not identical with the neurosecretory cells and are not connected in series with afferents from receptors for hemodynamic and other nonosmotic stimuli. Rather, the osmoreceptors must be totally separate neurons that connect in parallel with hemodynamic and other afferents to act on the same population of neurosecretory cells. Otherwise, it would not be possible to selectively ablate osmoreceptor function [15, 24] or to osmotically suppress VP in the presence of nonosmotic stimuli such as hypotension, hypoglycemia or nausea [26, 33, 36]. Secondly, VP secretion may be tonically controlled by a pacemaker or integrator neuron whose activity, in turn, is regulated by input from the osmoreceptors, baroreceptors and possibly other structures. It is also possible that many of the factors that 'reset the osmostat' do so by altering the tonic activity of the pacemaker interneuron. Finally, one could postulate that the osmoreceptor may be bimodal, i.e. contain both inhibitory and stimulatory components the activity of which would be determined by the level of plasma osmolality. This model would predict that certain lesions or conditions could block the capacity to osmotically suppress VP secretion without affecting osmotic stimulation.

VP is secreted into CSF probably by neurons separate from those that secrete the hormone into plasma. The regulation of VP secretion into CSF is presently unclear but at least some of the factors (osmolality) that influence release into plasma stimulate secretion into CSF. In addition, it is likely that other variables such as circadian rhythm may exert an effect on CSF but not plasma levels at least in animals.

References

1 Robertson, G.L.; Mahr, E.A.; Athar, S.; Sinha, T.: The development and clinical application of a new method for the radioimmunoassay of arginine vasopressin in human plasma. J. clin. Invest. *52:* 2340–2352 (1973).

2 Robertson, G.L.: The regulation of vasopressin function in health and disease. Recent Prog. Horm. Res. *33:* 333–385 (1977).

3 Robertson, G.L.: Vasopressin and water metabolism; in Ingbar, Contemporary endocrinology, 1982 (Plenum Publishing, New York, in press).

4 Robertson, G.L.: Vasopressin and water metabolism; in Ingbar, The year in endocrinology 1977, pp. 205–231 (Plenum Publishing, New York, 1978).
5 Robertson, G.L.; Athar, S.; Shelton, R.L.: Osmotic control of vasopressin function; in Grantham, Rector, Disturbances in body fluid osmolality, pp. 125–148 (Am. Physiological Society, Bethesda, 1977).
6 Zerbe, R.L.; Robertson, G.L.: A comparison of plasma vasopressin measurements with a standard indirect test in the differential diagnosis of polyuria. New Engl. J. Med. 305: 1539–1546 (1981).
7 Zerbe, R.L.; Miller, J.Z.; Robertson, G.L.: Reproducibility and heritability of vasopressin osmoregulation in humans. The Endocrine Society, Washington 1980. Endocrinology 106: suppl. 78, abstract 13 (1980).
8 Baylis, P.H.; Robertson, G.L.: Plasma vasopressin response to hypertonic saline infusion to assess posterior pituitary function. J. R. Soc. Med. 73: 255–260 (1980).
9 Zerbe, R.L.; Robertson, G.L.: Osmoregulation of thirst and vasopressin secretion in human subjects: effect of various solutes. Am. J. Physiol. 224: E607–E614 (1983).
10 Robertson, G.L.: Diseases of the posterior pituitary; in Felig, Baster, Broadus, Frohman, Endocrinology and metabolism, pp. 251–277 (McGraw-Hill, New York, 1981).
11 McKinley, M.J.; Denton, D.A.; Weisinger, R.S.: Sensors for antidiuresis and thirst-osmoreceptor or CSF sodium detector. Brain Res. 141: 89–103 (1978).
12 Thrasher, T.N.; Jones, R.G.; Keil, L.C.; Brown, C.J.; Ramsay, D.J.: Drinking and vasopressin release during ventricular infusion of hypertonic solutions. Am. J. Physiol. 238: R340–345 (1980).
13 Wang, B.C.; Share, L.; Crofton, J.T.; Kimura, T.: Effect of intravenous and intracerebroventricular infusion and hypertonic solutions on plasma and cerebrospinal fluid vasopressin concentrations. Neuroendocrinology 34: 215–221 (1982).
14 Wade, C.E.; Bie, P.; Keil, L.C.; Ramsey, D.J.: Osmotic control of plasma vasopressin in the dog. Am. J. Physiol. 243: E287–292 (1982).
15 Robertson, G.L.: Physiopathology of ADH secretion; in Tolis et al., Clinical neuroendocrinology: a pathophysiological approach, pp. 247–260 (Raven Press, New York, 1979).
16 Bealer, S.L.; Crofton, J.T.; Share, L.: Hypothalamic knife cuts alter fluid regulation, vasopressin secretion, and natriuresis during water deprivation. Neuroendocrinology 36, 364–370 (1983).
17 Mangiapane, M.L.; Thrasher, T.N.; Keil, L.C.; Simpson, J.B.; Ganong, W.F.: Deficits in drinking and vasopressin secretion after lesions of the nucleus medianus. Neuroendocrinology 37: 73–77 (1983).
18 Thrasher, T.N.; Keil, L.C.; Ramsey, D.J.: Lesions of the organum vasculosum of the lamina terminalis (OVLT) attenuate osmotically induced drinking and vasopressin secretion in the dog. Endocrinology 110: 1937 (1982).
19 Sklar, A.H.; Schrier, R.W.: Central nervous system mediators of vasopressin release. Physiol. Rev. 63: 1243–1280 (1983).
20 Brennan, L.A., Jr.; Malvin, R.L.; Jochim, K.E.; Roberts, D.E.: Influence of right and left atrial receptors on plasma concentrations of ADH and renin. Am. J. Physiol. 221: 273 (1971).
21 Fater, D.C.; Schultz, H.D.; Sundet, W.D.; Mapes, J.S.; Goetz, K.L.: Effects of left atrial stretch in cardiac-denervated and intact conscious dogs. Am. J. Physiol. 242: H1056 (1982).

22 Schrier,, R.W.; Berl, T.; Anderson, R.J.; McDonald, K.M.: Nonosmolar control of renal water excretion; in Andreoli, Disturbances in body fluid osmolality, pp. 149–178 (Am. Physiology Society, Bethesda, 1977).

23 Robertson, G.L.: The role of osmotic and hemodynamic variables in regulating vasopressin secretion; in James, Proc. 5th Int. Congr. of Endocrinology, Hamburg, 1976. Excerpta Med. Int. Congr. Ser., No. 402, pp. 126–130 (1977).

24 Robertson, G.L.: Thirst and vasopressin function in normal and disordered states of water balance. J. Lab. clin. Med. *101:* 351–371 (1983).

25 Sawchenko, P.E.; Swanson, L.W.: Central noradrenergic pathways for the integration of hypothalamic neuroendocrine and autonomic responses. Science *214:* 685–687 (1981).

26 Dunn, F.J.; Brennan, T.J.; Nelson, A.E.; Robertson, G.L.: The role of blood osmolality and volume in regulating vasopressin in the rat. J. Clin. Invest. *52:* 3212–3219 (1973).

27 Zerbe, R.L.; Henry, D.P.; Robertson, G.L.: Vasopressin response to orthostatic hypotension: etiological and clinical implications. Am. J. Med. *74:* 265–271 (1983).

28 Weitzman, R.E.; Fisher, D.A.; DiStefano, J.H., III; Bennett, C.M.: Episodic secretion of arginine vasopressin. Am. J. Physiol. *233:* E32–E36 (1977).

29 Hammer, M.; Engell, H.C.: Episodic secretion of vasopressin in man. Acta endocr., Copenh. *101:* 517 (1982).

30 Katz, F.H.; Loeffel, D.E.; Roper, E.F.; Lock, J.P.; Husain, M.: Circadian variation of plasma vasopressin in normal active humans; in Proc. 58th Annu. Meet. of the Endocrine Society, San Francisco, 1976. Endocrinology *98:* suppl. No. 322 (1976).

31 Repper, S.M.; Artman, M.G.; Swaminathan, S.; Fisher, D.A.: Vasopressin exhibits a rhythmic daily pattern in cerebrospinal fluid but not in blood. Science *213:* 1256 (1981).

32 Robertson, G.L.; Athar, S.: The interaction of blood osmolality and blood volume in regulating plasma vasopressin in man. J. clin. Endocr. Metab. *42:* 613–620 (1976).

33 Kannan, H.; Yagi, K.: Supraoptic neurosecretory neurons: evidence for the existence of converging inputs both from carotid baroreceptors and osmoreceptors. Brain Res. *145:* 385–390 (1978).

34 Dunn, F.L.; Brennan, T.J.; Nelson, A.E.; Robertson, G.L.: The role of blood osmolality and volume in regulating vasopressin in the rat. J. clin. Invest. *52:* 3212–3219 (1973).

35 Rowe, J.W.; Shelton, R.L.; Helderman, J.H.; Vestal, R.E.; Robertson, G.L.: Influence of the emetic reflex on vasopressin release in man. Kidney int. *16:* 729–735 (1979).

36 Baylis, P.H.; Robertson, G.L.: Rat vasopressin response to insulin-induced hypoglycemia. Endocrinology *107:* 1975–1979 (1980).

37 Baylis, P.H.; Zerbe, R.L.; Robertson, G.L.: Arginine vasopressin response to insulin-induced hypoglycemia in man. J. clin. Endocr. Metab. *53:* 935–940 (1981).

38 Baylis, P.H.; Heath, D.A.: Plasma-arginine-vasopressin response to insulin-induced hypoglycemia. Lancet *ii:* 428–430 (1977).

39 Baylis, P.H.; Robertson, S.L.: Vasopressin response to 2-deoxy-*D*-glucose in the rat. Endocrinology *107:* 1970–1974 (1980).

40 Thompson, D.A.; Campbell, R.G.; Lilavivat, U.; Welle, S.L.; Robertson, G.L.:

Increased thirst and plasma arginine vasopressin levels during 2-deoxy-D-glucose-induced glucoprivation in humans. J. clin. Invest. 67: 1083–1093 (1981).

41 Kimura, T.; Share, L.; Wang, B.C.; Crofton, J.T.: The role of central adrenoreceptors in the control of vasopressin release and blood pressure. Endocrinology 105: 1829–1836 (1981).

42 Roman, R.J.; Cowley, A.W.; Lechene, C.: Water diuretic and natriuretic effect of clonidine in the rat. Pharmac. exp. Ther. 211: 385–393 (1979).

43 Miller, T.; Handelman, W.A.; Arnold, P.E.; Mc Donald, K.M.; Molinoff, P.B.; Schrier, R.W.: Effect of central catecholamine depletion on the osmotic and non-osmotic stimulation of vasopressin (antidiuretic hormone) in the rat. J. clin. Invest. 64: 1599–1607 (1979).

44 Kamoi, K.; Henry, D.P.; Robertson, G.L.: Depletion of hypothalamic norepinephrine does not alter the osmoregulation of vasopressin in rats (accepted for publication).

45 Hoffman, W.E.; Phillips, M.I.; Schmid, P.: The role of catecholamines in central antidiuretic and pressor mechanisms. Neuropharmacology 16: 563–569 (1977).

46 Miller, M.: Inhibition of ADH release in the rat by narcotic antagonists. Neuroendocrinology 19: 241–251 (1975).

47 Bisset, G.W.; Chowdrey, H.S.; Feldberg, W.: Release of vasopressin by enkephalin. Br. J. Pharmacol. 62: 370–371 (1978).

48 Weitzman, R.E.; Fisher, D.A.; Minick, S.; Ling, N.; Guillemin, R.: β-Endorphin stimulates secretion of arginine vasopressin in vivo. Endocrinology 101: 1643–1646 (1977).

49 Bisset, G.W.; Chowdry, H.S.; Feldberg, W.: Release of vasopressin by enkephalin. Br. J. Pharmacol. 62: 370–371 (1978).

50 Tseng, L.F.; Loh, H.H.; Li, C.H.: Beta-endophrin: antidiuretic effects in rats. Int. J. Pept. Protein Res. 12: 173–176 (1978).

51 Summy-Long, J.V.; Keil, L.C.; Deen, K.; Severs, W.B.: Opiate regulation of angiotensin-induced drinking and vasopressin release. J. Pharmac. exp. Ther. 217: 630–637 (1981).

52 Greidanus, T.B. van W.; Thody, T.J.; Verspaget, H.; Rotte, G.A. de; Goedemans, H.J.H.; Croiset, G.; Ree, J.M. van: Effects of morphine and beta-endorphine on basal and elevated plasma levels of alpha-MSH and vasopressin. Life Sci. 24: 579–586 (1979).

53 Aziz, L.A.; Forsling, M.L.; Woolf, C.J.: The action of morphine on vasopressin release in the rat. J. Physiol., Lond. 399: 24P–25P (1980).

54 Brownell, J.; Pozo, E. del; Donatsch, P.: Inhibition of vasopressin secretion by a Met-enkephalin (FK 33-824) in humans. Acta endocr., Copenh. 94: 304–308 (1980).

55 Grossman, A.; Besser, G.M.; Milles, J.J.; Baylis, P.H.: Inhibition of vasopressin release in man by an opiate peptide. Lancet. i: 1108–1110 (1980).

56 Kamoi, K.; White, K.; Robertson, G.L.: Opiates elevate the osmotic threshold for vasopressin release in rats (Abstract). Clin. Res. 27: 254A (1979).

57 Zerbe, R.L.; Henry, D.P.; Robertson, G.L.: A new Met-enkephalin analogue suppresses plasma vasopressin in man. Peptides 1: 199–201 (1981).

58 Huidobro, F.; Huidobro-Toro, J.P.: Antidiuretic effect of morphine and vasopressin in morphine tolerant and non-tolerant rats, differential effects on urine composition. Eur. J. Pharmacol. 59: 55–64 (1979).

59 Huidobro-Toro, J.P.; Huidobro, F.; Croxatto, R.: Effects of β-endorphine and D-alanine enkephalin-amide on urine production and urinary electrolytes in the rat. Life Sci. *24:* 697–703 (1979).
60 Oiso, Y.; Robertson, G.L.: Role of endogenous opiates in mediating ethanol-induced suppression of vasopressin. Proc. 64th Annu. Meet. of the Endocrine Society, San Francisco, 1982, abstr. 751, p. 267.
61 Endogenous Opiates and Their Action, Editorial. Lancet *ii:* 305–307 (1982).
62 Keil, L.C.; Summy-Long, J.; Severs, W.B.: Release of vasopressin by angiotensin II. Endocrinology *96:* 1063–1065 (1975).
63 Gregg, C.M.; Malvin, R.L.: Localization of central sites of action of angiotensin II on ADH release in vitro. Am. J. Physiol. *234:* F135–F140 (1978).
64 Bealer, S.L.; Phillips, M.I.; Johnson, A.K.; Schmid, P.G.: Anteroventral third ventricle lesions reduce antidiuretic responses to angiotensin II. Am. J. Physiol. *236:* E610–E615 (1979).
65 Van Houten, M.; Schiffrin, E.L.; Mann, J.F.E.; Posner, B.I.; Bouchler, R.: Radioautographic localization of specific binding sites for bloodhorne angiotensin II in the rat brain. Brain Res. *186:* 480–485 (1980).
66 Hoffman, P.K.; Share, L.; Crofton, J.T.; Shade, R.E.: The effect of intracerebroventricular indomethacin on osmotically stimulated vasopressin release. Neuroendocrinology *34:* 132–139 (1982).
67 Walker, B.R.: Suppressed basal antidiuretic hormone release during cycloxygenase inhibition in conscious dogs. Am. J. Physiol. *244:* R487–R491 (1983).
68 Kamoi, K.; Cunningham, M.; Robertson, G.L.: On the failure of prostaglandin synthetase inhibitor indometacin to affect the osmoregulation of plasma arginine vasopressin in rats. Endocr. Jap. *30:* 121–126 (1983).
69 Wang, B.C.; Share, L.; Crofton, J.T.: Central infusion of vasopressin decreased plasma vasopressin concentration in dogs. Am. J. Physiol. *243:* E365–E369 (1982).
70 Engel, P.; Rowe, J.; Minaker, K.; Robertson, G.L.: Stimulation of vasopressin by exogenous vasopressin: effect of sodium intake and age. Am. J. Physiol. (in press, 1984).
71 Forsling, M.L.; Akerlung, M.; Stromberg, P.: Variations in plasma concentrations of vasopressin during the menstrual cycle. J. Endocr. *89:* 263–266 (1981).
72 Punnonen, R.; Viinamaki, O.; Multamaki, S.: Plasma vasopressin during normal menstrual cycle. Hormone Res. *17:* 90–92 (1983).
73 Vokes, T.; Gaskill, M.; Robertson, G.L.: Changes in the osmoregulation of thirst and vasopressin during the normal menstrual cycle. 7th Int. Congr. of Endocrinology, Quebec City, 1984.
74 Durr, J.A.; Stamoutsos, B.A.; Lindheimer, M.D.: Osmoregulation during pregnancy in the rat: evidence for resetting of the threshold for vasopressin secretion during gestation. J. clin. Invest. *68:* 337–346 (1981).
75 Davison, J.M.; Gilmore, E.A.; Durr, J.; Robertson, G.L.; Lindheimer, M.D.: Altered osmotic threshold for vasopressin secretion and thirst in human pregnancy. Am. J. Physiol. *246:* F105–F109 (1984).
76 Barron, W.M.; Stamoutsos, B.A.; Lindheimer, M.D.: Role of volume in the regulation of vasopressin secretion during pregnancy in the rat (in press).
77 Skowsky, W.R.; Swan, L.; Smith, P.: Effects of sex steroid hormones on arginine vasopressin in intact and castrated male and female rats. Endocrinology *104:* 105–108 (1979).

78 Miller, M.; Wilks, J.W.: Urinary antidiuretic hormone excretion during the menstrual cycle and pregnancy in the monkey. Neuroendocrinology *12:* 174–178 (1973).
79 Forsling, M.L.; Stromberg, P.; Akerlund, M.: Effect of ovarian steroids on vasopressin secretion. J. Endocr. *95:* 147–151 (1982).
80 Walsh, C.H.; Baylis, P.H.; Malins, J.M.: Plasma arginine vasopressin in diabetic ketoacidosis. Diabetologia *16:* 93–96 (1979).
81 Zerbe, R.L.; Vinicor, F.; Robertson, S.L.: Plasma vasopressin in uncontrolled diabetes mellitus. Diabetes *28:* 503–508 (1979).
82 Van Itallie, C.M.; Fernstrom, J.D.: Osmolal effects on vasopressin secretion in the streptozotocin-diabetic rat. Am. J. Physiol. *242:* E411–417 (1982).
83 Zerbe, R.L.; Vinicor, F.; Robertson, G.L.: The regulation of plasma vasopressin in insulin-dependent diabetes mellitus (submitted for publication).
84 Aycinena, P.R.; Robertson, G.L.: Effects of insulin on the osmoregulation of vasopressin in insulin-dependent diabetes mellitus (DM-I). Am. Soc. for Clinical Investigation Annu. Meet. Washington, 1983. Clin. Res. *31:* 469A (1983).
85 Gold, P.W.; Robertson, G.L.; Post, R.M.; Kaye, W.; Ballenger, J.; Rubinow, D.R.; Goodwin, F.K.: The effect of lithium on the osmoregulation of arginine vasopressin secretion. J. clin. Endocr. Metab. *562:* 295–299 (1983).
86 Gold, P.W.; Robertson, G.L.; Ballenger, J.C.; Kaye, W.; Chen, J.; Rubinow, D.R.; Goodwin, F.K.; Post, R.M.: Carbamazepine diminishes the sensitivity of the plasma arginine vasopressin response to osmotic stimulation. J. clin. Endocr. Metab. *57:* 952–957 (1983).
87 Rowe, J.W.; Kilgore, A.; Robertson, G.L.: Evidence in man that cigarette smoking induces vasopressin release via an airway-specific mechanism. J. clin. Endocr. Metab. *51:* 170–172 (1980).
88 Rydin, H.; Verney, E.B.: The inhibition of water diuresis by emotional stress and muscular exercise. Q. J. exp. Physiol. *223:* 343–374 (1938).
89 Brennan, T.C.; Shelton, R.L.; Robertson, G.L.: Effect of stress on plasma vasopressin and corticosterone in rats. Am. Fed. for Clinical Research, Atlantic City, 1975. Clin. Res. *23:* 234A (1975).
90 Husain, M.K.; Manger, W.M.; Rock, T.W.; Weiss, R.J.; Frantz, A.G.: Vasopressin release due to manual restraint in the rat: role of body comparison with other stressful stimuli. Endocrinology *104:* 641–644 (1979).
91 Helderman, J.H.; Vestal, R.E.; Rowe, J.W.; Tobin, J.D.; Andres, R.; Robertson, G.L.: The response to arginine in intravenous ethanol and hypertonic saline in man: the impact of aging. J. Geront. *33:* 39–47 (1978).
92 Robertson, G.L.; Rowe, J.W.: The effect of aging on neurohypophyseal function. Peptides *1:* suppl. 1, pp. 159–162 (1980).
93 Rowe, J.W.; Minaker, K.L.; Sparrow, D.; Robertson, G.L.: Age-related failure of volume-pressure-mediated vasopressin release. J. clin. Endocr. Metab. *54:* 661–664 (1982).
94 Granberg, P.O.: Effect of acute hypoxia on renal hemodynamics and water diuresis in man. Scand. J. Lab. Invest. *15:* suppl. 63, pp. 5–62 (1962).
95 Farber, M.O.; Weinberg, M.H.; Robertson, G.L.; Fineberg, N.S.; Manfredi, F.: Hormonal abnormalities affecting sodium and water balance in acute respiratory failure due to chronic obstructive lung disease. Chest *85:* 49–54 (1984).

96 Forsling, M.L.; Rees, M.: Effects of hypoxia and hypercapnia on plasma vasopressin concentration. J. Endocr. *67:* 62P–63P (1975).
97 Anderson, R.J.; Pluss, R.G.; Berns, A.S.; Jackson, J.I.; Arnold, P.E.; Schrier, R.W.; McDonald, K.M.: Mechanism of effect of hypoxia on renal water function. J. clin. Invest. *62:* 769–777 (1978).
98 Heyes, M.P.; Farber, M.O.; Manfredi, F.; Robertshaw, D.; Weinberger, M.; Fineberg, N.; Robertson, G.L.: Acute effects of hypoxia on renal and endocrine function in normal humans. Am. J. Physiol. *243:* R265–R270 (1982).
99 Baylis, P.H.; Stockley, R.A.; Heath, D.A.: Effect of acute hypoxaemia on plasma arginine vasopressin in conscious man. Clin. Sci. mol. Med. *53:* 401–404 (1977).
100 Claybaugh, J.R.; Hansen, J.E.; Wozniak, D.B.: Response of antidiuretic hormone to acute exposure to mild and severe hypoxia in man. J. Endocr. *77:* 157–160 (1978).
101 Bernabe, J.E.; Martiner, M.: Hypercalcemic nephropathy. Archs. intern. Med. *138:* 777–779 (1978).
102 Ayciena, P.R.; Robertson, G.L.: Effects of hypercalcemia on secretion of the antidiuretic hormone, arginine vasopressin. Proc. Endocrine Soc. Annu. Meet., San Antonio, 1983, abstr. 245, p. 142.
103 Rutecki, G.W.; Cox, J.W.; Robertson, G.L.; Francisco, L.L.; Ferris, T.F.: Urinary concentrating ability and antidiuretic hormone responsiveness in the potassium-depleted dog. J. Lab. clin. Med. *100:* 53–60 (1982).
104 Zusman, R.M.: Prostaglandins, vasopressin and renal water reabsorbtion. Med. Clin. Am. *65:* 915–925 (1981).
105 Luerssen, T.G.; Robertson, G.L.: Cerebrospinal fluid vasopressin and vasotocin in health and disease; in Wood, Neurobiology of cerebrospinal fluid, pp. 613–623 (Plenum Publishing, New York, 1980).
106 Gold, P.W.; Goodwin, F.K.; Ballenger, J.C.; Post, R.M.; Weingartner, H.; Robertson, G.L.: Central vasopressin function in affective illness. Int. J. ment. Health *9:* 91–107 (1981).
107 Dogterom, J.; Wimersma Greidanus, Tj.B. van; De Wied, D.: Vasopressin in cerebrospinal fluid and plasma of man, dog and rat. Am. J. Physiol. *234:* E463–E467 (1978).
108 Gold, P.W.; Kaye, W.; Robertson, G.L.; Ebert, M.: Abnormalities in plasma and cerebrospinal fluid arginine vasopressin in patients with anorexia nervosa. New Engl. J. Med. *308:* 1117–1123 (1983).
109 Wang, B.C.; Share, L.; Crofton, J.T.; Kimura, T.: Changes in vasopressin concentration in plasma and cerebrospinal fluid in response to hemorrhage in anesthetized dogs. Neuroendocrinology *33:* 61–66 (1981).
110 Szczepanska-Sadowska, E.; Gray, D.; Simon-Oppermann, C.: Vasopressin in blood and third ventricle CSF during dehydration, thirst, and hemorrhage. Am. J. Physiol. *245:* R549–R555 (1983).
111 Mens, W.B.J.; Andringa-Bakker, E.A.D.; Wimersma Greidanus, Tj.B. van: Changes in cerebrospinal fluid levels of vasopressin and oxytocin of the rat during various light-dark regimes. Neurosci. Lett. *34:* 51–56 (1982).
112 Reppert, S.M.; Artman, H.G.; Swaminathan, S.; Fisher, D.A.: Vasopressin exhibits a rhythmic daily pattern in cerebrospinal fluid but not in blood. Science *213:* 1256–1257 (1981).

113 Reppert, S.M.; Coleman, R.J.; Heath, H.W.; Keutmann, H.T.: Circadian properties of vasopressin and melatonin rhythms in cat cerebrospinal fluid. Am. J. Physiol. *6:* E489–E498 (1982).
114 Raichle, M.E.; Grubb, R.L.: Regulation of brain water permeability by centrally released vasopressin. Brain Res. *143:* 191–194 (1978).
115 Schmid, P.G.; Sharabi, F.M.; Guo, G.B.; Abbound, F.M.; Thames, M.D.: Vasopressin and oxytocin in the neural control of the circulation. Fed. Proc. *43:* 97–102 (1984).
116 De Wied, D.: Behavioral effects of intraventricularly administered vasopressin and vasopressin fragments. Life Sci. *19:* 685–690 (1976).
117 Wimersma Greidanus, Tj.B. van; De Wied, D.: Modulation of passive-avoidance behavior of rats by intracerebroventricular administration of antivasopressin serum. Behav. Biol. *18:* 325–333 (1976).
118 De Wied, D.; Bohus, B.: Long-term and short-term effects on retention of a conditioned avoidance response in rats by treatment with long acting pitressin and α-MSH. Nature, Lond. *212:* 1484–1486 (1966).

Tamara Vokes, MD, University of Chicago, 5841 S. Maryland Avenue, Box 131, Chicago, IL 60637 (USA)

Clinical and Laboratory Observations in the Adult with Diabetes insipidus and Related Syndromes[1]

Arnold M. Moses

VA Medical Center and State University Hospital, Syracuse, N.Y., USA

Clinical Features

Diabetes insipidus
Central or neurogenic diabetes insipidus (DI) is due to impaired renal conservation of water which results from deficient arginine vasopressin (AVP) release in response to normal physiological stimuli. Polyuria, excessive thirst and polydipsia are the major clinical features. Eneuresis may be an important feature in children. These symptoms are usually sudden in onset when the disorder first appears and whenever the effects of administered vasopressin disappear during therapy of patients with the disorder. In severe cases the urine volume may be as much as 8–12 l/day. More often, the urine volume is only moderately excessive (2.5–6 l/day). Urinary concentration is well below that of serum in severe cases, but may be higher (290–600 mosm/kg) in patients with partial DI. Patients who fail to empty their bladders adequately because of urethral obstruction or other reasons may develop enlarged atonic bladders, hydroureter, hydronephrosis and renal failure. These features are more common in patients with nephrogenic DI.

The author has accumulated detailed information on 119 cases of longstanding central DI (table I). These are patients whose DI has been present for more than 6 months or until the patient died. For the group as

---

[1] This work is supported by VA research funds and Grant RR 229, from the General Clinical Research Centers Program of the Division of Research Resources, National Institutes of Health, Bethesda, Md.

Table I. 119 cases of longstanding central diabetes insipidus diagnosed by the author at SUNY, Upstate Medical Center. Categories are arranged in order of increasing median age of onset

Cause	Age of onset, years		Males	Females	Number of cases	Percent
	median[1]	range				
Histiocytosis	2	1–20	2	2	4	3
Primary brain tumor – postoperative[2]	15	6–50	9	11	20	17
Idiopathic	16	<1–66	19	11	30	25
Primary brain tumor – preoperative[3]	17	7–58	15	1	16	13
Head trauma	23	5–48	14	6	20	17
Non-traumatic encephalomalacia	37	15–73	3	4	7	6
Ruptured cerebral aneurysm	39		1	0	1	1
Post-hypophysectomy	40	24–68	4	7	11	9
Sarcoidosis	42		0	1	1	1
Metastatic cancer	56	32–71	5	4	9	8
			(61%)	(39%)		

[1] Median age of entire group = 19 years.
[2] 16 cases were of craniopharyngioma.
[3] 5 cases were of glioma, 5 of germinoma and 4 of craniopharyngioma.

a whole the median age of onset was 19 years and 61% of the patients were male. The largest group was idiopathic (25%), followed closely in number by patients who developed DI before or after surgery for benign brain (extrapituitary) tumors (24%). Patients who developed DI following head trauma constituted the third largest group (17%). 9% of the patients developed DI following hypophysectomy for treatment of pituitary hypersecretion, diabetic retinopathy or cancer, and 8% developed the disease as a consequence of cancer metastatic to the brain (including multiple myeloma and leukemia). 6% of the cases resulted from malignant brain tumors, before or after surgery, and the same percent was due to non-traumatic encephalomalacia which resulted in brain death and DI. 3% had histiocytosis X and the remaining cases were due to sarcoidosis or to a ruptured cerebral aneurysm. Causes which have not been observed personally include thrombotic thrombocytopenic purpura [1], pituitary apoplexy [2], postpartum pituitary necrosis [3, 4], Wegener's granulomatosis [5] and systemic blastomycosis [6].

Our youngest group of patients developed DI from histiocytosis X (table I). The patterns of abnormal water metabolism in these patients have been well described [7]. The next youngest of our patients developed DI as the result of surgery for a primary brain tumor. Of the 20 cases, 16 were of craniopharyngioma. Two of the other 4 patients had benign lesions and 2 had malignant lesions. Eleven of the patients with craniopharyngioma also had anterior pituitary abnormalities, as did 4 other patients [see also *Czernichow* et al., this volume].

The median age of onset of apparent idiopathic DI was 16 years, and as in most of the groups, the majority of patients were male (63%). Thirteen of the patients who were classified in table I as having idiopathic DI had one or more additional abnormalities (table II). Eleven patients had evidence of anterior pituitary dysfunction which probably reflects more widespread hypothalamic disease. Of these 11 patients, 6 had 1 anterior pituitary abnormality, 2 patients had 2 abnormalities, 2 had 3, and 1 had 4 abnormalities. Two patients had thickened infundibular stalks on CT scan of the brain. This has previously been reported in patients with DI [8], but there is no information available pertaining to the relationship of this finding to the cause of the DI. Our own experience has revealed thickened infundibular stalks in 4 patients with DI (table III). Other findings in patients with apparent idiopathic DI are noted in table II. The author believes that patients with DI with any of the findings listed in table II should be considered as having DI, cause to be determined, and should be examined more frequently and carefully than patients who have no abnormalities other than DI.

Pathological studies in several patients with idiopathic DI including the inherited type have disclosed a striking depletion of the Nissl granules and in the number of neurons in the supraoptic and paraventricular nuclei [9]. Some patients with idiopathic DI have circulating antibodies to vasopressin producing cells of the hypothalamus [10] thus suggesting an immunological origin for the disease [see also *Scherbaum* et al., this volume]. Idiopathic DI is almost always permanent. However, we observed 1 patient with idiopathic DI who spontaneously reverted to normal after approximately 10 months of the disease.

Sixteen patients with the median age of 17 developed DI from primary brain tumors, before or without surgery. There were 5 cases of glioma, 5 of germinoma and 4 of craniopharyngioma. Three patients

Table II. 11 of the 30 patients classified as having idiopathic diabetes insipidus had 1 or more additional endocrinological, neurological or radiological findings as follows:

5 TSH deficiency
4 Increased prolactin levels
4 Growth hormone deficiency
4 ACTH deficiency
3 Gonadotrophin deficiency[1]
2 Thickened infundibular stalk on CT scan
1 Enlarged sella turcica
1 Lytic lesion of sphenoid bone
1 Abnormal cerebrospinal fluid
1 Narcolepsy
1 Cerebral palsy

[1] A total of 11 patients had some evidence of abnormal anterior pituitary function.

Table III. Evidence of thickened infundibular stalk with CT scan

	Normal stalk	Thickened stalk
Traumatic DI	6	1
Idiopathic DI	5	2
Primary polydipsia	2	1
Nephrogenic DI	3	0
Post-hypophysectomy	2	0
Non-traumatic encephalopathy	1	0
Post-surgery for benign brain tumor	1	0
Total	20	4

with germinoma, 3 with craniopharyngioma, 1 with glioma and 1 with diencephalic cyst also had anterior pituitary abnormalities.

DI is a well recognized complication of blunt head trauma. This type of DI is occurring with increasing frequency, probably due to improved medical care for head trauma victims which has resulted in survival of patients who previously would have died from severe neurological damage. We had 20 patients in this category. See the contribution by *Verbalis* et al. [this volume] for complete discussion.

DI was caused by anoxemic or non traumatic encephalomalacia in 7 patients with a median age of 37 years (table IV). This cause of DI is

Table IV. Clinical data on 7 patients who developed diabetes insipidus from non-traumatic encephalopathy. All patients had to be maintained on total life support systems and were brain dead at time diabetes insipidus developed

Patient data	Cause
M, 15	gunshot wound of chest causing shock
M, 23	watershed necrosis due to overaggressive treatment of hypertension
F, 33	embolus to left middle cerebral artery and uncal herniation
F, 37	hypertensive encephalopathy
M, 49	cardiopulmonary arrest
F, 72	methanol poisoning
F, 73	cardiopulmonary arrest

being described more frequently and at least 9 cases have already been reported [11–15]. The previously reported patients had brain death due to asphyxia, drug-induced respiratory failure, diabetic ketoacidosis, cardiac arrest, and shock from penetrating thoracic trauma. Our 7 patients (3 males, 4 females) ranged in age from 15 to 73 years and developed encephalopathy and DI from a variety of severe cerebral insults (table IV). They were all brain dead and had to be maintained on total life support systems. One of our patients who survived for 68 days after the criteria for brain death were satisfied had findings at autopsy which were characteristic of an advanced stage in the sequence of changes that follow brain infarction with prolonged preservation of systemic circulatory function [16]. Anterior pituitary function cannot be properly evaluated because of many factors including steroid therapy, and the effect of severe illness on thyroid and gonadal functions.

Eleven cases, with a median age of 40, developed DI following surgical hypophysectomy. Three of the patients were operated for Cushing's syndrome, and 2 each for acromegaly, chromophobe adenoma, and diabetic retinopathy, and 1 each for prolactinoma and breast cancer.

DI resulted from metastatic malignancy in 9 cases with a median age of 56 years. There were 2 cases each with carcinoma of lung and breast, and 1 case each from carcinoma of kidney and rectum, 1 carcinoma with an unknown primary site, 1 patient with multiple myeloma, and 1 patient with leukemia. DI as a complication of leukemia, though rare, has been previously reported [17, 18]. Thirty-nine cases of metastatic breast carcinoma causing DI have been reported [19]. Following radiotherapy to the hypothalamus there was substantial laboratory and clinical improvement in the patients with renal carcinoma and with multiple myeloma.

Only 1 of our patients had sarcoidosis. However, DI has been reported by others in patients with sarcoidosis [20, 21]. Sarcoidosis may also cause 2 other states of hypotonic polyuria: primarily polydipsia and acquired nephrogenic DI [22, 23]. We studied 1 patient with DI secondary to a ruptured cerebral aneurysm. He had substantial permanent neurological damage and was hypodipsic.

Diabetes insipidus with Inadequate Thirst

Normal function of the thirst center insures that polydipsia closely matches polyuria, so that dehydration is seldom detectable except by a slight elevation of serum sodium concentration.

However, when replenishment of water is inadequate, dehydration and hypernatremia may become severe, causing weakness, fever, psychic disturbances, vascular collapse and death. Five patients with central DI were unable to concentrate their urine even with very high levels of serum sodium, and were hypodipsic or adipsic. This combined problem was caused by head trauma in 2 patients, by germinoma in 2 patients and by a ruptured aneurysm in 1 patient. These patients were maintained in the hospital by careful adjustment of fluid intake on the basis of urine volume, estimated insensible and sensible fluid losses, with modifications made according to body weight and serum electrolyte and osmolality determinations. However, it has been very difficult to accomplish this properly at home. Although *Bode* et al. [24] reported that chlorpropamide may restore drinking behavior it has not been successful in our patients.

There are a variety of vascular, neoplastic and granulomatous causes of adipsic hypernatremia [25]. One form of this abnormality is asymptomatic or essential hypernatremia. This is generally associated with dilution as well as concentration of the urine around an elevated serum sodium concentration and may be equated with a high-set osmoreceptor mechanism. Normonatremia may be restored by the provision of an adequate water supply [26]. The majority of patients with myotonic dystrophy may have asymptomatic hypernatremia due to impaired release of AVP and impaired stimulation of thirst by hypertonicity in association with an intact volume control of AVP release [27]. We have found a similar situation in 2 patients with Kallmann's syndrome [28].

Table V. Clinical data on 7 patients with primary polydipsia

Patient data	Age of onset	Urine volume l/24 h	Response to therapy	Other comments
F, 30	very young	4.6		
F, 52	18	4.1		
F, 36	30	4.9		
F, 32	32	6.4	very good	thick infundibular stalk, TSH-deficient hypothyroidism, history of encephalitis
M, 40	38	4.5		
M, 45	40	5.3	very good	depressed skull fracture at age 20
F, 48	48	4.1		

Note: (1) These patients all had normal UAVP vs Uosm relationships, eliminating confusion with NDI. (2) All patients had Posm vs Uosm relationships that were normal or to the left of normal and no response to Pitressin at the plateau in Uosm. This is different from DI. (3) However, all patients had Posm vs UAVP in low normal or to right of normal as would be expected in mild DI.

Primary Polydipsia

Primary polydipsia is a state of hypotonic polyuria with suppressed AVP levels secondary to excessive fluid intake. Any tendency toward polydipsia can be aggravated by drugs such as thioridazine which causes dryness of the mouth [29]. Primary polydipsia is often psychogenic in origin with the symptoms being less consistent than in patients with central or nephrogenic DI. In fact, some patients with psychogenic polydipsia may have normal water balance for days or weeks, followed by periods of excessive water turnover. Increased thirst and polydipsia due to increased renin and/or angiotensin levels may occur in patients with renal artery stenosis [30].

Our 7 patients who appear to have primary polydipsia began their symptoms from a very early age through the age of 48 with a median age of 32 (table V). The duration of the problem at the time they were studied ranged from several months to almost an entire lifetime. Two of the patients had good responses to dDAVP nasal spray and insisted on continued therapy with this agent. These 2 patients had evidence of possible organic brain disease (table V). The differential diagnosis between primary polydipsia and DI can be difficult and perhaps impossible, particularly when the patients appear to have disease of the hypothalamus and respond to therapy with vasopressin. The response to therapy in our 2 patients was unlike that of our other patients with primary polydipsia, as those patients become bloated and uncomfortable when treated with ADH [31]. When improvement occurs it may be related to the effect of ADH on increasing salivary flow rates and ameliorating thirst in patients with DI [32]. This is apparently a peripheral action of ADH, since centrally administered vasopressin has a dipsogenic action in the dog [33]. Alternatively the vasopressin may produce a mild SIADH and hyponatremia which than 'satisfies' a lower thirst threshold.

Interestingly, of our 7 patients with congenital nephrogenic DI, 4 males and 3 females with severe polydipsia and polyuria, 2 of the females substantially increased Uosm in response to large amounts of injected dDAVP, and 1 patient has been treated with large amounts of dDAVP nasal spray for more than a year, with dramatic improvement in her polydipsia, polyuria, and overall sense of well-being.

Laboratory Features

DI is due to either a deficiency of AVP or impaired osmoreceptor function. The osmoreceptor mechanism may be totally inoperative. Alternatively, AVP release may be subnormal with increasing Posm, with a normal or elevated osmotic threshold [34]. Selective osmoreceptor failure has been recognized for many years, based on observations that some patients with DI have an antidiuretic response to nicotine but not to hypertonic saline [35, 36]. Some patients with defective osmoreceptor mechanisms are capable of releasing AVP in response to hypovolemia or hypotension. These patients may also release AVP and concentrate their urine in response to nausea and drugs such as chlorpropamide and clofibrate [37, 38]. These responses indicate that there is sufficient synthesis and storage of AVP to adequately concentrate the urine in response to appropriate non-osmotic stimuli.

Selective osmoreceptor failure may also be the cause of hypernatremia [39–42]. The osmotic threshold in hypernatremic patients may be absent or elevated [43]. Hypernatremia, instead of DI occurs in patients with impaired osmoreceptor function when they are hypodipsic [44]. With a normal or sensitive thirst mechanism patients with impaired osmoreceptor function drink before an effective osmotic stimulus to AVP release is attained and polyuria results.

The diagnosis of DI is based on evaluation of the hypothalamo-neurohypophyseal system including both the osmoreceptor mechanism and the ability of AVP to be released in consequence of the activation of this receptor. Even though nicotine, nausea, hypovolemia and hypotension may release AVP, and are useful in attempting to understand the underlying pathophysiological abnormality in individual patients, the results are often not helpful in explaining the degree of symptomatology or need for therapy.

The diagnostic procedures to be discussed in this section were developed in a clinical setting in an effort to utilize osmotic factors that influence AVP release. My approach to finding the cause of hypotonic polyuria is definitely pragmatic. The following description basically relates my attempt to develop tests that are readily available, rapidly performed, reliable, safe and inexpensive in order to allow the physician to establish the diagnosis and initiate therapy as quickly as possible.

Hypertonic NaCl

The early tests used for the diagnosis of DI were based on the presence or absence of a decrease in urine flow rate in patients being infused with hypertonic saline [45, 46]. In an effort to define the plasma osmolality at which the release of AVP was *initiated*, we modified the saline infusion tests and developed the concept of an *osmotic threshold* for AVP release by using 5% saline to raise the Posm in water-loaded subjects [47]. When a hydrated normal subject is infused with hypertonic saline at a constant rate there is a linear rise in Posm and after an interval of time depending on the infusion rate, concentration of saline and initial Posm, there is an abrupt progressive fall in free water clearance without a change in solute or creatinine excretion. We defined the Posm at the initiation of antidiuresis under these conditions as the osmotic threshold for AVP release and found that it occurred in normal subjects at a mean value ($\pm$ SD) of 287.3 $\pm$ 3.6 mosm/kg [47, 48]. Similar studies which have subsequently been done by relating direct assay of PAVP to Posm have given remarkably similar results [25, 49].

The first patients with DI whom we investigated had no attainable osmotic thresholds, but we soon recognized patients who had elevated osmotic thresholds (high-set osmoreceptor), and depending on the situation, mainly the sensitivity of the thirst mechanism, could be diagnosed as having either incomplete or partial DI or essential hypernatremia [34]. At the present time we rarely use the osmotic threshold test *clinically*, except where there is a problem in differentiating diabetes insipidus from primary polydipsia. In the latter condition the osmotic threshold is often low and slowly returns toward normal if fluids are restricted.

The osmotic threshold can be altered by numerous physiological and pharmacological factors. Hypervolemia inhibits the release of AVP and acts to raise the osmotic threshold for release of the hormone [50]. Cortisol and oxillorphan also raise the osmotic threshold for AVP release [34, 47]. In contrast, there is a lower osmotic threshold for AVP release when the osmotic stimulus is provided by dehydration than by infusion of 5% saline [51]. Pregnancy is another factor that may lower the osmotic threshold for AVP release [52].

Dehydration Test

We now attempt to determine *maximum* urinary concentrating ability and to relate this to normal and to degrees of AVP deficiency. The possibility of incomplete vasopressin deficiency had previously been claimed by *Lipsett* et al. [53, 54] and to overcome this problem we modified existing dehydration tests [55] to develop a test which was not based on weight loss or time of water deprivation, but on the premise that normal subjects were limited in their maximal Uosm by their renal concentrating mechanism. In contrast, maximum Uosm in patients with impaired neurohypophyseal function was limited by the amount of AVP released by water deprivation [56]. In normal subjects, the injection of ADH after a plateau in urine osmolality had been attained, produced no further rise in Uosm. In contrast, vasopressin administration after prolonged dehydration induced a further rise in Uosm in patients with neurogenic DI and even in patients with nephrogenic DI (fig. 1).

The maximal Uosm after dehydration may be 800–1400 mosm/kg in robust normal subjects, but may be only 450–800 mosm/kg in chronically debilitated or malnourished individuals with normal neurohypophyseal reserve. In these subjects, Uosm falls slightly or rises by less than 9% after administration of aqueous Pitressin [56]. Patients with primary polydipsia have values similar to those of the debilitated and malnourished subjects including failure to increase Uosm after injection of vasopressin (fig. 1). In severe DI, the Uosm reached before the Pitressin injection remains below Posm; while in those patients in whom DI is partial or moderately severe, Uosm rises above Posm during dehydration (fig. 1). A Pitressin-induced rise in Uosm of >50%, after the plateau in Uosm is reached, is found in patients with severe central DI, while in

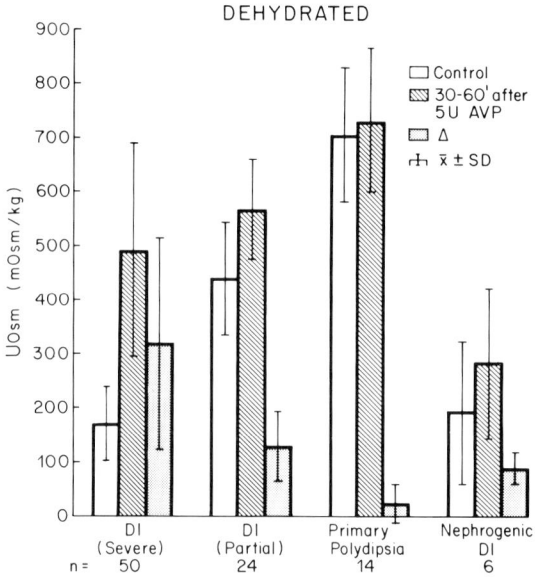

Fig. 1. Plateau in urine osmolality during dehydration (open bars), urine osmolality during the 30- to 60-min period after injection of 5 U AVP (cross-hatched bars), and increase in urine osmolality (stippled bars) in patients with neurogenic and congenital nephrogenic diabetes insipidus and primary polydipsia.

patients with partial neurogenic DI, Uosm increases 9–50% following injection of vasopressin [56] (fig. 1). Males with congenital nephrogenic DI have a plateau in Uosm below Posm, followed by a 0–45% rise in Uosm after the Pitressin injection. Females with severe symptomatic congenital nephrogenic DI may eventually concentrate urine to approximately 400–500 mosm/kg followed by an additional rise of about 100 mosm/kg after the injection of 5 U AVP or 1 μg dDAVP. The dehydration test is simple, generally available and safe even in patients with profound deficiency of AVP, if severe dehydration is prevented by not allowing body weight to fail by more than 5%. The principal value of this test is in differentiating patients without abnormality of vasopressin release or function from patients with neurogenic or nephrogenic DI. Patients with complete DI can usually be separated from partial neurogenic and nephrogenic DI but sometimes other means need to be employed to separate the last two.

The response to 5 U AVP when dehydrated (fig. 1) can be compared to the response when randomly hydrated (fig. 2). In the randomly

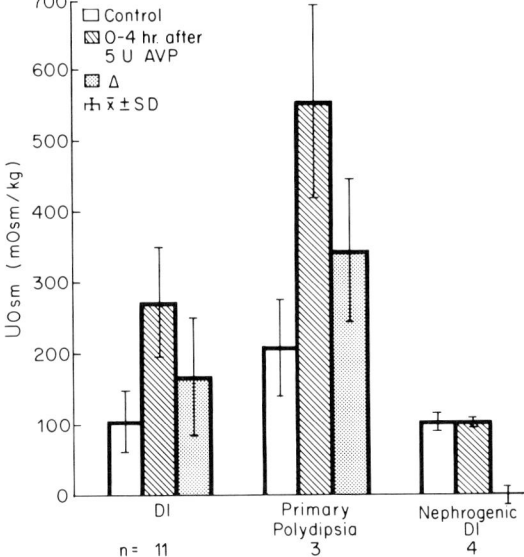

Fig. 2. Urine osmolality during random hydration (open bars), urine osmolality in the 4-hour period following injection of 5 U AVP (cross-hatched bars) and increase in urine osmolality (stippled bars) in patients with neurogenic and congenital nephrogenic diabetes insipidus and primary polydipsia.

hydrated condition, patients with congenital nephrogenic DI do not increase Uosm after 5 U AVP, while patients in the other 2 groups more than double Uosm (fig. 2).

Posm vs Uosm

The dehydration test as described above has provided much important information about the status of neurohypophyseal function in many patients but eventually it became obvious that the Uosm attained during dehydration had to be related carefully to Posm [34, 57, 58]. For instance, patients were found who had *apparently* normal dehydration tests, but with high Posm. Under random conditions, their thirst mechanisms might not allow them to get sufficiently dehydrated to concentrate the urine, so under random conditions they would have hypotonic polyuria. If their thirst mechanism was insensitive, they would be classified as having a high set osmoreceptor or essential hypernatremia. We collected Posm and Uosm data during water loading and dehydration and described a normal area indicated by the shaded areas in figure 3. This

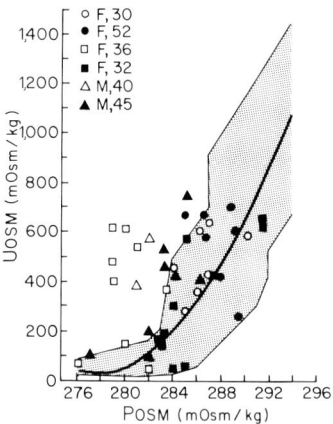

Fig. 3. Plasma vs urine osmolality relationships in normal subjects (shaded area) and in patients with primary polydipsia.

allowed an assessment of the range of Posm and Uosm relationships under conditions similar to those encountered in day-to-day variations of water balance. A similar relationship has been described for children [59]. As in all relationships between Posm and AVP release, the Posm reflects the presence of biologically active particles, essentially sodium and its associated anions. Increased amounts of urea or glucose have to be corrected down to normal amounts, and appropriate corrections have to be made for substances like ethanol which, like urea and glucose, have little or no effect on altering osmotic gradients across cell membranes.

The Posm vs Uosm coordinates can be divided into 4 areas. Points that fall in the shaded area where urine is hypotonic to plasma are indeterminate, while points that fall in the shaded area where urine is hypertonic to plasma are normal. Points that fall in the area of hypotonic urine with an elevated Posm are abnormal and indicate the presence of central or nephrogenic DI. To clarify the status of patients whose points fall in the indeterminate area, oral or intravenous fluids should be restricted and Posm and Uosm values obtained hourly to determine if subsequent coordinates fall into the areas that indicate normal or DI. If dehydration causes the urine to become concentrated but only at the expense of an elevated Posm, the patient probably has partial neurogenic or nephrogenic DI. With water restriction, patients with primary polydipsia develop coordinates which fall into the normal area or to the left of normal.

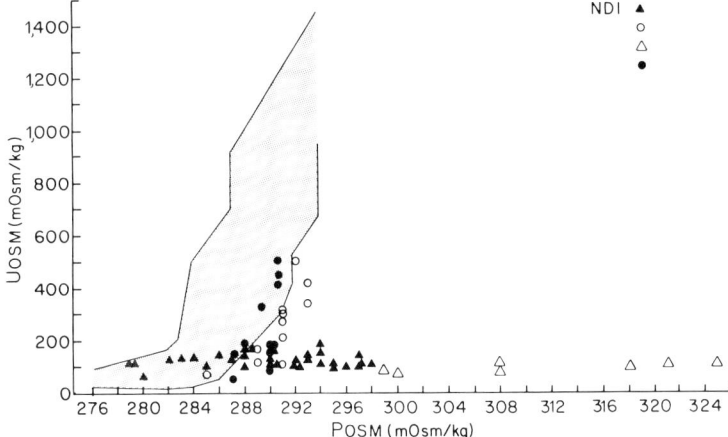

Fig. 4. Plasma vs urine osmolality relationships in normal subjects (shaded area), in 2 males (triangles), and 2 females (circles) with congenital nephrogenic diabetes insipidus.

When the pattern becomes clear, the test may be terminated before a plateau in Uosm is reached, or alternatively a formal dehydration test (plateau in Uosm followed by injection of ADH) may be conducted as described above. Another aspect of this formulation is that when initial data fall into the area of hypertonicity or hypernatremia with concentrated urine, the gentle forcing of hypotonic fluids will cause subsequent points to fall into areas that indicate normality or DI, thus generally allowing a definitive diagnosis to be established. Posm vs Uosm coordinates are most useful in differentiating patients with primary polydipsia from patients with neurogenic and nephrogenic DI (compare fig. 3, 4).

Measurement of Urinary AVP

When a definitive diagnosis is not possible by the use of the above methods, measurement of plasma [25, 41, 60] or urinary AVP [61] allows the diagnosis to be established. In patients with renal disease, plasma not urinary AVP measurements have to be used. We have obtained data on urinary AVP excretion relative to Posm in adults with normal renal function (shaded area fig. 5, 6). Urinary AVP is normal or supranormal relative to Posm in patients with nephrogenic DI (fig. 5). In contrast, patients with central DI have low levels of urinary AVP in relation to Posm (fig. 5). This relationship, as well as the Uosm in relation to UAVP (vide infra), allows the separation of the two types of DI but is

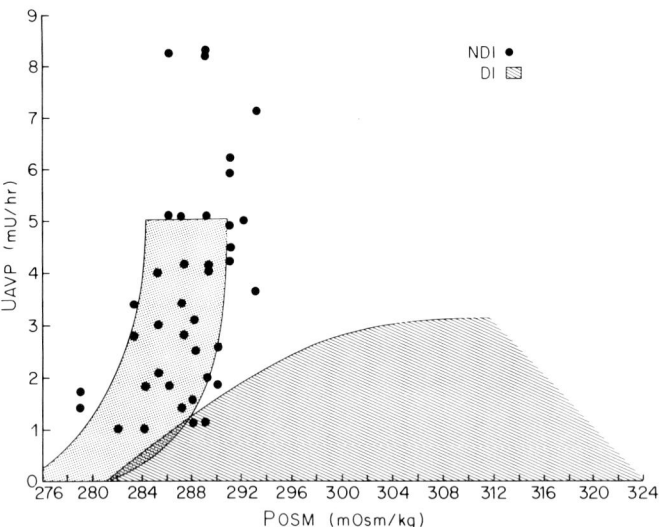

Fig. 5. Relationship between plasma osmolality and urine AVP in normal subjects (shaded area) and in patients with congenital nephrogenic diabetes insipidus. Range of values in patients with neurogenic diabetes insipidus is indicated by shading to right of normal.

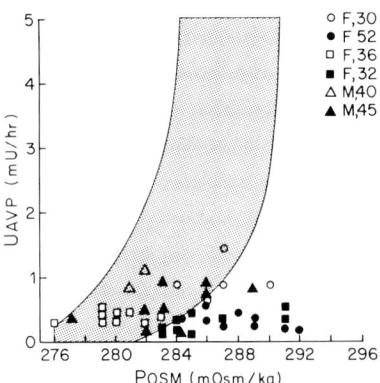

Fig. 6. Relationship between plasma osmolality and urine AVP in normal subjects (shaded area) and in patients with primary polydipsia.

of direct clinical value only when simpler means of differentiating the types of DI give unclear results. Urinary AVP tends to be low in patients with primary polydipsia (fig. 6) but, as indicated above, the differentiation of primary polydipsia from DI should be clear from simpler tests. Studies are now being conducted to determine whether the low UAVP in

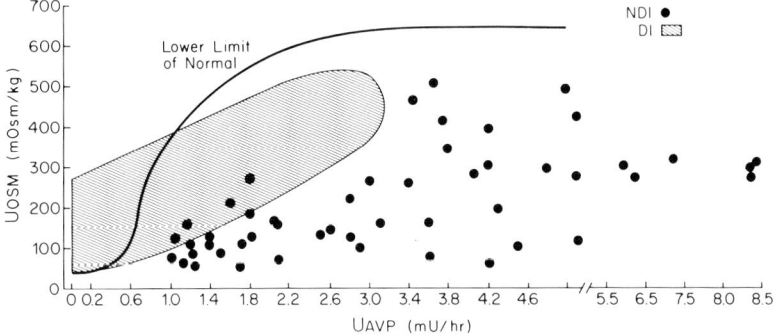

Fig. 7. Relationship between urine AVP excretion and urine osmolality in normal subjects (lower limit depicted by line), in patients with congenital nephrogenic diabetes insipidus, and in patients with central diabetes insipidus (shaded area).

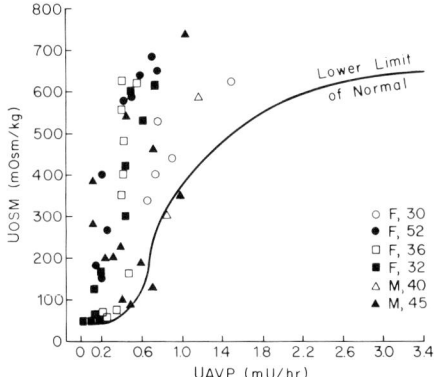

Fig. 8. Relationship between urine AVP and osmolality in normal subjects (lower limit depicted by line) and in patients with primary polydipsia.

patients with primary polydipsia reflects decreased neurohypophyseal release of hormone or decreased renal clearance.

The third way of comparing Posm, Uosm and UAVP is the relationship between UAVP and Uosm. This formulation has the advantage of having integrated values of both measurements over the same period of time. The lower limit of the normal relationship is indicated in figures 7 and 8. Patients with nephrogenic DI have points which fall to the right and below normal (fig. 7), while patients with primary polydipsia have coordinates in the normal range (fig. 8). In patients with complete neurogenic DI both Uosm and UAVP are low and meaningful renal resistance data cannot be obtained this way. Patients with partial neurogenic DI show some renal resistance to AVP (fig. 7).

Evaluation

The clinically practical differentiation of the 3 major types of hypotonic polyuria can be readily accomplished. History and physical examination reveal important clues such as age of onset and evidence of systemic disease. Clearly depressed or elevated plasma sodium or Posm can be of obvious value. However, in our experience these values are normal in the vast majority of patients with hypotonic polyuria. My next step is usually the collection of 4-h urines during random hydration, and the determination of the Uosm response to 5 U aqueous AVP. Patients with DI and primary polydipsia respond, while patients with congenital nephrogenic DI do not increase Uosm (fig. 2). The differentiation between primary polydipsia and neurogenic DI can then be made by dehydration test, including the response to 5 U aqueous AVP at the plateau in Uosm. If results are unclear further studies with AVP measurements can be conducted. In my experience nephrogenic DI can always be distinguished from the other causes of hypotonic polyuria by use of the above criteria, including measurement of AVP levels when necessary. Complete and partial neurogenic DI represent a continuum and are not clearly separable. The only clinical problem is whether or not to treat the patient. There may be difficulty at times differentiating between mild forms of neurogenic DI and primary polydipsia. The response to therapy cannot be used as an ultimate diagnostic criterion because some patients with primary polydipsia may markedly improve when treated with ADH, the improvement being related to an increased salivary flow rate and amelioration of thirst [32]. There is no single definitive diagnostic standard for the differential diagnosis of the 3 major hypotonic polyurias, but the diagnosis can be established by the use of a combination of factors.

References

1. Van Slyck, E.; Jurgensen, J.; Cargill, J.: Diabetes insipidus complicating thrombotic thrombocytopenic purpura. J. Am. med. Ass. *209:* 768–770 (1969).
2. Veldhuis, J.; Hammond, J.: Endocrine function after spontaneous infarction of the human pituitary: report, review, and reappraisal. Endocr. Rev. *1:* 100–107 (1980).

3   Schwartz, A.; Leddy, A.: Recognition of diabetes insipidus in postpartum hypopituitarism. Obstet. Gynec. N.Y. *59:* 394–398 (1982).
4   Baylis, P.; Milles, J.; London, D.; Butt, W.: Postpartum cranial diabetes insipidus. Br. med. J. *280:* 20 (1980).
5   Haynes, B.; Fauci, A.: Diabetes insipidus associated with Wegener's granulomatosis successfully treated with cyclophosphamide. New Engl. J. Med. *299:* 764 (1978).
6   Kelly, P.: Systemic blastomycosis with associated diabetes insipidus. Ann. intern. Med. *96:* 66–67 (1982).
7   Helbock, H.; Krivit, W.; Nesbit, M., Jr.: Patterns of antidiuretic function in diabetes insipidus caused by histiocytosis, X. J. Lab. clin. Med. *78:* 194–202 (1971).
8   Manelfe, C.; Louvet, J.-P.: Computed tomography in diabetes insipidus. J. Comput. assit. Tomogr. *3:* 309–316 (1979).
9   Braverman, L.; Mancini, J.; McGoldrick, D.: Hereditary idiopathic diabetes insipidus. A case report with autopsy findings. Ann. intern. Med. *63:* 503–508 (1965).
10 Scherbaum, W.; Bottazzo, G.: Autoantibodies to vasopressin cells in idiopathic diabetes insipidus: evidence for an autoimmune varient. Lancet *i:* 897–901 (1983).
11 Taubin, H.; Matz, R.: Cerebral edema, diabetes insipidus, and sudden death during the treatment of diabetic ketoacidosis. Diabetes *17:* 108–109 (1968).
12 Machiedo, G.; Bolanowski, P.; Bauer, J.; Neville, W.: Diabetes insipidus secondary to penetrating thoracic trauma. Ann. Surg. *181:* 31–34 (1975).
13 Moyson, F.: Anoxie cérébrale et diabète insipide après inhalation d'un corps étranger chez l'enfant. Ann. Chir. infant. *16:* 241–248 (1975).
14 Glauser, F.: Diabetes insipidus in hypoxemic encephalopathy. J. Am. med. Ass. *235:* 932–933 (1976).
15 Rothschild, M.; Shenkman, L.: Diabetes insipidus following cardiorespiratory arrest. J. Am. med. Ass. *238:* 620–621 (1977).
16 Parisi, J.; Kim, R.; Collins, G.; Hilfinger, M.: Brain death with prolonged somatic survival. New Engl. J. Med. *306:* 14–16 (1982).
17 Betkerur, U.; Shende, A.; Lanzkowsky, P.: Acute myeloblastic leukemia presenting with diabetes insipidus. Am. J. med. Sci. *273:* 325–327 (1977).
18 Kornberg, A.; Zimmerman, J.; Matzner, Y.; Polliack, A.: Acute lymphoblastic leukemia: association with vasopressin-responsive diabetes insipidus. Archs. intern. Med. *140:* 1236 (1980).
19 Yap, H-Y.; Tashima, C.; Blumenschein, G.; Eckles, N.: Diabetes insipidus and breast cancer. Archs. intern. Med. *139:* 1009–1011 (1979).
20 Winnacker, J.; Becker, K.; Katz, S.: Endocrine aspects of sarcoidosis. New Engl. J. Med. *278:* 427–434, 483–492 (1968).
21 Bruning, P.; Koster, H.; Hekster, R.; Luyendijk, W.: Sarcoidosis presenting with diabetes insipidus followed by acute cranial nerve syndrome. Acta med. scand. *205:* 441–444 (1979).
22 Panitz, F.; Shinaberger, J.: Nephrogenic diabetes insipidus due to sarcoidosis without hypercalcemia. Ann. intern. Med. *62:* 113–120 (1965).
23 Stuart, C.; Neelon, F.; Lebovitz, H.: Disordered control of thirst in hypothalamic-pituitary sarcoidosis. New Engl. J. Med. *303:* 1078–1082 (1980).
24 Bode, H.; Harley, B.; Crawford, J.: Restoration of normal drinking behavior by chlorpropamide in patients with hypodipsia and diabetes insipidus. Am. J. med. *51:* 304–313 (1971).

25 Robertson, G.; Aycinena, P.; Zerbe, R.: Neurogenic disorders of osmoregulation. Am. J. med. 72: 339–353 (1982).
26 Smitz, S.; Legros, J.: Hypodipsie et dysfonction sélective des osmorecepteurs dans le syndrome d'hypernatrémie. C. r. Séanc. Soc. Biol. 175: 874–881 (1981).
27 Smals, A.; Kloppenborg, P.: Hypernatraemia in human myotonic dystrophy. Neth. J. med. 23: 95–103 (1980).
28 Hochberg, Z.; Moses, A.; Miller, M.; Benderli, A.; Richman, R.: Altered osmotic threshold for vasopressin release and impaired thirst sensation: additional abnormalities in Kallmann's syndrome. J. clin. Endocr. Metab. 55: 779–782 (1982).
29 Rao, K.; Miller, M.; Moses, A.: Water intoxication and thioridazine (Mellaril). Ann. intern. Med. 82: 61 (1975).
30 Rogers, P.; Kurtzman, N.: Renal failure, uncontrollable thirst, and hyperreninemia. J. Am. med. Ass. 225: 1236–1238 (1973).
31 Barlow, E.; De Wardener, H.: Compulsive water drinking. Q. Jl. Med. 28: 235–257 (1959).
32 Pasqualini, R.; Codevilla, A.: Thirst-suppressing (antidipsetic) effect of Pitressin in diabetes insipidus. Acta endocr., Copenh. 30: 37–41 (1959).
33 Szczepańska-Sadowska, E.; Sobocińska, J.; Sadowski, B.: Central dipsogenic effect of vasopressin. Am. J. Physiol. 242: R372–R379 (1982).
34 Miller, M.; Moses, A.: Clinical states due to alteration of ADH release and action; in Moses, Share, Neurohypophysis, pp. 153–166 (Karger, Basel 1977).
35 Cates, J.; Garrod, O.: The effect of nicotine on urinary flow in diabetes insipidus. Clin. Sci. 10: 145–160 (1951).
36 Dingman, J.; Benirschke, K.; Thorn, G.: Studies of neurophyophyseal function in man. Diabetes insipidus and psychogenic polydipsia. Am. J. med. 23: 226–238 (1957).
37 Moses, A.; Numann, P.; Miller, M.: Mechanism of chlorpropamide-induced antidiuresis in man: Evidence for release of ADH and enhancement of peripheral action. Metabolism 22: 59–66 (1973).
38 Moses, A.; Howanitz, J.; Gemert van, M.; Miller, M.: Clofibrate-induced antidiuresis. J. clin. Invest. 52: 535–542 (1973).
39 DeRubertis, F.; Michelis, M.; Beck, N.; Field, J.; Davis, B.: Essential hypernatremia due to ineffective osmotic and intact volume regulation of vasopressin secretion. J. clin. Invest. 50: 97–111 (1971).
40 Halter, J.; Goldberg, A.; Robertson, G.; Porte, D.; Jr.: Selective osmoreceptor dysfunction in the syndrome of chronic hypernatremia. J. clin. Endocr. Metab. 44: 609–616 (1977).
41 Kimura, T.; Matsui, K.; Sato, T.; Yoshinaga, K.; Hoshi, T.: Relationship between plasma osmolality and plasma concentration of antidiuretic hormone in normal subjects, patients with chronic renal diseases, and patients with central diabetes insipidus. Tohoku J. exp. Med. 113: 77–88 (1974).
42 Brezis, M.; Weiler-Ravell, D.: Hypernatremia, hypodipsia and partial diabetes insipidus: a model for defective osmoregulation. Am. J. med. Sci. 279: 37–45 (1980).
43 Mahoney, J.; Goodman, A.: Hypernatremia due to hypodipsia and elevated threshold for vasopressin release. Effects of treatment with hydrochlorothiazide, chlorpropamide and tolbutamide. New Engl. J. Med. 279: 1191–1196 (1968).
44 Sridhar, C.; Calvert, G.; Ibbertson, H.: Syndrome of hypernatremia, hypodipsia and

partial diabetes insipidus: a new interpretation. J. clin. Endocr. Metab. *38:* 890–901 (1974).

45 Hickey, R.; Hare, K.: The renal excretion of chloride and water in diabetes insipidus. J. clin. Invest. *23:* 768–775 (1944).

46 Carter, A.; Robbins, J.: The use of hypertonic saline infusions in the differential diagnosis of diabetes insipidus and psychogenic polydipsia. J. clin. Endocr. Metab. *7:* 753–766 (1947).

47 Aubry, R.; Nankin, H.; Moses, A.; Streeten, D.: Measurement of the osmotic threshold for vasopressin release in human subjects and its modification by cortisol. J. clin. Endocr. Metab. *25:* 1481–1492 (1965).

48 Moses, A.; Streeten, D.: Differentiation of polyuric states by measurement of responses to changes in plasma osmolality induced by hypertonic saline infusions. Am. J. med. *42:* 368–377 (1967).

49 Gold, P.; Robertson, G.; Ballenger, J.; Kaye, W.; Chen, J.; Rubinow, D.; Goodwin, F.; Post, R.: Carbamazepine diminishes the sensitivity of the plasma arginine vasopressin response to osmotic stimulation. J. clin. Endocr. Metab. *57:* 952–957 (1983).

50 Moses, A.; Miller, M.; Streeten, D.: Quantitative influence of blood volume expansion on the osmotic threshold for vasopressin release. J. clin. Endocr. Metab. *27:* 655–662 (1967).

51 Moses, A.; Miller, M.: Osmotic threshold for vasopressin release as determined by saline infusion and by dehydration. Neuroendocrinology *7:* 219–226 (1971).

52 Durr, J.; Stamoutsos, B.; Lindheimer, M.: Osmoregulation during pregnancy in the rat. Evidence for resetting of the threshold for vasopressin secretion during gestation. J. clin. Invest. *68:* 337–346 (1981).

53 Lipsett, M.; MacLean, J.; West, C.; Li, M.; Pearson, O.: An analysis of the polyuria induced by hypophysectomy in man. J. clin. Endocr. Metab. *16:* 183–195 (1956).

54 Lipsett, M.; Pearson, O.: Further studies of diabetes insipidus following hypophysectomy in man. J. Lab. clin. Med. *49:* 190–199 (1957).

55 Dashe, A.; Cramm, R.; Crist, C.; Habener, J.; Solomon, D.: A water deprivation test for the differential diagnosis of polyuria. J. Am. med. Ass. *185:* 699–703 (1963).

56 Miller, M.; Dalakos, T.; Moses, A.; Fellerman, H.; Streeten, D.: Recognition of partial defects in antidiuretic hormone secretion. Ann. intern. Med. *73:* 721–729 (1970).

57 Notman, D.; Mortek, M.; Moses, A.: Permanent diabetes insipidus following head trauma: Observations on ten patients and an approach to diagnosis. J. Trauma *20:* 599–602 (1980).

58 Moses, A.; Notman, D.: Diabetes insipidus and syndrome of inappropriate antidiuretic hormone secretion (SIADH). Adv. intern. Med. *27:* 73–100 (1982).

59 Richman, R.; Post, E.; Notman, D.; Hochberg, Z.; Moses, A.: Simplifying the diagnosis of diabetes insipidus in children. Am. J. Dis. Child. *135:* 839–841 (1981).

60 Skowsky, W.; Rosenbloom, A.; Fisher, D.: Radioimmunoassay measurement of arginine vasopressin in serum: development and application. J. clin. Endocr. Metab. *38:* 278–287 (1974).

61 Miller, M.; Moses, A.: Urinary antidiuretic hormone in polyuric disorders and in inappropriate ADH syndrome. Ann. intern. Med. *77:* 715–721 (1972).

Dr. Arnold M. Moses, State University Hospital, 750 East Adams Street, Syracuse, NY 13210 (USA)

Diagnosis of Diabetes insipidus

Gary L. Robertson

University of Chicago, Chicago, Ill., USA

Diabetes insipidus (DI) is a disorder characterized by the excretion of abnormally large volumes (>30 ml/kg/day) of dilute urine (<250 mosm/l)[1]. This combination of abnormalities can result from any of 3 basic defects. Probably the most common is deficient secretion of the antidiuretic hormone, arginine vasopressin (AVP). This disorder is variously referred to as neurogenic, cranial or hypothalamic DI. It is usually if not always due to loss of the neurosecretory neurons of the neurohypophysis and can be caused by a variety of pathologic processes[1]. Diabetes insipidus can also result from renal insensitivity to the antidiuretic effects of AVP. This disorder is usually referred to as nephrogenic DI and it can result from a lesion at any of several different sites in the chain of biochemical reactions that mediate the antidiuretic effects of AVP[2]. These defects can be caused by a variety of diseases or drugs[1]. Symptomatic polydipsia and polyuria can also result from excessive intake of water. This condition is usually referred to as primary polydipsia. It can result from an abnormality in the thirst mechanism, in which case the patients usually report thirst as the reason for drinking; or it can be psychogenic, in which case the patients usually give other often fanciful explanations for their compulsion to drink[3]. In patients with inappropriate thirst, a variety of pathologic processes can be responsible. In most cases, the cause is idiopathic and there is some evidence that it may occasionally occur on a familial basis[3]. Recent studies have shown that it can also be caused by sarcoidosis[4, 5] or lithium[6, 7] and, in animal models at least, may also be produced by surgical lesions in the septum and other midline structures[8–10].

In theory, differentiating between neurogenic DI, nephrogenic DI, and primary polydipsia should be possible with a few relatively simple and widely available clinical tests. In practice, however, these distinctions may not be as clear and straight forward as one might expect. For

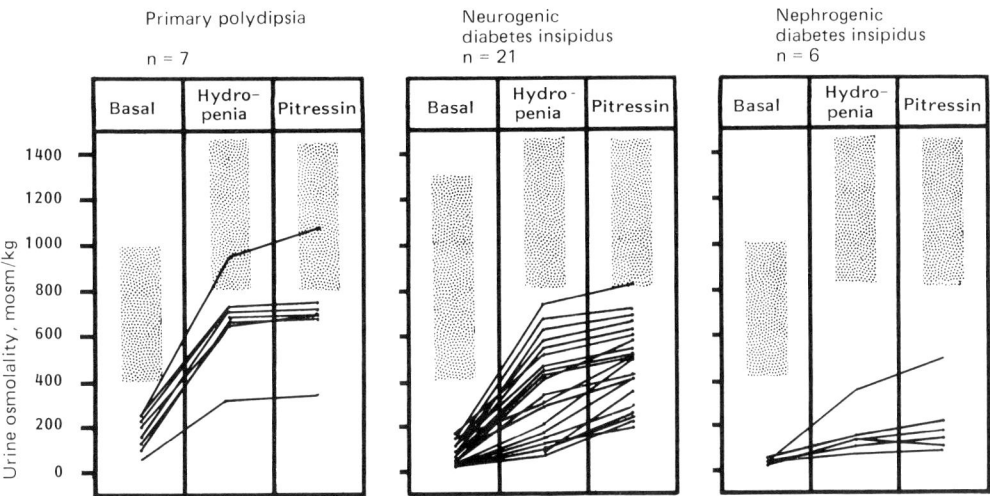

Fig. 1. The effect of dehydration and exogenous antidiuretic hormone on urinary concentration in patients with polyuria of diverse etiology. The shaded areas indicate the range of values in healthy adults under identical conditions. From [3] and unpublished observations.

example, basal plasma osmolality and sodium ought to be depressed in primary polydipsia and increased in neurogenic DI or nephrogenic DI. These differences are demonstrable when the mean values for the 3 groups are compared and may even be diagnostically useful in a few individuals whose values are at one extreme or the other [3, 11]. More often, however, the individual values for basal plasma osmolality are diagnostically ambiguous because of extensive overlap between the groups. This overlap is due largely to individual differences in the 'set' of the osmoregulatory mechanisms for thirst and AVP secretion [12].

The response of urine osmolality to fluid deprivation and administration of antidiuretic hormone may be similarly ambiguous (fig. 1). Although fluid deprivation in patients with primary polydipsia almost always produces concentration of the urine as expected, the absolute levels of urine osmolality achieved are appreciably and sometimes markedly less than normal. This abnormality is not due to a deficiency of circulating AVP because a relatively low osmolality of the urine persists following the injection of aqueous pitressin. Moreover, many patients with the two forms of DI respond to these procedures in essentially the same way (fig. 1). Thus, fluid deprivation causes concentration of the ur-

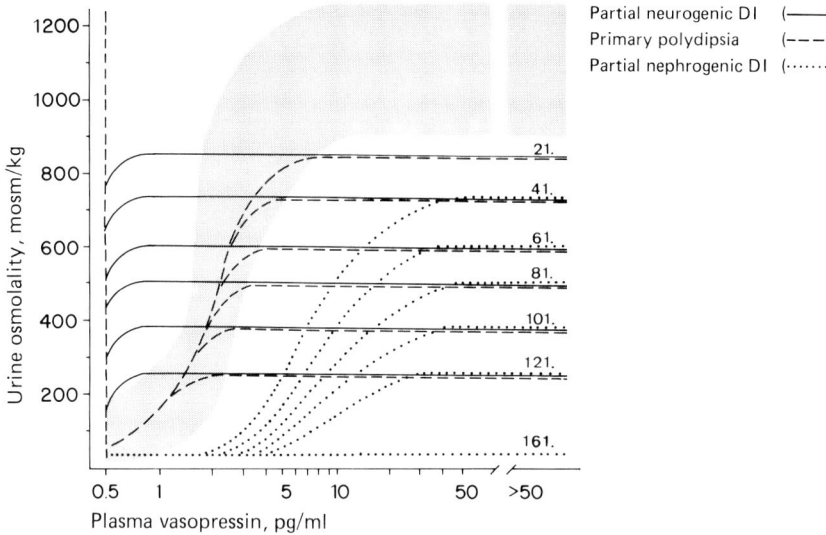

Fig. 2. The relationship of urine osmolality to plasma vasopressin in patients with polyuria of diverse etiology and severity. Note that, for each of the 3 categories of polyuria (neurogenic DI, nephrogenic DI, and primary polydipsia), the relationship is described by a family of sigmoid curves that differ in height. These differences in height reflect differences in maximum concentrating capacity due to 'washout' of the medullary concentration gradient. They are proportional to the severity of the underlying polyuria (indicated in l/day at the right end of each plateau) and are largely independent of the etiology. Thus, the 3 categories of DI differ principally in the submaximal or ascending portion of the dose response curve. In patients with partial neurogenic DI, this part of the curve lies to the left of normal, reflecting increased sensitivity to the antidiuretic effects of very low concentrations of plasma AVP. In contrast, in patients with partial nephrogenic DI, this part of the curve lies to the right of normal, reflecting decreased sensitivity to the antidiuretic effects of normal concentrations of plasma AVP. In primary polydipsia, this relationship is relatively normal. Schematic representation of data from [3].

ine in many patients with neurogenic DI or nephrogenic DI but the absolute levels of urine osmolality achieved are usually less than normal and remain so even after the injection of pitressin.

The failure of these simple clinical tests to differentiate reliability between the various types of DI is due to several factors. First, chronic polyuria of any etiology interferes with the maintenance of the medullary concentration gradient [13–16]. Because of this 'washout' effect, the maximum concentrating capacity of the kidney is diminished. The extent of this blunting varies in direct proportion with the severity of the

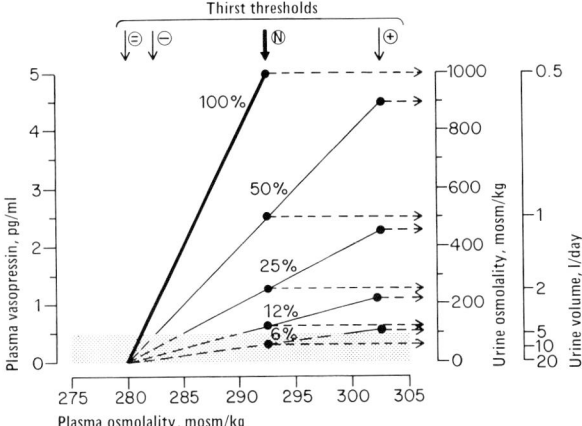

Fig. 3. Plasma AVP, urine osmolality and urine flow as a function of the osmoregulation of thirst and AVP secretion. Each oblique line depicts schematically the relationship between plasma AVP and plasma osmolality when secretory capacity is reduced to a specific percentage of normal. Each vertical arrow indicates the osmotic threshold for thirst as it occurs normally (N) or when it is abnormally high (+) or low (−), (=). The closed circles on each oblique line indicate the highest level to which plasma osmolality and AVP are normally allowed to rise at each thirst setting. The broken, horizontal arrows indicate the daily urine osmolalities and volumes that result when plasma AVP is limited to the specified levels. These figures assume a solute load of 500 mosm/day and normal renal sensitivity to AVP. Note the degree of impairment of AVP secretory capacity that is required to produce significant polyuria (>2 l/day) depends on the 'set' of the thirst mechanism. If it is normal, polyuria does not begin until AVP secretory capacity falls to 25% of normal. However, if the thirst threshold is 'set' 10 mosm/kg higher, secretory capacity must be reduced to 12% of normal for the same degree of polyuria to occur. Conversely, if the thirst threshold is 'set' 10 mosm/kg below normal, polyuria occurs even if AVP secretory capacity is normal. Schematic representation of data from [1, 12, 17].

polyuria but is independent of its cause [3]. Hence, for any given level of basal urine output, the maximum urine osmolality achieved in the presence of saturating levels of AVP is depressed to the same extent in patients with primary polydipsia, neurogenic DI and nephrogenic DI (fig. 2).

The second cause of the diagnostic confusion is that many patients with neurogenic or nephrogenic DI have an incomplete defect in AVP secretion (fig. 3) or action (fig. 2). As a consequence, the rise in plasma osmolality produced by fluid deprivation may force the secretion of enough additional AVP to permit concentration of the urine. Normally,

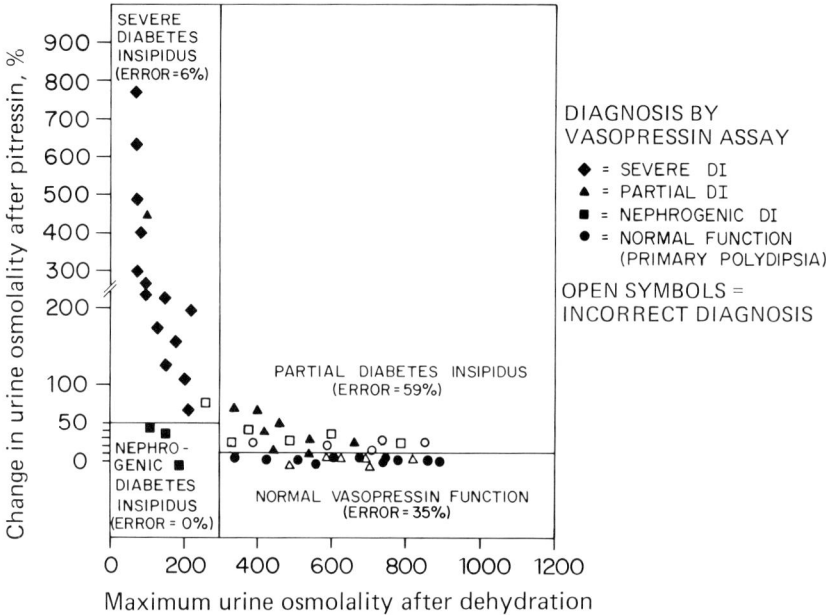

Fig. 4. The results of direct and indirect tests of AVP function in patients with polyuria of diverse etiology. The indirect tests were performed and interpreted as described by *Miller* et al. [18, 19]. The direct tests, which are based on an immunoassay of plasma AVP were performed and interpreted as described by *Zerbe and Robertson* [3]. The error rates indicate the percentage of patients in each of the 4 categories that were misdiagnosed by the indirect tests. From [3] and unpublished observations.

when water intake is unrestricted, these patients do not concentrate their urine because their thirst mechanisms do not allow their plasma osmolalities to rise high enough to stimulate AVP secretion to a level sufficient to maintain antidiuresis. Only when the physician overrides this barrier by instructing the patient not to drink during the dehydration test does plasma AVP reach a level that produces urinary concentration. Depending on the severity of the underlying impairment in AVP secretion or action and the duration of the test, the levels of urine osmolality achieved during fluid deprivation may or may not reach the maximum permitted by washout of the medullary concentration gradient. Therefore, further elevating plasma AVP by giving pitressin, may or may not produce a significant further increase in urine osmolality *in either type of patient.* This problem is further complicated by the fact that many patients with neurogenic DI appear to be supersensitive to the antidiuretic effect of very low levels of plasma AVP (fig. 2). Hence, even though their secretory capacity may be markedly diminished, a relatively mild stimulus like water deprivation may raise their plasma AVP and urine osmolality to the maximum permitted by washout of their medullary concentration gradient.

We found that the diagnoses provided by a widely used indirect test [18, 19] were reliable only in patients with severe defects in AVP secretion or action. In this category, i.e. in patients who fail to concentrate their urine during a standard dehydration test, a rise in urine osmolality of more than 50% after the administration of pitressin accurately distinguished most (95%) patients with severe neurogenic DI from those with severe nephrogenic DI (fig. 4). However, these indirect criteria are unable to differentiate between partial neurogenic DI, partial nephrogenic DI, or primary polydipsia (fig. 4). In these patients, i.e. in those who are able to concentrate their urine to varying degrees when water deprived, the error rate of the indirect test approximates 50%. In our series, the most common mistake was to misdiagnose partial nephrogenic DI as partial neurogenic DI. However, the indirect test also tended to confuse partial neurogenic DI and primary polydipsia with the incidence of false positive and false negative results being about equal in the two diagnostic groups. In almost every patient in this series, the diagnosis provided by direct assay of AVP was independently confirmed by closely monitoring the patient's response to a 2 to 6 day therapeutic trial with desmopressin [3].

Given the serious limitations of the conventional indirect methods for differentiating between the three forms of polyuria, what can be done? One possibility might be to modify the conditions of the indirect tests so as to correct the abnormalities in renal responsiveness that are largely responsible for the confusion. Such modifications may be possible since a considerable amount is known about factors that influence receptor function as well as the medullary concentration gradient. However, efforts along these lines probably would require many years of painstaking trial and error and, even then, would not be assured of success. Therefore, it would seem to be much more rational to favor direct assay of AVP in plasma or urine.

In our experience, the measurement of plasma AVP and plasma osmolality (or sodium) after a suitable period of fluid deprivation pro-

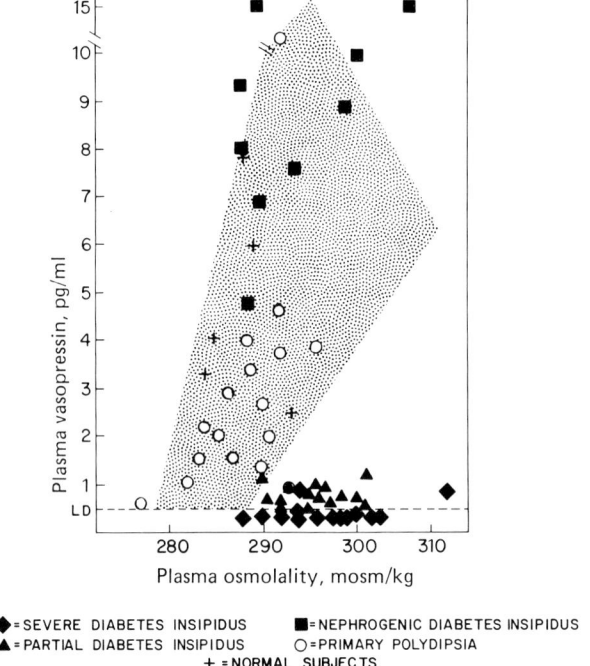

Fig. 5. Plasma AVP as a function of the concurrent plasma osmolality in patients with polyuria of diverse etiology. All samples were obtained at the end of a standard dehydration test. From [3] and unpublished observations.

vides a simple, safe and reliable way of differentiating neurogenic DI from nephrogenic DI and from primary polydipsia (fig. 5). For correct interpretation, the plasma AVP values *must* be expressed as a function of the concurrent level of plasma osmolality (or sodium). If the results are analyzed in this way, most if not all of the values in patients with neurogenic DI fall clearly below the normal range. An occasional value will be ambiguous usually because the level of plasma osmolality achieved during water deprivation is not high enough to clearly separate it from normal. In such a situation, repeating the measurements of plasma AVP and osmolality after a short infusion of hypertonic saline invariably suffices to clarify the diagnosis.

In sharp contrast to the results obtained in neurogenic DI, the plasma AVP values from patients with nephrogenic DI or primary polydip-

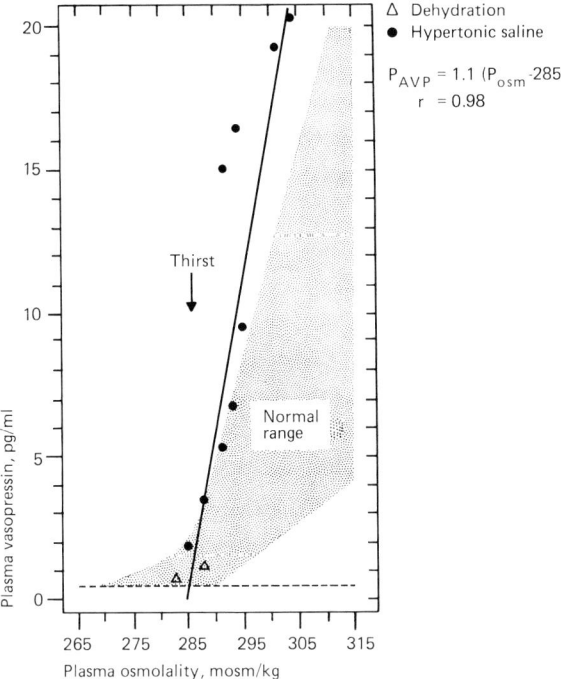

Fig. 6. Thirst and plasma AVP as a function of plasma osmolality in a woman with dipsogenic DI. Note that her osmotic threshold for thirst approximates that for AVP release. In healthy adults, it is usually 'set' 10–15 mosm/kg higher [12].

sia are invariably normal or even supranormal relative to the concurrent plasma osmolality (fig. 5). In theory, patients with primary polydipsia might secrete subnormal amounts of AVP due to the chronic suppressive effects of overhydration. This possibility is difficult to test conclusively because healthy volunteers will not drink excessively for periods long enough to faithfully reproduce the state of mild chronic overhydration characteristic of patients with primary polydipsia. However, none of our patients with psychogenic or other forms of severe polydipsia have ever shown any evidence of pituitary suppression (fig. 5). If anything, their plasma AVP response to osmotic stimuli tends to be slightly supranormal (fig. 6). Hence, it seems unlikely that chronic polydipsia suppresses posterior pituitary function to the extent that it interferes with the use of plasma AVP measurements for diagnosing neurogenic DI.

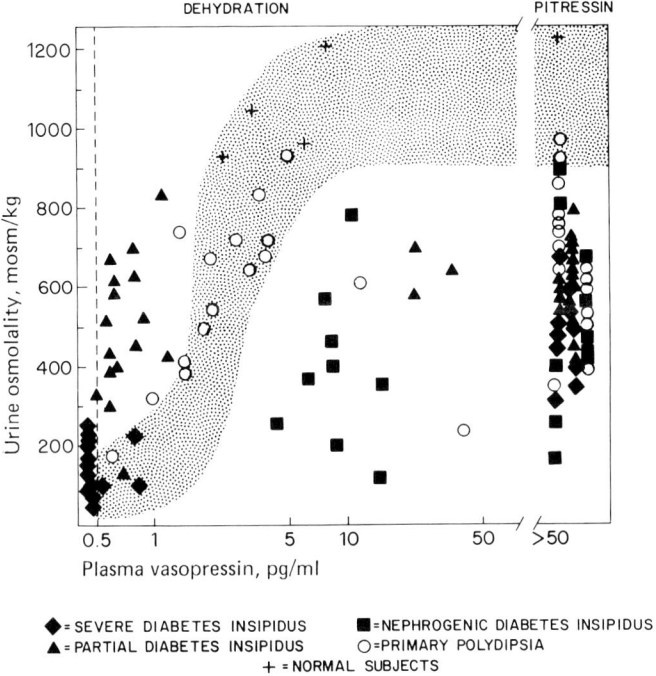

Fig. 7. Urine osmolality as a function of the concurrent plasma AVP in patients with polyuria of diverse etiology. All samples were obtained at the end of a standard dehydration test. From [3] and unpublished observations.

Those patients who exhibit a normal or supranormal AVP response to osmotic stimulation can be further differentiated into nephrogenic DI and primary polydipsia by analyzing the relationship of urine osmolality to plasma AVP during fluid deprivation (fig. 7). However, the proper interpretation of this relationship is complicated by the marked blunting of maximum concentrating capacity that results from chronic polyuria per se (fig. 2). Because of this blunting, the relationship of urine osmolality to plasma AVP during fluid deprivation in a patient with severe primary polydipsia may be more subnormal than in a patient with partial nephrogenic DI (fig. 7). This kind of problem can be obviated to some extent if the urine osmolality-plasma AVP values are plotted on a nomogram composed of a family of lines which reflect the reduction in urinary concentrating capacity that would be expected for a given level of polyuria (fig. 2). In many cases, however, even this approach is insufficient to distinguish primary polydipsia from partial nephrogenic DI because the confidence limits for the relationship of maximum concentrating capacity to basal urine output are actually quite wide [3]. Moreover, by the end of the standard dehydration test, plasma AVP is usually high enough to cause maximum urinary concentration and, as noted (fig. 2), values on this part of the dose response curve do not permit differentiation between nephrogenic DI and primary polydipsia. To make this distinction, it is necessary to examine the relationship when plasma AVP approximates normal basal levels (1 to 3 pg/ml). This requirement can usually be met by measuring plasma AVP and urine osmolality sequentially throughout the dehydration test although it is sometimes also necessary to perform a water load-pitressin infusion test.

In using a nomogram like figure 5 to diagnose DI, considerable care should be taken to avoid interference by nonosmotic stimuli. Nausea or hypotension are particularly troublesome in this regard because they are extremely potent stimuli and can induce significant elevations in plasma AVP even in patients with severe neurogenic DI [20]. Moreover, nausea and/or hypotension often accompany the vasovagal reactions that are caused in some patients by minor stresses such as venipuncture. Other, less common causes include adrenal insufficiency and/or orthostatic hypotension. Cigarette smoking without nausea or hypotension can also elevate plasma AVP [21] and in some patients with partial neurogenic DI this effect is great enough to totally abolish their polyuria [unpublished observations]. Even though modest, the increases in plasma AVP induced by these superimposed nonosmotic stimuli occasionally are sufficient to give the appearance of a normal hormone response to osmotic stimuli in patients with neurogenic DI. To avoid the kind of diagnostic error that would otherwise result from failure to recognize such an effect, the dehydration tests are best performed with frequent monitoring of symptoms and activity as well as vital signs.

Several other precautions are also necessary to obtain consistently reliable diagnostic results with direct tests of AVP function. First, the AVP assay employed must be extremely sensitive and largely if not totally free from interference by other components of plasma. Many assays which are suitable for other investigative purposes do not meet these requirements. Second, the plasma AVP values must be interpreted in relation to the concurrent level of plasma osmolality. The plasma AVP values alone are usually meaningless no matter how accurate they are because they often reach 'normal' basal levels during dehydration in patients with partial neurogenic DI. Third, the measurements of plasma osmolality must also be quite accurate and, if necessary, corrected for the contribution of abnormal elevations of ineffective solutes like glucose and urea [22]. Because the margin for error is small (fig. 5), the osmometry should have an accuracy of 1%. This requirement cannot be met by most hospital based laboratories and can never be achieved on serum or frozen plasma [unpublished observations]. Thus, the osmometry must be specially arranged and accompanied by simultaneous measurements of plasma glucose and urea. An alternative and possibly better approach might be to relate plasma AVP values to plasma sodium since it is the major if not the only effective osmotic stimulus in man and is measured more accurately by most clinical laboratories.

Finally, the possibility that plasma AVP levels may be spuriously high due to the presence of antibodies to AVP should always be remembered although systematic studies of this problem indicate that AVP antibodies are unlikely to be a significant cause of misdiagnosis because they occur infrequently [23].

Given the fastidious requirements of diagnostic methods based on quantitating the relationship of plasma AVP to plasma and urine osmolality, other approaches need to be considered. The use of nonosmotic stimuli like apomorphine or hypotension has much to recommend it since these stimuli are extremely potent and might, therefore, obviate the need for very sensitive and specific measurements of AVP and osmolality. However, apomorphine is extremely unpleasant, producing hypotension is cumbersome and requires close monitoring, and insulin-induced hypoglycemia is too weak and inconsistent a stimulus to AVP secretion to be of any diagnostic value [20]. Moreover, all of these non-osmotic methods have the potential disadvantage of missing a physiologically significant deficiency of AVP secretion caused by selective loss of the osmoregulation of AVP [20].

In our view, the only satisfactory alternative to this direct method is to conduct a closely monitored therapeutic trial with desmopressin (DDAVP). Our studies indicate that standard doses of this drug (25 µg intranasally twice a day) produce the critical response needed to differentiate partial nephrogenic DI from neurogenic DI or primary polydipsia (fig. 2). Thus, if the drug produces a significant antidiuretic effect, nephrogenic DI is effectively excluded and the differential can be narrowed to neurogenic DI vs primary polydipsia. These two conditions can then be distinguished by monitoring changes in water balance during therapy. If polydipsia as well as polyuria are abolished and plasma osmolality or sodium do not fall significantly below the normal range, the patient probably has neurogenic DI. Conversely, if the DDAVP reduces urine output without reducing water intake and body water increases abnormally as evidenced by the development of hyponatremia, then the patient probably has primary polydipsia. Needless to say, the hazards of inducing serious and possibly fatal water intoxication in such patients require that the test be performed under closely monitored conditions using a drug such as DDAVP that has an intermediate duration of action.

It should be emphasized, however, that the results of a therapeutic trial can also be diagnostically misleading even if the test is conducted properly. Some patients with severe neurogenic DI of many years duration will develop water intoxication when first treated with antidiuretic hormone. Their failure to appropriately reduce water intake has not been clearly explained but may be due to habits acquired during long experience with the disease. Moreover, in certain patients with primary polydipsia due to resetting of the thirst osmostat, antidiuretic therapy abolishes thirst and water intake without producing water intoxication. This response occurs because the small rise in body water induced by the antidiuresis lowers plasma osmolality below the threshold for thirst. This effect is well exemplified by the patient shown in figure 6. When she was treated with DDAVP, water intake and excretion fell promptly to normal and her plasma osmolality stabilized around the lower limit of normal (280 mosm/kg). This response was clinically indistinguishable from that exhibited by many patients with partial neurogenic DI even though she had no evidence of deficient AVP secretion by direct assay. The patient shown in figure 6 developed polydipsia and polyuria acutely during pregnancy in association with eosinophilia and cuts in the *nasal* half of her visual fields. Nevertheless, her AVP secretion and action were normal and the only detectable abnormality in osmoregulation was resetting of the thirst osmostat. These findings as well as other studies in animals [8–10] show clearly that intracranial pathology can cause abnormally increased thirst as well as impaired AVP secretion. At present, therefore, direct measurement of plasma AVP as well as plasma and urine osmolality during a dehydration test provides a consistently reliable way of differentiating between neurogenic DI, nephrogenic DI and primary polydipsia.

References

1 Robertson, G.L.: Diseases of the posterior pituitary; in Felig, Baxter, Broadus, Frohman, Endocrinology and metabolism, pp. 251–277 (McGraw-Hill, New York 1981).
2 Singer, I.; Forrest, J.N., Jr.: Drug-induced states of nephrogenic diabetes insipidus. Kidney int. *10:* 82–95 (1976).
3 Zerbe, R.L.; Robertson, G.L.: A comparison of plasma vasopressin measurements with a standard indirect test in the differential diagnosis of polyuria. New Engl. J. Med. *305:* 1539–1546 (1981).
4 Stuart, C.A.; Neelon, F.A.; Lebovitz, H.E.: Disordered control of thirst in hypothalamic-pituitary sarcoidosis. New Engl. J. Med. *303:* 1078–1082 (1980).
5 Kirkland, J.L.; Pearson, D.J.; Goddard, C.; Davies, I.: Polyuria and inappropriate secretion of arginine vasopressin in hypothalamic sarcoidosis. J. clin. Endocr. Metab. *56:* 269–272 (1983).
6 Cox, M.; Singer, I.: Lithium and water metabolism. Am. J. Med. *59:* 153–157 (1975).
7 Weiss, N.; Robertson, G.L.: Lithium induced polyuria. Clin. Res. *32:* 536A (1984).
8 Grossman, S.P.; Grossman, L.; Halaris, A.: Effects of hypothalamic and telencephalic NE and 5-HT of tegmental knife cuts that produce hyperphagia and hyperdipsia in the rat. Pharmacol. Biochem. Behav. *6:* 101–106 (1977).
9 Hennessy, J.W.; Grossman, S.P.; Kanner, M.: A study of the etiology of the hyperdipsia produced by coronal knife cuts in the posterior hypothalamus. Physiol. Behav. *18:* 73–80 (1977).
10 Grossman, S.P.; Grossman, L.: Food and water intake in rats after transections of fibers en passage in the tegmentum. Physiol. Behav. *18:* 647–658 (1977).
11 Barlow, E.D.; De Wardener, H.E.: Compulsive water drinking. Q. Jl Med. *28:* 235–258 (1959).
12 Robertson, G.L.: Abnormalities of thirst regulation. Kidney Int. (in press 1984).
13 Epstein, F.H.; Kleeman, C.R.; Hendrikx, A.: A influence of bodily hydration on the renal process. J. clin. Invest. *36:* 629–634 (1957).
14 De Wardener, H.E.; Herxheimer, A.: The effect of high water intake on the kidney's ability to concentrate the urine in man. J. Physiol. Lond. *139:* 42–52 (1957).
15 Alexander, C.S.; Filbin, D.M.; Fruchtman, S.A.: Failure of vasopressin to produce normal urine concentration in patients with diabetes insipidus. J. Lab. clin. Med. *54:* 566–571 (1959).
16 Harrington, A.R.; Valtin, H.: Impaired urinary concentration after vasopressin and its gradual correction in hypothalamic diabetes insipidus. J. clin. Invest. *47:* 502–510 (1968).
17 Robertson, G.L.; Aycinena, P.R.; Zerbe, R.L.: Neurogenic disorders of osmoregulation. Am. J. Med. *72:* 339–353 (1982).
18 Miller, M.; Dalakos, T.; Moses, A.M.; Fellerman, H.; Streeten, D.H.P.: Recognition of partial defects in antidiuretic hormone secretion. Ann. intern. Med. *73:* 721–729 (1970).
19 Streeten, D.H.P; Moses, A.M.; Miller, M.: Disorders of the neurohypophysis; in Isselbacher, Adams, Braunwald, Petersdorf, Wilson, Harrison's principles of internal medicine, 9th ed., pp. 1684–1694 (McGraw-Hill, New York 1980).

20 Zerbe, R.L.; Baylis, P.H.; Robertson, G.L.: Vasopressin function in clinical disorders of water balance; in Robertson, Beardwell, Clinical endocrinology, Butterworth's Int. Med. Review Series, vol. 1, pp. 297–329 (Butterworths, London 1981).
21 Rowe, J.W.; Kilgore, A.; Robertson, G.L.: Evidence in man that cigarette-smoking induces vasopressin release via an airway-specific mechanism. J. clin. Endocr. Metab. *51:* 170–172 (1980).
22 Zerbe, R.L.; Robertson, G.L.: Osmoregulation of thirst and vasopressin secretion in human subjects: effect of various solutes. Am. J. Physiol. E *224:* 607–614 (1983).
23 Vokes, T.; Gaskill, M.; Robertson, G.L.: Antivasopressin antibodies in neurogenic diabetes insipidus. Clin. Res. *32:* 275A (1984).

G.L. Robertson, MD, University of Chicago, 5841 S. Maryland Avenue, Box 131, Chicago, IL 60637 (USA)

Neurogenic Diabetes insipidus in Children

P. Czernichow, R. Pomarede, R. Brauner, R. Rappaport
with the technical assistance of A. Basmaciogullari

Département de Pédiatrie, Service d'Endocrinologie Pédiatrique et Diabète, Clinique Robert Debré, Hôpital des Enfants-Malades, Paris, France

Abnormal urinary concentration observed in hypothalamic diabetes insipidus is most often due to a destructive process along the hypothalamopituitary axis. Some familial cases have been described, and recent progress in our understanding of the processing of neurohormones may allow future discovery of inherited diseases due to an abnormal vasopressin molecule [*Robinson and Verbalis, Richter and Schmale,* this volume]. Inadequate secretion of arginine vasopressin (AVP) results in polyuria and secondary polydipsia. In some rare cases polyuria is primary due to polydipsia. In this situation excessive urine output is due to abnormally high water consumption. Primary polydipsia is extremely rare in children and will be discussed at the end of the present chapter. Besides the well-characterized patients in whom secretion of AVP is deficient, there are some patients with chronic hypernatremia and deficient thirst in whom the control of AVP secretion by the osmostat seems to be abnormal. This syndrome is not well understood but will be discussed in the light of available data on plasma AVP obtained under various condition.

Clinical Manifestations

Diabetes insipidus in children affects both sexes equally and may begin at any age from a few months to adolescence. Interestingly, even in familial cases, occurrence of polyuria is rare in neonates. Very young children usually present with chronic dehydration, unexplained fever, vomiting, constipation and failure to thrive. These may lead to serious neurologic disturbances and may produce long-term neurologic deficits. In older children, polyuria is the main symptom and often has a dramatic onset. The intensity of the polyuria is variable but may reach 10–15 liters a day if children have free access to water intake and urine output does not correlate well with the severity of the vasopressin deficiency. The intensity of polyuria is similar in patients with both partial and complete diabetes insipidus. Enuresis is less common in children under 10 years, while anorexia, sleep disturbances and difficulty with school work are common presenting complaints. Growth is usually normal and, if abnormal, indicates either decreased growth hormone secretion or inadequate water intake. In the latter, catch-up growth will be observed after appropriate treatment but the former will require treatment with growth hormone (GH) for proper growth.

Diagnosis

Differentiation of hypothalamic diabetes insipidus from other causes of polyuria is usually not difficult. As shown by *Aaronson and Svenningsen* [1], administration of 1-desamino-8-*D*-arginine-vasopressin (DDAVP) intranasally, 20 µg in children and 10 µg in infants, will cause concentration of the urine similar to values obtained after 22 h of water deprivation. In some rare cases with longstanding polyuria, maximum urinary osmolality is not obtained after the first administration of DDAVP and the test should be repeated several times. As discussed by *Harrington and Valtin* [2], patients with hypothalamic diabetes insipidus may have impaired urinary concentration in response to administered vasopressin because of hypotonicity of the renal medulla. In very young children, the maximum ability to concentrate the urine is less than in older children, and the DDAVP test will not differentiate between partial diabetes insipidus and partial unresponsiveness of the renal tubules to the administered vasopressin. In this situation, measurement of vasopressin in plasma will be diagnostic (see below).

Several tests have been described in the literature to assess the renal concentrating ability in children with polyuria [3–5]. The potential danger of testing children with severe diabetes insipidus should be noted. Blood pressure and weight loss must also be monitored during any test involving dehydration and the test should be interrupted if weight loss is

Table I. Plasma and urinary osmolality, plasma AVP at the end of a 14-hour dehydration test

	AVP pg/ml ± SD	Osmolality, mosm/kg ± SD	
		plasma	urine
Normal children n = 22	1.9 ± 0.2	283 ± 1	1,056 ± 47
Partial DI n = 18	2.0 ± 1.8	298 ± 9	511 ± 117
Complete DI n = 23	1.3 ± 0.8	312 ± 15	150 ± 70

5% of body weight or thirst is intolerable. We have for 5 years used a short overnight dehydration test in which plasma and urine osmolality are measured and interpreted relative to measurement of AVP [6]. The ability to distinguish between partial and total vasopressin deficiency was evaluated by comparing results obtained after 14 versus 18 h dehydration. In this study, only 2 of 21 patients failed to reach their individual maximum urinary osmolality after 14 h hydropenia (starting at 6 p.m. the previous day). We, therefore, routinely use a 14-hour overnight dehydration test, under close medical supervision. However, in children with excessive polyuria or in children under 4 years of age we start the test at midnight to shorten the time of dehydration. Our experience with this test is summarized in table I and figure 1. In normal children, maximum urinary osmolality is not different from that in adults. The urinary osmolality was greater than 800 mosm after dehydration. In this group of subjects, plasma osmolality was 283 mosm, a value not different from random values in children.

Patients were classified as partial or complete diabetes insipidus according to urinary osmolality above or below 300 mosm, respectively. As shown in figure 1, in 41 children with diabetes insipidus half were complete and half were partial. Plasma osmolality at the end of the dehydration period was very high. In patients with partial defects urinary osmolality was between 300 and 750 mosm, a value significantly different from that observed in complete diabetes insipidus. Mean plasma osmolality was 295 mosm. All of these patients responded to DDAVP by increasing their urinary osmolality to above 800 mosm.

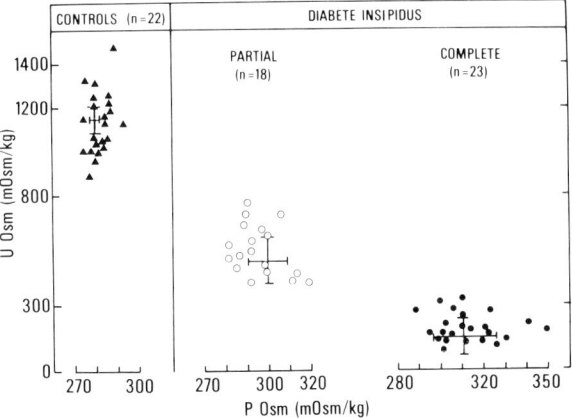

Fig. 1. Urinary osmolality (UOsm) in relation to plasma osmolality (POsm) in normal children and in patients with partial and complete diabetes insipidus.

It is clear from figure 1 that there is a continuum between normal and total defects. In most patients with severe polyuria a short period of dehydration with urine and plasma collection in the morning will establish inadequate urinary concentration with elevated plasma osmolality. In partial defects, a well-controlled dehydration test must be compared to normal values for the same age group. As shown in figures 1 and 2 several patients with partial defects had urinary concentrations close to normal at the end of this test. All of them, however, at that time, had elevated plasma osmolality, confirming inadequate secretion of vasopressin. In some patients, e.g. those with histiocytosis-X or dysembryoma who frequently have mild defects, prolongation of dehydration until hyperosmolality is present may be necessary to reveal inadequate secretion of vasopressin. As stated earlier, we found no difference in the urinary output between patients with both types of diabetes insipidus; their thirst mechanism must not allow them to become dehydrated. *De Wardener and Herscheimer* [7] showed that prolonged water consumption induced a transient state of inability to concentrate the urine and *Royer* et al. [8] showed that progressive water restriction will, after several days, increase urinary concentration even when a patient is unable to concentrate his urine after a short test. This is demonstrated in 2 patients in figure 3. Both patients complained of intolerable thirst during the short dehydration test and were considered as having complete dia-

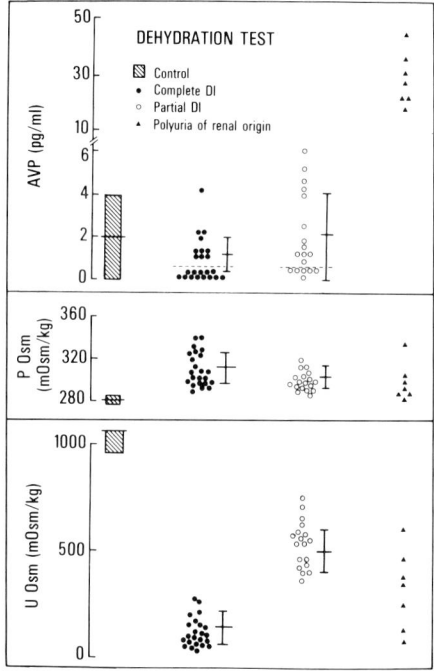

Fig. 2. Plasma AVP, POsm and UOsm in control children and in patients with neurogenic and nephrogenic diabetes insipidus.

betes insipidus. After a progressive decrease in water intake 1 patient still had complete diabetes insipidus but in the other patient the urinary osmolality was 705 at the end of the test. In both plasma osmolality was elevated. These examples illustrate the limitation of the dehydration test as an absolute measure of vasopressin deficiency, but the longer procedure is expensive and time-consuming and may only document the limited secretion of vasopressin with levels of plasma osmolality which are clinically undesirable.

Plasma AVP was measured during the short dehydration test. As shown in figures 2 and 4, most patients with complete diabetes insipidus have plasma AVP levels less than 1 pg/ml. Patients with partial defects tended to have plasma AVP levels higher than controls, but these values need to be correlated with the measured plasma osmolality. In normal children, as shown previously in adults [*Vokes and Robertson*, this volume; 9], there is a correlation between plasma AVP and plasma osmo-

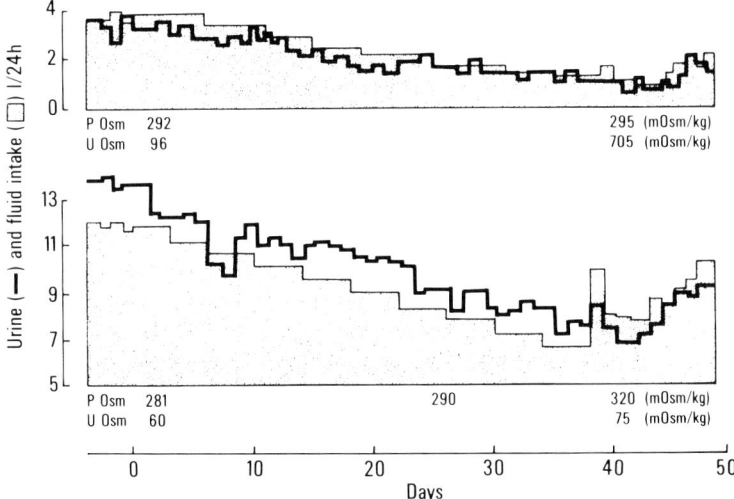

Fig. 3. Effect of progressive restriction of water intake in 2 patients classified as having complete diabetes insipidus on the basis of the short dehydration test. After reduction of water intake 1 patient (lower panel) still has complete diabetes insipidus although the other case (upper panel) was able to concentrate to 705 mosm.

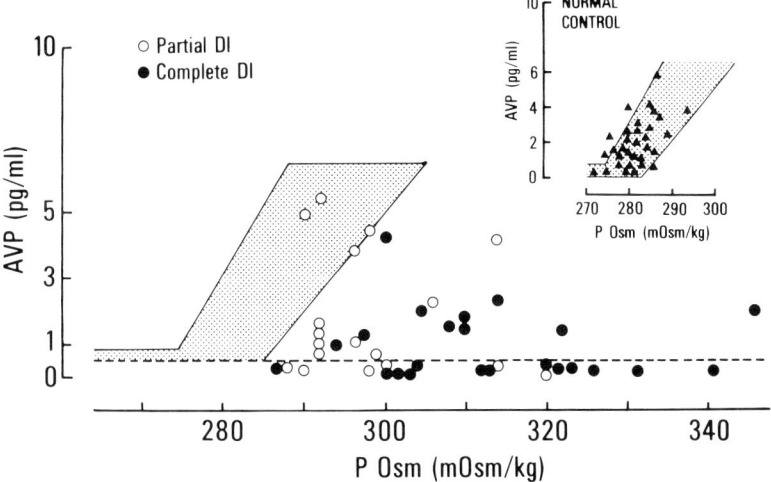

Fig. 4. Plasma AVP in relation to POsm in complete and partial diabetes insipidus. The grey zone indicates normal value. Lower limit of the AVP assay is indicated by the dotted line. The inset in the upper right part indicate individual values in normal control subjects.

lality (fig. 4). However, there are great individual variations which limit the diagnostic value to a single determination unless the plasma osmolality is greatly elevated.

In patients with diabetes insipidus the plasma AVP is inappropriate for the plasma osmolality. As shown in figure 4, patients with severe diabetes insipidus have higher plasma osmolality and lower plasma AVP than patients with partial defects. Five patients with diabetes insipidus had plasma AVP levels close to or even in the normal range. We have no explanation for the normal levels. *Weitzman* et al. [10] demonstrated in dogs and sheep an episodic secretion of AVP and the normal values might represent an acute single burst of AVP which does not reflect basal release. However, measurements of plasma AVP every 5 min for 1 h in 5 patients demonstrated stable values without significant variation. We have assumed that in these few patients, the sensitivity of the kidney is different from normal children in whom urinary osmolality is above 800 mosm with similar levels of AVP.

In patients with polyuria of renal origin (fig. 2) plasma AVP levels have great diagnostic importance and are from 10 to 50 pg/ml. In very young children in whom the dehydration test is not easy to perform and in whom the normal range of urinary osmolality response is broad, we found AVP measurement useful. A single determination of AVP will confirm the correct diagnosis. However, it should be emphasized that plasma AVP has to be measured during hypernatremia. As shown in table II, plasma AVP is close to normal when measured in normally hydrated patients. When measured at the end of a dehydration period plasma AVP is diagnostic in this type of polyuria.

Associated Anterior Pituitary Dysfunction

The frequency of anterior pituitary insufficiency in association with diabetes insipidus depends on the etiology of the diabetes insipidus (table III). GH deficiency is the most common accompanying anterior pituitary deficit, but deficiency of thyroid-stimulating hormone (TSH) and adrenocorticotropic hormone (ACTH) may also occur. ACTH deficiency requires special attention because it may mask the polyuria of diabetes insipidus. When ACTH deficiency is suspected, the dehydration test should not be performed until at least 3 days of replacement therapy with hydrocortisone.

Table II. Plasma osmolality (mosm/kg) and AVP (pg/ml) obtained in 4 patients with nephrogenic diabetes insipidus at the end of a dehydration test (left) and at random (right)

Patient	POsm	PAVP	POsm	PAVP
I	312	10.2	280	2.0
	290	7.5	280	2.0
II	330	8.6	264	4.3
III	302	6.8	275	2.1
IV	295	6.3	284	2.3

Table III. Anterior pituitary deficiency in neurogenic diabetes insipidus

	STH	TSH	ACTH	STH+TSH	STH+ACTH	TSH+ACTH	STH ACTH TSH	None
Post neurosurgery (23)	0	0	0	1	0		22	0
Intracranial lesion (26)	6	1	0	3	2	0	2+7*	5
Idiopathic and familial cases (31)	6	2+2*	0	1	3	1+1*	0	15

Asterisk indicates prolonged TSH secretion after TRH stimulation.

Etiology of Hypothalamic Diabetes insipidus in Children

The etiologies of diabetes insipidus in a large series of 118 patients is presented table IV.

Postoperative Diabetes insipidus

Diabetes insipidus that occurs after intracranial surgery may be the first manifestation of water imbalance in patients who had normal posterior pituitary function before operation. Patients should be adequately tested prior to surgery, but this is sometimes difficult for patients with tumor and associated intracranial hypertension. As shown in figure 5 and table IV, GH deficiency is common in such cases and often deficiencies of other anterior pituitary hormones is present. Some of these patients may have abnormal thirst. This should be noted during the

Table IV. Etiology of 118 cases of diabetes insipidus[1]

Causes	n	Total number of cases	Percent
Post surgery			
Craniopharyngioma	22		
Other	2	24	20.3
Central lesion			
Tumor			
Dysgerminoma	14		
Craniopharyngioma	8		
Floor of the 3rd ventricle	6	32	27.1
Von Recklinghausen	2		
Glioma	1		
Chordoma	1		
Histiocytosis	18	18	15.3
Other causes			
Meningitis	2		
Cranial trauma	2		
Cerebral malformation	1		
Aneurysm (carotid)	1	10	8.5
Associated neurological symptoms	1		
With empty sella	3		
Idiopathic			
Isolated cases	27	27	22.9
Familial	7	7	5.9

[1] From the Hôpital des Enfants-Malades, Paris, 1955 through 1983.

dehydration test since it has important clinical consequences for treatment.

Intracranial Lesions

Diabetes insipidus may indicate the presence of an intracranial lesion. However, even with appropriate X-ray studies, the diagnosis may not be apparent for several years. In our experience, 49% of cases resulting from intracranial lesions were not diagnosed at the onset of polyuria. In 18% of the cases, in spite of thorough evaluation, the diagnosis was not established for up to 2 years, and in an additional 6% not until 4 years (fig. 6). After 4 years of following, the likelihood of finding an intracranial tumor as the cause of diabetes insipidus is extremely small. An-

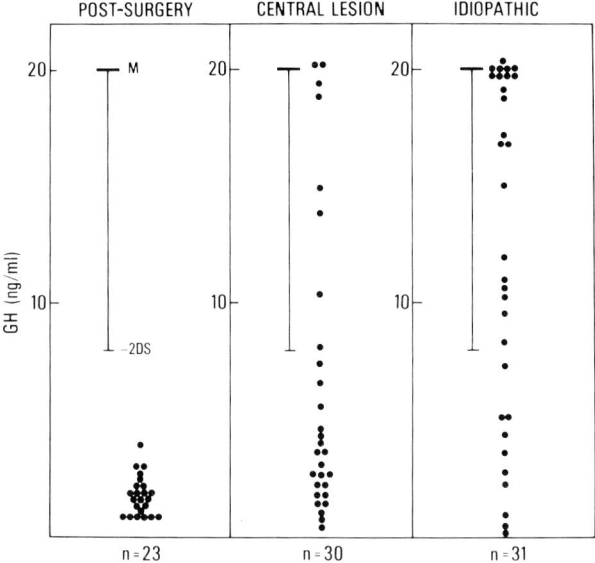

Fig. 5. Growth hormone (GH) secretion in different types of diabetes insipidus. Familial cases are included in the idiopathic type.

terior pituitary deficiency is commonly associated with diabetes insipidus when the diabetes insipidus is caused by an intracranial lesion (table III, fig. 5). GH deficiency was found in 80%, and combined deficiencies of GH, ACTH, and TSH were found in 36% of cases.

Dysgerminoma was the most common intracranial tumor in our series of children [*Pomarede* et al., this volume; 11]. Craniopharyngioma is the second most common cause. 10–35% of patients with craniopharyngioma have been reported to have diabetes insipidus [12, 13]. In our experience the majority of patients have inadequate urine concentration, but this may reflect selection of more severe cases in our series or the possibility that ACTH deficiency masked the polyuria in less carefully evaluated cases.

Diabetes insipidus is also frequent in children with histiocytosis-X; 40% according to a recent series by *Nezelof* et al. [14]. Polyuria may be moderate and may disappear during the course of the disease [15]. Severe head trauma with basal skull fracture is also a cause of diabetes insipidus [*Verbalis* et al., this volume].

We have seen in the last 5 years 3 cases of diabetes insipidus associated with empty sella of unknown etiology, a condition rarely diagnos-

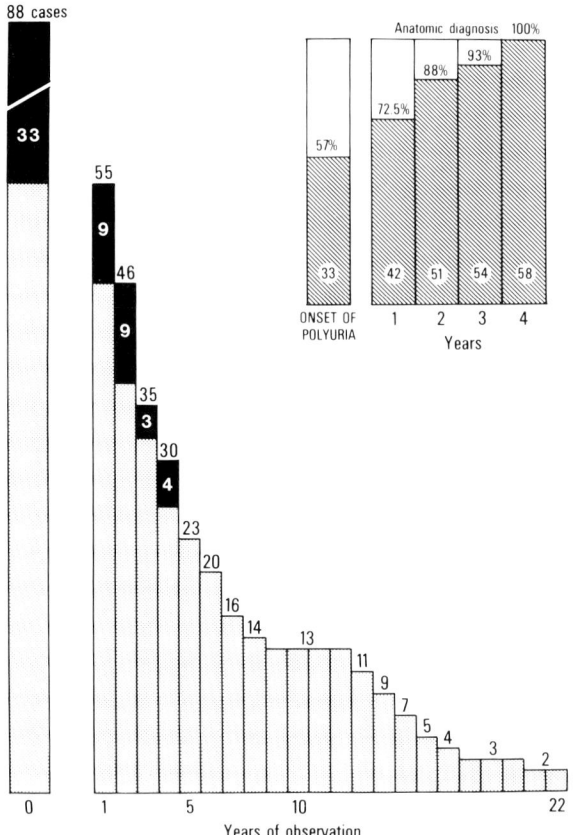

Fig. 6. The diagnosis of idiopathic diabetes insipidus is a temporary classification and depends on the years of observation. Thirty-three patients with intracranial lesion (ICL in black) were recognized at onset of polyuria. Among the 55 patients, 9 had an intracranial lesion diagnosed during the first year. After 4 years no new cases of intracranial lesion were found. As shown in the inset 57% of all cases of diabetes insipidus recognized in our clinic were diagnosed at onset of polyuria and 100% of all known cases in the 4 following years.

ed before the advent of CT scanning. Finally, polyuria with associated pyramidal symptoms was found in an 8-year-old boy with a degenerative disorder of unknown cause.

Idiopathic

About 30% of all cases of diabetes insipidus in children are classified as idiopathic. 17 cases of isolated idiopathic diabetes insipidus were fol-

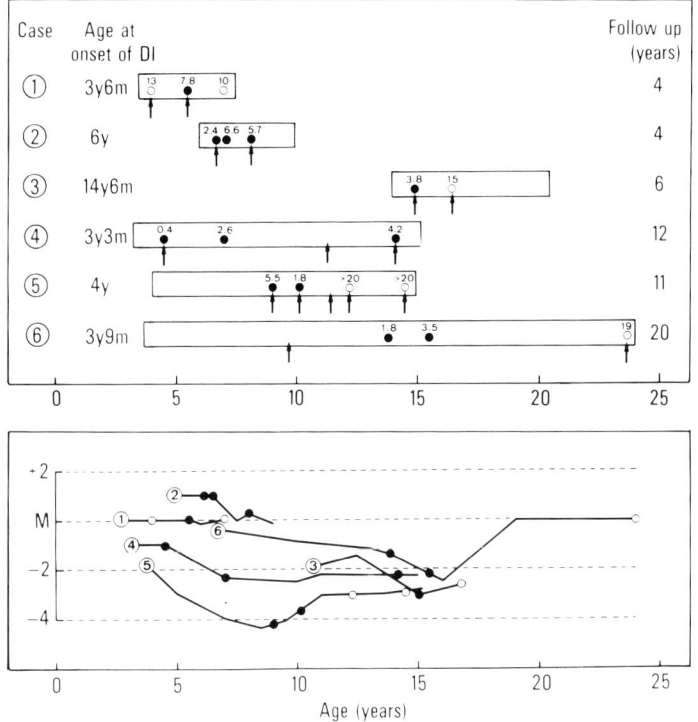

Fig. 7. Clinical survey of 6 patients with idiopathic diabetes insipidus. Open bars indicate the length of observation and arrows the moment of neuroradiological examination. Dots indicated normal (open) or low (black) GH secretion. Growth of these patients is represented in standard deviation score.

lowed in our clinic from 4 up to 26 years. The diagnosis of idiopathic diabetes insipidus was based on routine clinical tests and repeated neuroradiological investigations which were negative for at least 4 years. Interestingly, some of these cases of idiopathic diabetes insipidus were associated with an anterior pituitary dysfunction [16]. As shown in figure 5, GH deficiency was present in 6 children. Insufficient TSH secretion after thyrotropin releasing hormone (TRH) stimulation was demonstrated in 1 case and abnormal response to metopyrone test in 2 cases. Elevated prolactin and TSH were also present in 3 and 2 cases, respectively. Some of these anomalies were transient. Data relevant to the 6 patients with GH deficiency are summarized figure 7. In this group, 4 patients had a transient GH deficiency. Despite long-term follow-up and repeated evaluation with arginine insulin tolerance tests, GH secretion remained deficient in 2 patients (cases 2 and 4). Case 1 had a normal growth rate. Case 2 had a normal height but a decreased growth rate. Cases 3–6 had short stature with a decreased growth rate. Patient 6, however, had a pubertal growth spurt and reached a normal final height. There may be anterior pituitary deficiencies in some patients with permanent idiopathic diabetes insipidus. Occasionally, these deficiencies are transient and corrected with treatment of diabetes insipidus and normalization of water balance.

The presence of anterior pituitary deficiency always raises the suspicion of an associated tumor. CAT scans were normal in all patients with GH deficiency with the single exception of patient 5 (fig. 7) who had a slightly enlarged pituitary stalk. This finding has been described by others in idiopathic diabetes insipidus [*Manelfe* et al., this volume; 17]. In our patient, it remained stable for 2 years.

There are few studies of anatomy in patients with idiopathic diabetes insipidus. In adult patients, it has been reported that the neurohypophysis as well as the paraventricular and supraoptic nuclei are atrophic [18, 19]. Because of the anterior pituitary dysfunction in our patients we speculate that the destructive process may not only involve vasopressin-synthesizing cells but also other cells of the hypothalamus. The etiology of idiopathic diabetes insipidus remains unexplained. This group may be heterogenous. Despite extensive investigation some known causes may be missed. Histiocytosis-X, for example, cannot be excluded with certainty. Antisera which react with vasopressin-synthesizing cells were found in some adult patients with non-tumoral diabetes insipidus [20]. In preliminary data in our group of patients, some children with idiopathic diabetes insipidus also had antibodies which reacted with vasopressin cells [*Scherbaum* et al., this volume].

Familial Diabetes insipidus

Familial diabetes insipidus is rare. Early manifestations are usually present at the end of the first year of life but may begin later and may be only partial. In various families, transmission has been found to be autosomal dominant [21–24] or X-linked recessive [25]. Decreased numbers of cells in the supraoptic and paraventricular nuclei have been reported in familial diabetes insipidus [26]. In the usual case, the diabetes insipidus is clinically similar to idiopathic diabetes insipidus, but in one more severe familial syndrome, diabetes insipidus may be associated with diabetes mellitus, optic atrophy and deafness (DIDMOAD) [27].

Diabetes insipidus in Infants

Central diabetes insipidus is infrequently recognized in the neonatal period. Most cases reported in newborn are usually in association with severe trauma or infectious disease of the central nervous system [28–30]. In cases of familial diabetes insipidus, polyuria does not seem to occur in the early days of life. This observation is parallel to what is observed in the Brattleboro rats in which the polydipsia does not occur before 3 or 4 weeks of age [31].

The treatment of diabetes insipidus in infants is particularly difficult. Use of subcutaneous injection of DDAVP may be necessary. Despite careful adjustment of water, episodes of hyper- or hyponatremia may occur. As the infants grow and can control their water intake, the treatment is usually easier.

Primary Polydipsia

Primary polydipsia in adults is discussed chapters 9 and 10. This syndrome occurs rarely in children. Probably because of the close association of the neurogenic centers of thirst with the centers that control synthesis and secretion of antidiuretic hormone, some cases of primary polydipsia have been reported with diseases that are usually associated with diabetes insipidus: dysgerminomas [*Pomarede* et al., this volume] and histiocytosis-X. The possibility of hypothalamic disease should always be considered in these patients. In some cases the brain dysfunction is not well categorized and the etiology may be purely psychological [32]. Polydipsia may also be mother-induced and be of such an intensity as to produce water intoxication [6, 32–38]. Reports of such cases are summarized in table V.

Chronic Hypernatremia

Chronic hypernatremia is a complex syndrome usually described in conjunction with recognized pathology of the hypothalamus [39–42], but occasionally without demonstrable lesion of the central nervous system [43]. The prominent feature is a chronic plasma hyperosmolality with adipsia or oligodipsia, sometimes with a marked aversion for fluids. In contrast to the hypernatremia associated with diabetes insipidus, here there is no excessive urinary output and usually the urine is concentrat-

Table V. Primary polydipsia

Authors	Reported cases	Maximum urine osmolality mosm/kg H_2O
Frazier et al. [33]	5 cases with psychiatric disorders	614–942
Dugan and Holliday [34]	2 cases (4 and 5 months old) of mother-induced polydipsia	300
Stevko et al. [35]	2 cases (9 and 12 years old)	500–514
Nickman et al. [36]	2 cases of mother-induced polydipsia	
Royer et al. [8]	1 case (8 years) with psychiatric symptoms; 3 cases (5, 7 and 8 years) of mother-induced polydipsia	
Linshaw et al. [37]	3 cases of induced polydipsia (1 month, 6 months, and 2.5 years); signs of water intoxication	900–962
Kohn et al. [38]	2 cases (3 and 5 years; water intoxication but associated nephropathy)	408–719

Table VI. Plasma and urinary osmolality and plasma AVP, renin and aldosterone in 4 patients with chronic hypernatremia. Normal values per hour for renin and aldosterone are 17.4 pg/100 ml and 6.6 ng/ml/h, respectively

Patient		Osmolality, mosm/kg		AVP pg/ml	Renin ng/ml/h	Aldosterone pg/100 ml
No.	age years	plasma	urine			
I	4 1/12	331	805	3.6	40	57
		349	1,133		128	160
II	3 10/12	370	1,200	9.7	176	99
III	2 5/12	339	1,075	<1	44	12.2
		313	900	23,7		
IV	1 2/12	335	510	<1	40	140

ed. Behavioral and neuromuscular disorders are frequently present and may regress when hypernatremia is corrected. The hypernatremia appears to result from chronic deficiency of body water; renin is elevated and (when measured) total body water is decreased. Finally the hypernatremia is corrected by forced fluid intake. Table VI describes a series

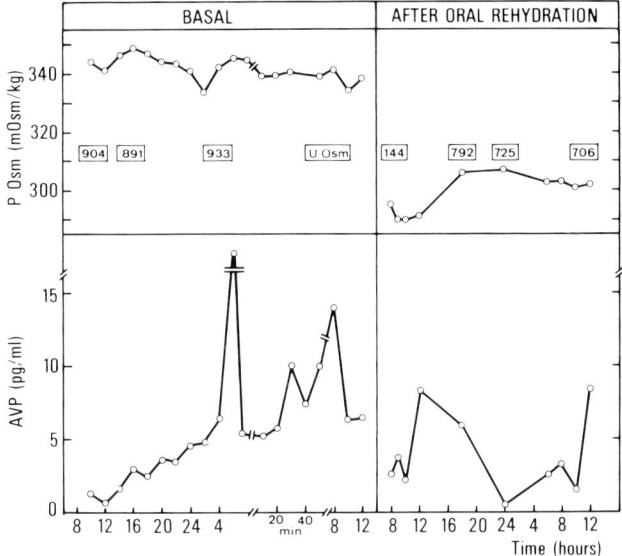

Fig. 8. Plasma AVP and osmolality during frequent sampling in a patient with chronic hypernatremia.

of 4 patients with chronic hypernatremia observed in our department. Patients 1, 2 and 4 had abnormal mid-brain structure with hypoplastic corpus callosum and microventricules. Patient 3 had no demonstrable brain lesion. When measured at random in such patients, renin was elevated and total body water was decreased.

Frequent blood sampling was performed in patient 1 during hypernatremia (fig. 8); plasma osmolality was elevated and rather constant while vasopressin was variable with a range from undetectable values to 26 pg/ml. These large variations in levels of vasopressin may explain why the urine was usually concentrated. After 10 days of careful rehydration with fluids orally, plasma osmolality decreased to values close to normal but without significant variation of plasma AVP. Urinary osmolality decreased and was in the range of total or partial diabetes insipidus when compared to plasma osmolality. Plasma AVP and plasma osmolality under various conditions are plotted in figure 9. When compared to normal children and patients with severe or partial diabetes insipidus,

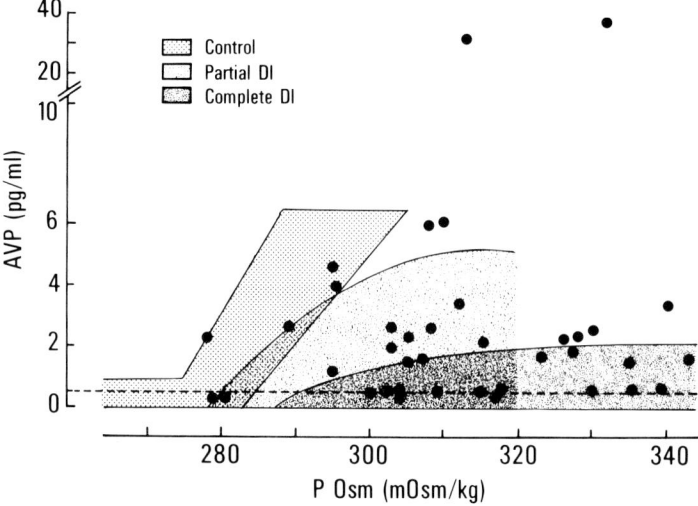

Fig. 9. Plasma AVP in relation to POsm in 4 patients with chronic hypernatremia. Correlations observed in normal and in children with diabetes insipidus are represented.

the observed values may correspond to either normal children or to patients with diabetes insipidus.

Several hypotheses have been proposed to explain this syndrome. Initially, it was proposed that the syndrome was due to a resetting of osmoreceptor around a higher level of plasma osmolality. This theory is not supported by recent observation of plasma AVP in adult patients and in our own patients. A reset receptor should have normal release of AVP when the new osmotic threshold is exceeded. On the basis of indirect studies of AVP release *De Rubertis* et al. [40] postulated that baroreceptor release was intact in such patients and that this may be the only effective stimulus for AVP secretion, i.e. the osmoreceptor was completely ineffective. This hypothesis was substantiated in their patients as well as in others by stimulation of AVP release during a hypotensive [42] or emetic [44] stimulus but no response to infusion of hypertonic saline. Our experience is consistent with this hypothesis. When chronic rehydration and volume replacement is accomplished these patients usually dilute their urines but the plasma osmolality may remain in the hypertonic range. The erratic variations of plasma AVP observed in our patient must therefore be explained by an absent or abnormal osmoreceptor input to the AVP releasing system. These observations have therapeutic consequences. Although these patients are able to synthesize and secrete AVP the clinical picture is more like diabetes insipidus. They respond well to DDAVP administration and fixed water intake.

References

1 Aaronson, A.S.; Svenningsen, H.W.: DDAVP test for estimation of renal concentrating capacity in infants and children. Archs Dis. Childh. *49:* 654–659 (1974).
2 Harrington, A.R.; Valtin, H.: Impaired urinary concentration after vasopressin and its gradual correction in hypothalamic diabetes insipidus. J. clin. Invest. *47:* 502–512 (1968).
3 Frazier, S.D.; Kutnik, L.A.; Achmidt, R.T.; Smith, F.G., Jr.: A water deprivation test for the diagnosis of diabetes insipidus in children. Am. J. Dis. Child. *114:* 157–160 (1967).
4 Edelman, C.J., Jr.; Barnett, H.L.; Stark, H.; Boikis, H.; Soriano, J.R.: A standardized test of renal concentrating capacity in children. Am. J. Dis. Child. *114:* 639–644 (1967).
5 Richman, R.A.; Post, F.M.; Notman, D.D.; Hochberg, L.; Moses, A.M.: Symplifying the diagnosis of diabetes insipidus in children. Am. J. Dis. Child. *135:* 839–841 (1981).
6 Czernichow, P.; Pomarede, R.; Basmaciogullari, A.; Rappaport, R.: Diabetes insipidus in children. I. Arginine-vasopressin determination in plasma during short dehydration test. Acta paediat. scand. suppl. 277, pp. 64–68 (1979).
7 De Wardener, E.; Herscheimer, A.: The effect of a high water intake on the kidney's ability to concentrate the urine in man. J. Physiol., Lond. *139:* 43–52 (1957).
8 Royer, P.; Balsan, S.; Loirat, C.; LeDeunff, M.J.: Etude sur les diabètes insipides de l'enfant. II. Consommation spontanée et restriction progressive des boissons. Archs fr. Pédiat. *28:* 365–380 (1978).
9 Robertson, G.L.; Mahr, E.A.; Athar, S.; Sinha, T.: Development and clinical application of a new method for the radioimmunoassay of arginine vasopressin in human plasma. J. clin. Invest. *52:* 2340–2352 (1973).
10 Weitzman, R.F.; Fisher, D.A.; Di Stefano, J.J., III; Bennet, C.M.: Episodic secretion of arginine vasopressin. Am. J. Physiol. *233:* E32–E36 (1977).
11 Pomarede, R.; Czernichow, P.; Finidori, J.; Pfister, A.; Roger, M.; Kalifa, C.; Zucker, J.M.; Kahn, A.P.; Rappaport, R.: Endocrine aspects and tumoral markers in intracranial germinoma. An attempt to delineate the diagnosis procedure in 14 patients. J. Pediat *101:* 374–378 (1982).
12 Lambertz, J.: Craniopharyngiome de l'enfant. Archs fr. Pédiat. *24:* 561–582 (1967).

13 Jenkins, J.S.; Gilbert, C.J.; Ang, V.: Hypothalamic pituitary function in patients with craniopharyngiomas. J. clin. Endocr. *43:* 394–399 (1976).
14 Nezelof, C.; Frilevx-Herbet, F.; Cronier-Sachot, J.: Disseminated histiocytosis X. Analysis of prognostic factors based on a retrospective study of 50 cases. Cancer *44:* 1824–1838 (1979).
15 Sims, D.G.: Histiocytosis X, follow-up of 43 cases. Archs Dis. Childh. *52:* 433–440 (1977).
16 Czernichow, P.; Pomarede, R.; Basmaciogullari, A.; Brauner, R.; Rappaport, R.: Diabetes insipidus in children: III. Anterior pituitary dysfunctions in idiopathic types. J. Pediat. *105* (1984).
17 Aubin, M.L.; Bentson, J.; Vignaud, J.: CT of the pituitary stalk. J. Neuroradiol. *5:* 153–160 (1978).
18 Blotner, H.: Primary or idiopathic diabetes insipidus: a system disease. Metabolism *7:* 191–200 (1958).
19 Green, J.R.; Buchan, T.L.; Alvord, E.; Swanson, A.G.: Hereditary and idiopathic types of diabetes insipidus. Brain *90:* 707–713 (1967).
20 Scherbaum, W.A.; Bottazzo, G.F.: Autoantibodies to vasopressin cells in idiopathic diabetes insipidus: evidence for an autoimmune variant. Lancet *i:* 897–901 (1983).
21 Levinger, E.L.; Escamilla, R.F.: Hereditary diabetes insipidus. Report of 20 cases in seven generations. J. clin. Endocr. Metab. *15:* 547–553 (1955).
22 Moehling, R.C.; Schultz, R.L.: Familial diabetes insipidus. Report of one of fourteen cases in four generations. J. Am. med. Ass. *158:* 725–727 (1955).
23 Pender, C.B.; Fraser, F.C.: Dominant inheritance of diabetes insipidus: a family study. Pediatrics, Springfield *11:* 246–254 (1953).
24 Brugnier, A.; Poisson, D.; Lestradet, H.; Labrune, B.: Diabète insipide familial d'origine centrale. Nouv. Presse méd. *10:* 897–899 (1981).
25 Forssman, H.: Two different mutations of the X-chromosome causing diabetes insipidus. Am. J. hum. Genet. *7:* 21–27 (1955).
26 Braverman, L.E.; Mancini, J.P.; McGoldrick, D.M.: Hereditary idiopathic diabetes insipidus. A case report with autopsy findings. Ann. intern. Med. *63:* 503–508 (1965).
27 Page, M.; Asmal, A.C.; Edwards, C.R.W.: Recessive inheritance of diabetes. The syndrome of diabetes insipidus, diabetes mellitus, optic atrophy and deafness. Q. Jl Med. *45:* 505–520 (1976).
28 Fenton, J.; Kleinman, A.: Transient diabetes insipidus in a newborn infant. J. Pediat. *85:* 79–81 (1974).
29 Adams, J.M.; Kenny, J.D.; Rudolph, A.J.: Central diabetes insipidus following intraventricular hemorrhage. J. Pediat. *88:* 292–294 (1976).
30 Pai, K.G.; Rubin, H.M.; Wedemeyer, P.P.; Linarelli, L.G.: Hypothalamic-pituitary dysfunction following group B beta-hemolytic streptococcal meningitis in a neonate. J. Pediat. *88:* 289–291 (1976).
31 Dlouhai, H.; Krecek, J.; Zicha, J.: In Valtin, Sokol, Postnatal development and diabetes insipidus in Brattleboro rats. Ann. N.Y. Acad. Sci. *394:* 10–20 (1982).
32 Mecklenburg, R.S.; Loriaux, D.L.; Thompson, R.H.; Andersen, A.E.; Lipsett, M.B.: Hypothalamic dysfunction in patients with anorexia nervosa. Medicine, Baltimore *53:* 681–690 (1976).
33 Frazier, S.D.; Kutnik, L.A.; Achmidt, R.T.; Smith, F.G., Jr.: A water deprivation test

for the diagnosis of diabetes insipidus in children. Am. J. Dis. Child. *114:* 157–160 (1967).
34 Dugan, S.; Holliday, M.A.: Water intoxication after voluntary fluid ingestion. Pediatrics, Springfield *39:* 418–420 (1967).
35 Stevko, R.M.; Basley, M.; Segar, W.E.: Primary polydipsia. Compulsive water drinking. J. Pediat. *73:* 845–851 (1968).
36 Nickman, S.L.; Buckler, J.H.M.; Weiner, L.B.: Further experiences with water intoxication. Pediatrics, Springfield *41:* 149–151 (1968).
37 Linshaw, M.A.; Hipp, T.; Gruskin, A.: Infantile psychogenic water drinking. Pediatrics, Springfield *85:* 520–522 (1974).
38 Kohn, B.; Norman, M.E.; Feldman, H.; Thier, S.O.; Singer, I.: Hysterical polydipsia (compulsive water drinking) in children. Am. J. Dis. Child. *130:* 210–212 (1976).
39 Christie, S.B.M.; Ross, E.J.: Ectopic pinealoma with adipsia and hypernatremia. Br. med. J. *ii:* 669–670 (1968).
40 De Rubertis, F.R.; Michelis, M.F.; Beck, N.; Field, J.B.; Davis, B.D.: 'Essential' hypernatremia due to ineffective osmotic and intact volume regulation of vasopressin secretion. J. clin. Invest. *50:* 97–111 (1971).
41 Halter, J.B.; Goldberg, A.P.; Robertson, G.L.; Porte, D.: Selective osmoreceptor dysfunction in the syndrome of chronic hypernatremia. J. clin. Endocr. Metab. *44:* 609–616 (1977).
42 Perelman, R.; Czernichow, P.; Danis, F.; Nathanson, M.; Gaudelus, J.; Cheritat, C.: Hypernatremie chronique neureogène avec arhinencéphalie. Annls Pédiat. *26:* 176–178 (1979).
43 Blank, M.S.; Farnsworth, P.B.: Idiopathic symptomatic hypernatremia in a 9-year-old boy: a clinical and physiologic evaluation. J. Pediat. *85:* 215–219 (1976).
44 Schaff, E.; Robertson, G.L.; Rosenfield, R.L.: Chronic hypernatremia from a congenital defect in osmoregulation of thirst and vasopressin. J. Pediat. *102:* 703–708 (1983).

P. Czernichow, Hôpital des Enfants Malades, 149, rue de Sèvres,
F-75743 Paris, Cedex 15 (France)

Computed Tomography in Diabetes insipidus

C. Manelfe[a], M.O. Balliana[a], J.P. Louvet[b], A. Sevely[a], J. Prere[a], P. Rochiccioli[c], A. Bonafe[a,1]

[a] Department of Neuroradiology, Centre Hospitalier Universitaire de Purpan, Toulouse; [b] Department of Endocrinology, Centre Hospitalier Universitaire de Purpan, Toulouse; [c] Department of Pediatric Endocrinology, Centre Hospitalier Universitaire de Rangueil, Toulouse, France

Introduction

In a previous report computed tomography (CT) appeared to represent a valuable method of investigating patients with diabetes insipidus [1].

We studied 51 patients presenting with diabetes insipidus diagnosed 7 months to 23 years earlier. All had CT between May 1977 and December 1983. In all patients, diabetes insipidus was documented by lack of ability to concentrate urine in response to osmotic stimulation, and by increased urine osmolality in response to exogenous hormone. No patient had a history of cranial trauma and/or cranial surgery.

Material and Methods

The clinical findings associated with diabetes insipidus are detailed on tables I and II. 28 were *children* (18 boys; 10 girls), aged 1 month to 15 years. In 7 out of 28 cases diabetes insipidus was isolated, and in 21 cases diabetes insipidus was associated with other conditions: endocrine disease (9 cases), histiocytosis X (5 cases) psychomotor retardation (3

---

[1] The authors wish to thank Drs. *Francis Bayard, Jean-Paul Carrière, Antoine Dalous, Robert Fedou, Bernard Guiraud, Claude Regnier* (Toulouse), *Emile Raynaud* (Clermont-Ferrand), and *Christian Seigneuric* (Montauban), for supplying clinical data; *Nicole Falga, Gérard Colas, Jean-Claude Trouis,* and *Richard Muraro* for technical assistance; *Janine Galien* and *Josette Dayde* for secretarial assistance; and *Anita Bories* for reviewing the manuscript.

cases), prematurity (2 cases), and headaches (2 cases). 23 were *adults* (14 males; 9 females) aged from 20 to 75 years. Diabetes insipidus was isolated in 5 cases, and associated with other conditions in 18: endocrine disease (14 cases), histiocytosis X (2 cases), hypertension (2 cases).

All patients were investigated with CT and 50 out of 51 with skull X-ray films and polytomograms of the sella. Pneumoencephalography was used in 13 patients, mainly at the beginning of our study, and angiography in 11. Patients were examined with a CT head scanner (Delta 25, 256 × 256 matrix), and from June 1982 with a total body scanner (CE 10,000 CGR, 512 × 512 matrix). Axial transverse sections were performed at $-10\,°C$ to the orbitomeatal line. The sellar and suprasellar regions were studied using either a 5-mm collimation overlapping with the head scanner, or a 1.5-mm collimation with the body scanner; above this level 8-mm sections were used. Direct or reconstructed coronal sections were performed in all cases and reconstructed sagittal sections in 21 cases. All patients were investigated before and after rapid intravenous contrast injection of 30% meglumine iothalamate (2 ml/kg body weight in adult; 3 ml/kg body weight in infant). Particular attention was paid to the hypothalamic-pituitary region and expecially to the size of the pituitary stalk and infundibulum. The diameter of the pituitary stalk was measured after intravenous contrast medium injection on axial transverse sections at the level of the dorsum sellae and on the section immediately above at the level of the chiasmatic cistern. Measurements were made with the cursor method on the scanner console. Direct or reconstructed coronal sections were also useful to appreciate the size and shape of the pituitary stalk and the infundibulum region. CT cisternography with water-soluble contrast medium (Metrizamide or Iopamidol) was used in 9 patients. 34 (19 children and 15 adults) out of 51 patients had follow-up CT scans over periods ranging from 6 months to 6 years. Follow-up CT scans were performed directly after intravenous injection of contrast medium.

10 patients (3 children, 7 adults) were operated on and histopathological specimens of the hypothalamic-pituitary lesion obtained.

Normal CT Anatomy of the Hypothalamic-Pituitary Region

The hypothalamic-pituitary region is well appreciated on thin axial and coronal CT sections and sagittal reformations. The pituitary gland presents an homogeneous density after contrast injection similar to that of the adjacent dura and cavernous sinuses. Accurate outer limits are often difficult to differentiate from adjacent bony and venous structures. Even with a high resolution CT scanner the normal pituitary gland can present, on direct coronal and reformatted sagittal sections, ill-defined hypodensities less than 3 mm in diameter; they may correspond to small pars intermedia cysts or – even in patients with no known pituitary disease – to small microadenomas as demonstrated by autopsy. Intrasellar hypodensities can also be related to artifacts when the petrous bones and the orbital walls are in the same plane.

On coronal sections the upper border of the pituitary gland is limited by a flat or slightly concave superiorly sellar diaphragma. The pituitary stalk is medial and at this point of attachment the diaphragm may be convex superiorly. The size of the pituitary gland ranges from 2 to 7 mm in adult and is usually smaller in male than in female. The size of the pituitary gland increases with pregnancy and can be considered as enlarged when it exceeds 10 mm in height. The suprasellar cistern appears as a five- or six-pointed 'star' or 'crown' and is bounded by the frontal lobes anteriorly, the temporal lobes laterally, and the cerebral peduncles posteriorly. The optic chiasm lies within the anterior portion of the cis-

Table I. Clinical and radiological findings in 28 children with DI

Patient No.	Sex and age (y = years; m = months)	Clinical data	Etiology	X-rays skull	PEG	angio	CT PS diameter mm	contrast enhancement PS	miscellaneous
1	F 12 y	hypercortisolism	large cystic craniopharyngioma	A	O	A	not seen		
2	F 14 y	delayed puberty	calcified craniopharyngioma	A	O	A	not seen		
3	M 7 y	precocious puberty	germinoma	N	N	A	not seen		ventricular enlargement
4	M 13 y	headaches		A	A	N	7–8	+++	
5	F 15 y			N	N	O	3–6	±	
6	F 1 m	prematurity	Listeria meningitis	N	O	O	4–5	++	
7	F 4 y			N	O	O	4–13	++	
8	F 11 y	short stature		N	O	O	4	+	
9	M 11 y	dwarfism	HGH deficiency	N	N	O	2+ inf. enl.	+	
10	M 5 y	short stature	Montgomery syndrome	N	A	O	10	++	contrast-enhancing masses in both temporal regions
11	M 4 y	hypothyroidism		N	O	O	3–5	++	empty sella
12	M 3 y	skin, bone, lung and liver localizations	histiocytosis X	A	O	O	6–8	±	
13	M 4 y	right exophthalmos	histiocytosis X	A	O	O	4–5	+	orbital infiltration and frontal lytic lesion

Computed Tomography in Diabetes insipidus

#	Sex/Age	Clinical features	Diagnosis						CT findings
14	M 2 y	skin lesions	histiocytosis X	A	O	O	+	5	
15	M 10 y			N	O	O	+	2	empty sella
16	F 7 y	bone localizations	histiocytosis X	A	O	O	+	1–2	
17	M 10 y		histiocytosis X	N	O	O	+	1–2	
18	M 3 y	dwarfism	septooptic, dysplasia	N	N	N	+	not seen	septum pellucidum present
19	F 12 y	dwarfism, optic atrophy	septo-optic dysplasia	N	O	O	++	3	agenesis of the septum pellucidum
20	F 1 y	psychomotor retardation		N	A	O	+	3	ventricular dilatation cortical atrophy
21	M 7 y	dwarfism	Wolfram syndrome	N	O	O	±	1–2.5	
22	M 3 y	prematurity		N	N	N	+	not seen	
23	M 1 m			O	O	O	+	1	moderate ventricular dilatation
24	M 4 y	psychomotor retardation		N	O	O	+	1	
25	M 7 y			N	O	O	+	1	
26	F 5 y			N	O	O	+	1	
27	M 5 y	short stature		N	O	O	+	1	
28	M 8 y	short stature hypothyroidism		N	O	O	+	1	

PEG = Pneumoencephalography; angio = angiography; PS = pituitary stalk; N = normal; A = abnormal; O = not performed.

Table II. Clinical and radiological findings in 23 adults with DI

Patient No.	Sex and age, years	Clinical data	Etiology	X-rays skull	PEG	angio	CT PS diameter mm	contrast enhancement PS	miscellaneous
29	F 57	panhypopituitarism	craniopharyngioma	A	O	O	3–4	++	
30	M 25	panhypopituitarism	craniopharyngioma	A	O	O	not seen		
31	F 25	galactorrhea	prolactinoma	A	O	A	not seen		
32	M 25	hypogonadism	germinoma	N	O	A	not seen		
33	F 41	diabetes mellitus	ganglioglioma	N	A	A	10–24	+++	
34	F 41	amenorrhea galactorrhea	breast cancer metastasis	N	O	O	6–8	++	
35	M 45		metastasis lung carcinoma	A	O	O	6–7	++	
36	M 20	gynecomastia		N	O	O	6–7	+++	
37	M 24	hypogonadism		N	O	O	5–7	+++	
38	M 50	sexual impotence		N	A	N	5.5–7	+++	
39	F 27	menstrual disorders		N	O	O	4–5	++	
40	F 56			N	O	O	4–5	++	
41	F 30	amenorrhea galactorrhea		A	A	N	5–6	+++	empty sella
42	M 67	gynecomastia		A	O	O	6–7	+++	empty sella
43	M 33	lung disease	histiocytosis X	A	O	O	7–8	+++	empty sella
44	M 42	skin lesion	histiocytosis X	A	O	O	2	+	empty sella
45	M 27	HTA		N	O	O	1.5–2	+	
46	M 59	HTA		N	N	O	3–4	++	cortical atrophy
47	M 75	galactorrhea		N	O	O	2	+	cortical atrophy
48	F 26			N	O	O	2	+	
49	M 20	menstrual disorders		N	O	O	1.5–2	+	
50	F 20	hyperthyroidism		N	O	O	2.5	+	
51	F 22			N	O	O	2.5	+	

PEG = Pneumoencephalography; angio = angiography; PS = pituitary stalk; N = normal; A = abnormal; O = not performed.

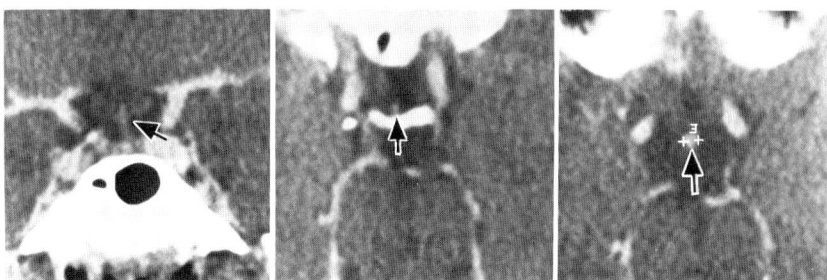

Fig. 1. Normal pituitary stalk and infundibulum after contrast enhancement. *a* Direct coronal section: the pituitary stalk (arrow) is medial, below the optic chiasma and above the sellar diaphragm. *b, c* Axial transverse sections at the level of the dorsum sellae *b* and chiasmatic cistern *c:* the pituitary stalk is well defined and its diameter approximately 3 mm (arrow).

tern forming a 'V'-shaped tissue mass. The circle of Willis with its main components, i.e. basilar artery and supraclinoid carotid arteries, is visualized after administration of contrast agent. As previously reported [1] the pituitary stalk is routinely demonstrated on 5-mm thickness axial sections after intravenous contrast injection: it appears as a small round, enhancing area in the center of the chiasmatic cistern, and just anterior to the dorsum sellae on the section below. The diameter of the pituitary stalk was considered as normal when inferior or equal to 3 mm in children and 4 mm in adults (fig. 1).

On upper sections the third ventricle is demonstrated on the midline as a slit between the hypothalamic region laterally, the plane of the frontal horns anteriorly and the quadrigeminal plate cistern posteriorly. On coronal sections the optic chiasma and pituitary stalk form a well-defined 'T'-shaped density in the chiasmatic cistern. The slit of the third ventricle is located immediately above the attachment of the pituitary stalk on the hypothalamus.

Results

The results of CT findings are separated for children and adults, and are summarized in tables I and II.

Children (table I)

3 patients had *tumors:* 1 large cystic craniopharyngioma (case 1), 1 calcified craniopharyngioma with suprasellar extension (case 2), and 1 germinoma (case 3) involving the floor of the third ventricle and the hypothalamus (fig. 2). These 3 patients were operated on and the lesion confirmed by pathology.

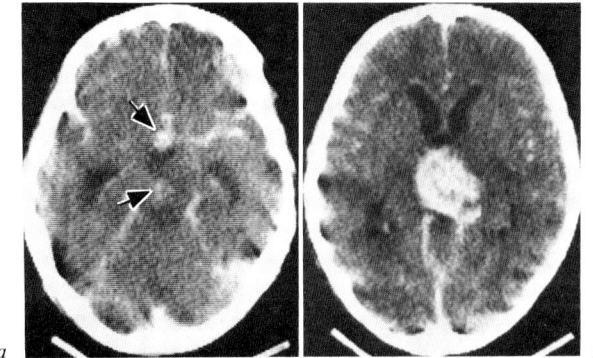

Fig. 2. Germinoma in a 7-year-old boy (case 3). Contrast-enhancing lesion involving: *a* the hypothalamic region and posterior inferior portion of the 3rd ventricle (arrows), and *b* the pineal region.

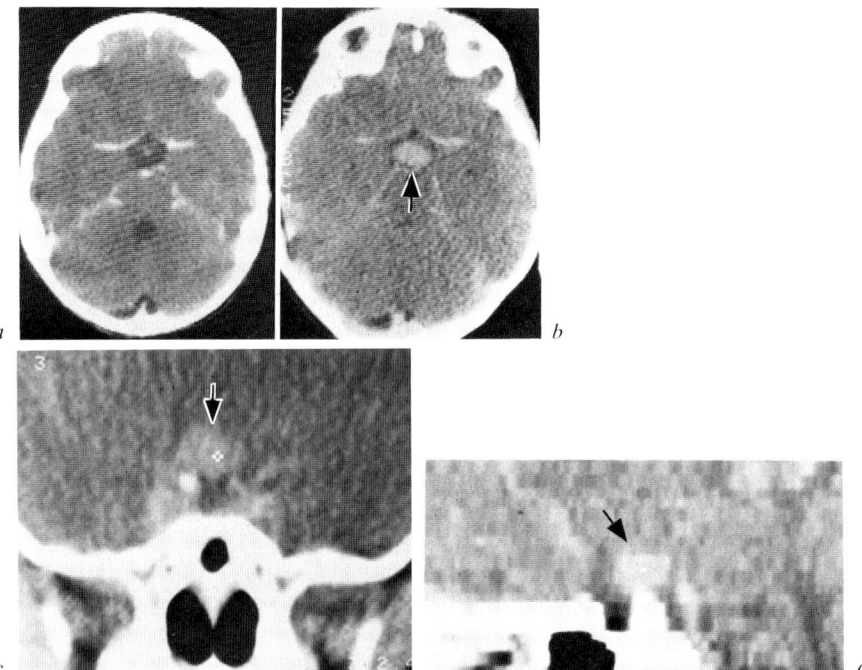

Fig. 3. Pituitary stalk and infundibulum enlargement in a 4-year-old girl (case 7). *a* First CT demonstrating a slight enlargement of the pituitary stalk (4 mm). 2 years later the size of the pituitary stalk (arrow) was considerably increased (approximately 13 mm) and well delineated on axial *(b)*, coronal *(c)*, and sagittal *(d)* sections.

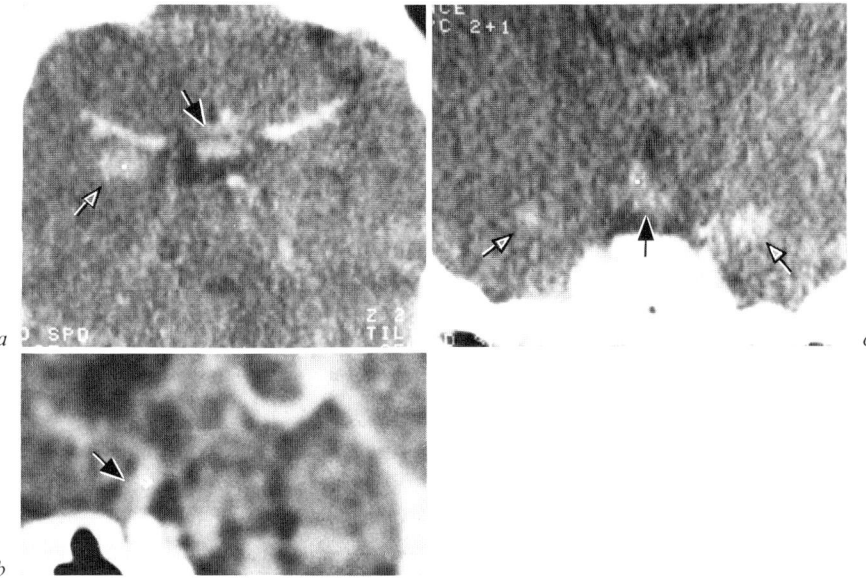

Fig. 4. a–c Pituitary stalk and infundibulum enlargement in 5-year-old boy with Montgomery syndrome (case 10). *a* Axial, *b* coronal and *c* sagittal CT sections. Note: associated with the lesion of the hypothalamic pituitary region (arrow), similar enhancing masses in both temporal regions (open arrows).

In 1 patient (case 15) a *primary empty sella* was demonstrated on CT.

13 patients had *no abnormality* in the hypothalamic-pituitary region (cases 16–28). Among them 2 patients (cases 18, 19) presented with septo-optic dysplasia but the absence of the septum pellucidum was only demonstrated on CT in case 18. In 2 patients ventricular dilatation was observed (cases 20, 23).

11 patients had *pituitary stalk and/or infundibulum enlargement* (cases 4–14). The size of the pituitary stalk ranged from 4 to 13 mm in diameter (fig. 3). Pituitary stalk and/or infundibulum enlargement was the only abnormality on CT in 6 patients (cases 4–9). This abnormality was associated with a Montgomery syndrome (case 10) (fig. 4), a primary empty sella (case 11), and histiocytosis X in 3 (cases 12–14) (fig. 5).

Among the 28 patients, 9 had only one CT examination, and 19 had two or more CT scans over periods ranging from 6 months to 6 years. 10 patients were followed by serial CT for 4–6 years, and 9 patients for 1–4

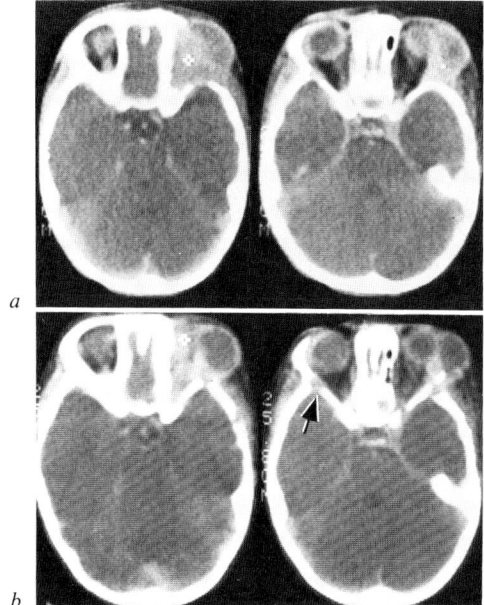

Fig. 5. Histiocytosis X in a 4-year-old boy (case 13). *a* Osteolytic lesion involving the right fronto-orbital region with contrast enhancing mass in the orbit. The size of the pituitary stalk was considered as normal (2–3 mm). *b* 1 year later there was some bony reconstruction on the right orbit but a more obvious lytic lesion was seen on the left (arrow). Moreover, the pituitary stalk appeared enlarged (4–5 mm).

years. In 3 cases (cases 4, 7 and 13) the enlargement of the pituitary stalk increased with time on the follow-up studies (fig. 3, 5); in 3 other cases (cases 8, 12 and 14) it decreased.

Adults (table II)

7 patients had *tumors* (cases 29–35) and 6 were operated on (cases 29–34) (fig. 6, 7). Case 35 was a 45-year-old male with primary lung carcinoma: CT demonstrated a contrast-enhancing mass at the level of the pituitary stalk and infundibulum consistent with a metastasis. The patient died before surgery.

Case 33, a 41-year-old woman, has been published [1, 2] and sarcoidosis was suspected on the basis of clinical and laboratory data. CT, pneumoencephalography, and angiography demonstrated a large enhancing mass obliterating the suprasellar cistern, and the hypothalamic

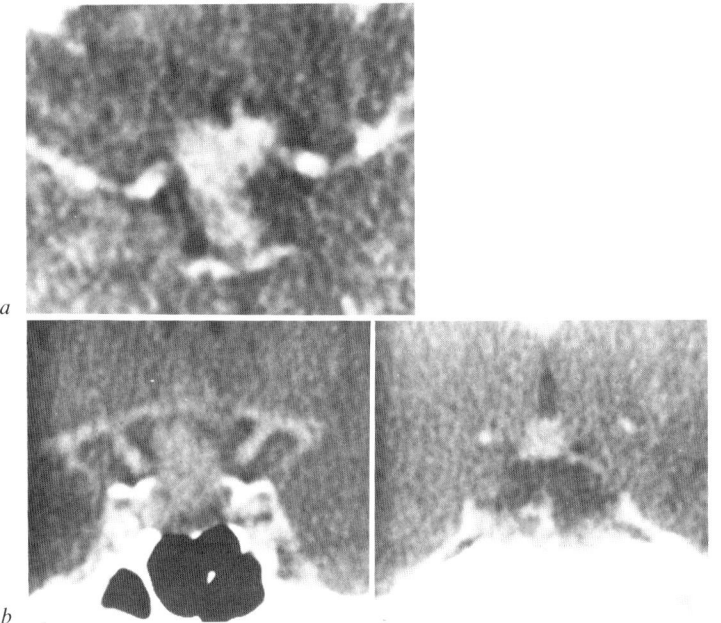

Fig. 6. Sellar germinoma in a 25-year-old male (case 32). Axial *a* and coronal *b, c* sections after contrast injection showing a large enhancing mass infiltrating the sella, the optic chiasma and the hypothalamic region.

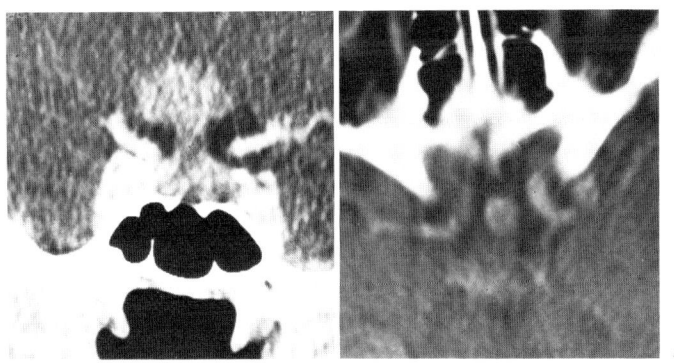

Fig. 7. Metastasis of breast carcinoma in a 41-year-old female (case 34). *a* Axial and *b* coronal CT sections demonstrating a contrast-enhancing lesion involving the hypophysis, the pituitary stalk, and the hypothalamus.

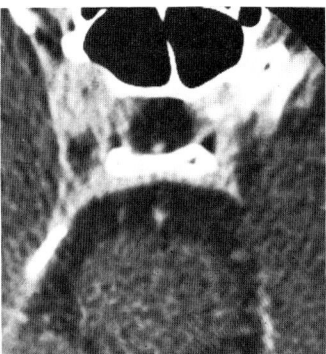

Fig. 8. Histiocytosis X in a 42-year-old male (case 44). Axial CT section showing a typical empty sella with a normal pituitary stalk.

region. Following steroid and vasopressin therapy, there was a marked decrease of the enhancing hypothalamic mass, but a few months later there was recurrence of the hypothalamic mass on CT and the patient was operated on. Pathologic examination demonstrated a ganglioglioma.

In 5 patients (cases 36–40) CT showed an *isolated pituitary stalk and/or infundibulum enlargement* ranging from 5 to 7 mm. In 4 cases a *primary empty sella* was seen on CT (cases 41–44). In case 44 the empty sella was the only abnormality on CT (fig. 8); in this patient, histiocytosis X was diagnosed by lung biopsy 4 years after the onset of the diabetes insipidus. In cases 41–43, the primary empty sella was associated with a pituitary stalk enlargement. In 7 cases (cases 45–51) CT was *normal* at the level of the hypothalamic and sellar regions.

Among the 23 adults, 8 patients had only one CT examination and 15 had two or more CT scans over periods ranging from 7 months to 6 years. 5 patients were followed by serial CT for more than 4 years, and 7 for 1–4 years. The lesions increased in size in 2 (cases 34, 38) and decreased in 4 (cases 36, 37, 39, 43).

Discussion

CT is now considered the most efficient diagnostic tool in neuroradiology and has replaced pneumoencephalography in the diagnosis of sellar and suprasellar areas [3]. Excellent spatial and contrast resolution provided by the new CT scanners allows diagnosis of intrasellar lesions often no larger than 3–4 mm [4, 5]. On thin sections small anatomical structures such as the pituitary stalk are routinely demonstrated after intravenous contrast medium injection [1, 6]. Our recent experience with a high-resolution CT scanner has confirmed the upper normal values of the pituitary stalk diameter in adults (4 mm) and in children (3 mm) [1].

Tumors of the hypothalamic-pituitary region were present in 10 of our patients. Craniopharyngiomas represent one of the most frequent causes of secondary diabetes insipidus in young patients and the CT appearance is well known [3, 7]. In 4 of our patients (cases 1, 2, 29, 30) an accurate preoperative diagnosis of craniopharyngioma was made using CT.

Suprasellar germinomas (cases 3, 32) are also common tumors of the hypothalamic region associated with diabetes insipidus, but the exact determination of the tumor type is usually possible only by pathological examination. Differential diagnosis includes meningiomas, optic chiasmal, and hypothalamic gliomas.

Metastatic involvement of the neurohypophysial system as a cause of diabetes insipidus has been reported in 5–20% [8] of patients with diabetes insipidus. Breast cancer and bronchogenic carcinoma were the most common site of origin of a primary tumor which metastasized to the pituitary stalk [8–11]. In 2 of our patients (cases 34, 35) CT showed a contrast-enhancing lesion involving the pituitary stalk and gland, and the hypothalamus (fig. 7). The development of diabetes insipidus secondary to metastases to the neurohypophysial system tends to occur late in the course of the primary tumor [8, 11].

The association of a primary empty sella with abnormal posterior pituitary function is rare, and few cases have been investigated with CT [2, 12, 13]. This association is observed more often in adults than in children [13]. In our series, CT demonstrated such an abnormality in 6 patients, (2 children, 4 adults). The empty sella was the only CT finding in 2 patients: a 10-year-old boy (case 15), and a 42-year-old man (case 44); in the latter histiocytosis X was proven by lung biopsy 4 years after the patient developed diabetes insipidus (fig. 8). In 4 patients the empty sella was associated with a pituitary stalk enlargement: a 4-year-old boy (case 11), and 3 adults (case 41–43). The significance of an isolated primary empty sella in diabetes insipidus is unclear and might represent a fortuitous anatomical variation since intrasellar herniation is present in 10% of autopsy cases without recognized pituitary disease [14].

The meaning of the enlargement of the pituitary stalk demonstrated by CT in 19 patients with diabetes insipidus is puzzling. In 11 patients (6 children, 5 adults) this enlargement represented the only CT abnormality (cases 4–9, and cases 36–40). Does it represent an infiltrative lesion such as a manifestation of histiocytosis X? Does it signify the presence of a small tumor? Or is it the manifestation of hypertrophied neurons in the hypothalamo-neurohypophysial system similar to lesions observed in Brattleboro rats [15]?

The diagnosis of histocytosis is easy when lesions are associated (skin, bone, liver, lung...) and lead to biopsy (cases 12–14), but when the lesion is isolated in the hypothalamic-pituitary area it is impossible to differentiate from glioma or teratoma [7]. As emphasized by *Pressman* et al. [16], radiotherapy is considered the treatment of choice for hypothalamic glioma and for some osseous lesions of histiocytosis X. Unfortunately, radiotherapy is of little value in visceral lesions of histiocytosis X, particularly those of the central nervous system, including the pituitary region. Good to excellent results with chemotherapy have been reported in disseminated histiocytosis X [16]. When there is a solitary hypothalamic mass in the absence of other signs of histiocytosis X, radiotherapy should not be undertaken without biopsy. Surprisingly in children, 2 out of 3 of the cases with histiocytosis X (cases 12, 14) showed a decrease in size of the lesion on follow-up CT. This may explain why in many patients, even with proven reticulosis, we are unable to demonstrate abnormalities by CT.

CT studies in some instances may correct the initial diagnosis. An example is case 33: the hypothalamic infiltration was initially considered to be sarcoidosis, but was in fact a ganglioglioma.

The group of patients without abnormality in the hypothalamic-pituitary region (20 out of 51) represents about 40%. The significance of the enhancement we observed in the hypothalamic-pituitary region must await more extensive studies. Perhaps the use of nuclear magnetic resonance imaging will tell us if these lesions are specific to diabetes insipidus.

References

1 Manelfe, C.; Louvet, J.P.: Computed tomography in diabetes insipidus. J. Comput. assist. Tomogr. *3:* 309–316 (1979).

2   Manelfe, C.; Louvet, J.P.; Boulard, C.; Regnier, C.; Rochiccioli, P.; Bayard, F.: Hypothalamic-pituitary changes in diabetes insipidus demonstrated by computerized tomography. Lancet *ii:* 1379–1380 (1978).
3   Naidich, T.P.; Pinto, R.S.; Kushner, M.J.; Lin, L.P.; Kricheff, I.I.; Leeds, N.E.; Chase, N.E.: Evaluation of sellar and parasellar masses by computed tomography. Radiology *120:* 91–99 (1976).
4   Syversten, A.; Haughton, V.M.; Williams, A.L.; Cusick, J.F.: The computed tomographic appearance of the normal pituitary gland and pituitary microadenomas. Radiology *133:* 385–391 (1979).
5   Taylor, S.: High resolution computed tomography of the sella. Radiol. Clin. N. Am. *20:* 207–236 (1982).
6   Aubin, M.L.; Bentson, J.; Vignaud, J.: Tomodensitometrie de la tige pituitaire. J. Neuroradiol. *5:* 153–160 (1978).
7   Miller, J.H.; Pena, A.M.; Segall, H.D.: Radiological investigation of sellar region masses in children. Radiology *134:* 81–87 (1980).
8   Yap, H.Y.; Tashima, C.K.; Blumenschein, G.R.; Eckles, N.: Diabetes insipidus and breast cancer. Archs intern. Med. *139:* 1009–1011 (1979).
9   Teears, R.J.; Silverman, E.M.: Clinicopathologic review of 88 cases of carcinoma metastatic to the pituitary gland. Cancer *36:* 216–224 (1975).
10 Krol, T.C.; Wood, W.S.: Bronchogenic carcinoma and diabetes insipidus: case report and review. Cancer *49:* 596–599 (1982).
11 Ozanne, P.; Jedynak, C.P.; Charbonnel, B.; Derome, P.J.: Metastases hypophysaires et hypothalamiques. Etude anatomo-clinique de cinq observations. Annls Méd. int. *133:* 92–96 (1982).
12 Marano, G.D.; Horton, J.A.; Vazquez, A.M.: Computed tomography in diabetes insipidus: posterior empty sella. Br. J. Radiol. *54:* 263–265 (1981).
13 Petrus, M.; Bonafe, A.; Dutau, G.; Manelfe, C.; Rochiccioli, P.: Syndrome de selle turcique vide primitive. Etude de deux cas. Archs fr. Pédiat. *38:* 581–586 (1981).
14 Bergland, R.M.; Ray, B.S.; Torack, R.M.: Anatomical variations in pituitary glands and adjacent structures in 225 human autopsy cases. J. Neurosurg. *28:* 93–99 (1968).
15 Valtin, H.: Hereditary hypothalamic diabetes insipidus. Am. J. Path. *83:* 633–636 (1976).
16 Pressman, B.D.; Waldron, R.L.; Wood, E.H.: Histiocytosis X of hypothalamus. Br. J. Radiol. *48:* 176–178 (1975).

C. Manelfe, Department of Neuroradiology, Centre Hospitalier Universitaire de Purpan, F-31059 Toulouse (France)

Nephrogenic Diabetes insipidus in Children

Patrick Niaudet[a], Michèle Dechaux[b], Danielle Leroy[b], Michel Broyer[a]

[a] Department of Pediatric Nephrology (Prof. *M. Broyer*) and
[b] Department of Physiology (Prof. *C. Sachs*), Hôpital Necker Enfants-Malades, Paris, France

Congenital nephrogenic diabetes insipidus is characterized by an insensitivity of the collecting duct to both endogeneous and exogeneous antidiuretic hormone. Clinically, polyuria develops shortly after birth and leads to severe volume contraction with plasma hyperosmolality. The pathogenesis of the disease is still unclear. It has been suggested that a defective generation of cyclic adenosine monophosphate (AMP) in response to vasopressin may be responsible for the concentrating defect. Several reports indicate that the situation is probably more complicated.

Various substances, such as phosphodiesterase, calcium, prostaglandins, can modulate the effects of vasopressin. Prostaglandin synthetase inhibitors have been used for several years in nephrogenic diabetes insipidus. They decrease urine output by mechanisms which are still debated.

Clinical Course

From 1957 to 1983, we observed 39 cases of nephrogenic diabetes insipidus. There were 31 boys and 8 girls. The first symptoms appeared during the first 3 months of life in 87% of the cases. Fever of unexplained origin (69%), anorexia, vomiting and constipation (45%) and growth retardation (67%) were the manifestations of chronic underhydration. Episodes of severe acute dehydration with hypernatremia occurred in 70% of the cases. Polyuria was constant. Urine osmolality did not exceed 100 mosm/kg and was not modified by 1-desamino-8-*D*-arginine-vasopressin (DDAVP) administration. During acute dehydration in 18 children aged less than 1 year, serum Na⁺ was increased to 169 ± 12.8 mmol/l whereas urine osmolality was low (170 ± 75.5 mosm/kg).

In 10 children, plasma vasopressin was measured by radioimmunoassay (Dr. P. Czernichow) and the results were analyzed relative to plasma osmolality. In 5 children with plasma osmolality >285 mosm/kg, vasopressin levels were elevated (5.2–11.5 pg/ml). In 9 children with plasma osmolality <280 mosm/kg, vasopressin levels were within the normal range (1–4 pg/ml) including 4 patients in whom elevated levels had been found previously during dehydration.

Acute dehydration episodes dominated the immediate prognosis. It induced subdural hematoma or intraventricular hemorrhage in 4 children, 3 of them died from this complication. Mental retardation secondary to cerebral lesions caused by hypernatremia was seen in 3 children. In 7 children, mental retardation regressed after 2–3 years when water equilibration was corrected. A complete arrest or severe delay of statural growth was observed in children with volume contraction and hypernatremia. When good equilibration of water was obtained, there was catch-up growth in most children after 2 years.

Genetics

Most family studies indicate that nephrogenic diabetes insipidus is a hereditary disease transmitted as a sex-linked recessive trait with variable penetrance in females. A genetic analysis could be performed in 29 cases: 16 were familial in 12 families and 13 cases were apparently sporadic. In 2 families, we observed an autosomal dominant inherence with a transmission from father to son. In one family of affected girls, a complete unresponsiveness to DDAVP was observed. We conclude that nephrogenic diabetes insipidus might be transmitted as an autosomal dominant trait with epistasia or as a sex-linked recessive trait.

Treatment

The aim of treatment in nephrogenic diabetes insipidus is to reduce urine output. The first step is to give a diet with low osmotic activity particularly in infants. This implies a strict limitation of sodium intake and a

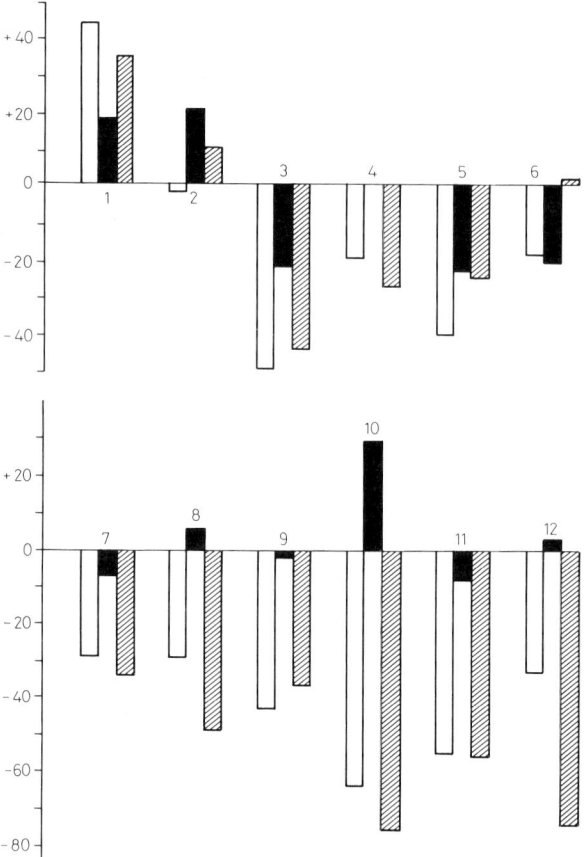

Fig. 1. Effects of Indomethacin in 12 children with nephrogenic diabetes insipidus. Percent increase (+) or decrease (−) of urine output, open bars, glomerular filtration rate, solid bars, and Na delivery, stippled bars, after 5 days Indomethacin.

low potassium and phosphorus intake. The best regimen is human milk or so-called 'humanized' milk in newborns. Diuretics are also effective in reducing polyuria. Hydrochlorothiazide by inducing a negative salt balance decreases extracellular volume which in turn increases proximal reabsorption. This leads to a decrease of distal water delivery. Hydrochlorothiazide induced a reduction of urine flow of 38 ± 8% in 21 of 31 cases.

The amount of water supplementation necessary for maintaining a normal water balance may be very high. It is necessary to give water con-

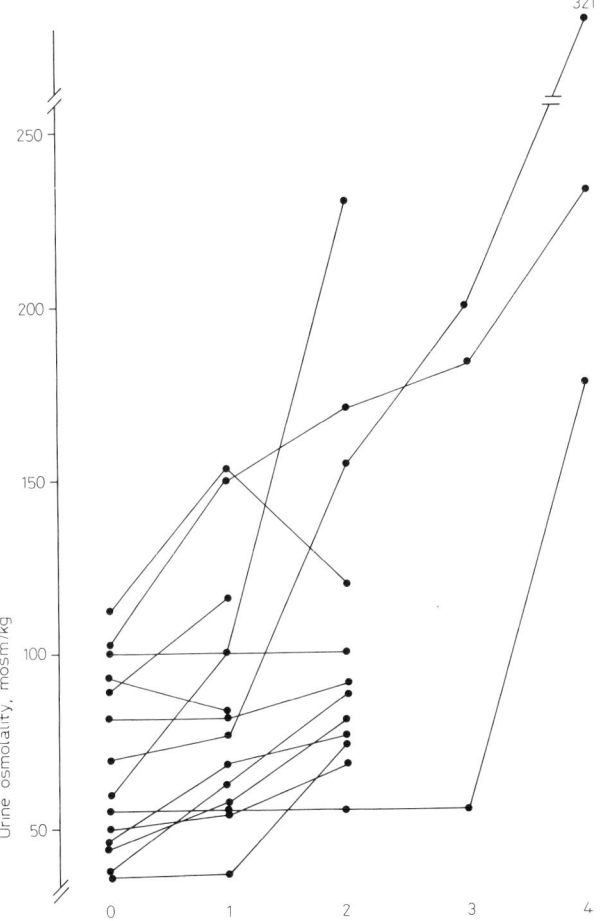

Fig. 2. Variations of urine osmolality in nephrogenic diabetes insipidus. 0 = No treatment; 1 = Indomethacin; 2 = Indomethacin + DDAVP; 3 = Indomethacin + thiazides; 4 = Indomethacin + thiazides + DDAVP.

tinuously by gastric tube on a 24-hour basis during the first months and during the nights hereafter. We have not observed severe dehydration or growth retardation with this technic during the past 7 years.

Recent reports have shown the therapeutic effect of treatment with prostaglandin synthetase inhibitors in nephrogenic diabetes insipidus. We studied the effects of Indomethacin in 12 children, 9 boys and 3 girls aged 1 month to 14 years, and investigated several parameters in order to clarify the mode of action of the drug. Glomerular filtration rate (GFR) was normal in 11 and diminished in 1 (56 ml/min/1.73 m²). Urine output varied between 1.2 and 9.2 l/day. Urine osmolality, after DDAVP, was between 60 and 126 mosm/kg. These children were maintained on a low osmotic diet without hydrochlorothiazide. Indomethacin, 1.5–3 mg/kg BW, was given for 5 days with low doses when the GFR was low.

Several parameters were studied before and after Indomethacin in water-loaded patients: creatinine clearance, urine osmolality, sodium and free water clearances, urinary cyclic AMP by radioimmunoassay. The sum $C_{H_2O} + C_{Na}$ was assumed to represent the Na delivery at the distal tubule and the ratio $C_{H_2O}/C_{H_2O} + C_{Na}$ the distal reabsorption of sodium.

The results are summarized in figure 1. GFR sightly decreased in 3 children and was unchanged or even higher in 9 children after Indomethacin. A significant reduction of diuresis >30% was observed in 8 children whereas in 3 children the reduction was 10–20% and in 1 child it increased. There was a reduction of the $C_{H_2O} + C_{Na}$ index of 25–76% in the 8 children with a significant reduction of urine output. In the other children, the index was reduced by 28% in 1 and was not changed in 3. The distal reabsorption of sodium as measured by $C_{H_2O}/C_{H_2O} + C_{Na}$ was unchanged after Indomethacin.

Urine osmolality increased after Indomethacin in 5 children with a further increase with Indomethacin and DDAVP. 4 children received Indomethacin and Hydrochlorothiazide with excellent clinical results. The decrease of diuresis varied from 50 to 80% compared to the period without treatment. Urine osmolality increased to more than 180 mosm/kg in 3 and reached 320 mosm/kg after DDAVP in 1 (fig. 2).

Urinary cyclic AMP varied from 4.3 to 28. It decreased in 3 patients, increased in 3 and was not modified in 5 after Indomethacin.

Discussion

It has been postulated that nephrogenic diabetes insipidus is a state where the absence of response to vasopressin is due to the absence of cyclic AMP formation in the epithelial cell of the collecting duct. *Fichman and Brooker* [4] showed that urinary cyclic AMP levels were the same in primary diabetes insipidus and in normal controls during water loading. Cyclic AMP increases after vasopressin infusion in both conditions. Urinary cyclic AMP was the same in nephrogenic diabetes insipidus and in controls but did not increase after vasopressin infusion in nephrogenic diabetes insipidus. It was concluded that there is a defect of cyclic AMP production in nephrogenic diabetes insipidus. Other authors found the same results. On the other hand, *Monn* [7] found that urinary cyclic AMP response to vasopressin was variable, and in some patients an increased urinary cyclic AMP excretion was observed. Thus, *Zimmerman and Green* [12] described 2 types of nephrogenic diabetes insipidus: type I with no increase of urinary cyclic AMP after vasopressin infusion and type II with an increased urinary cyclic AMP excretion in response to vasopressin. We measured urinary cyclic AMP levels before and after DDAVP administration in 12 cases and did not observe significant differences. However, the mean urinary cyclic AMP level before DDAVP was higher in the patients than in controls. These studies as well as those of *Jackson* et al. [5] and *Dousa and Valtin* [3] working on a strain of mice with nephrogenic diabetes insipidus do not allow the conclusion that the defect of cyclic AMP production in the epithelial cell is the only factor responsible for the concentrating defect.

The biological effects of renal prostaglandins will be briefly summarized [6, 8, 11] before discussing the mechanisms which could explain the action of Indomethacin in nephrogenic diabetes insipidus. Prostaglandins increase the renal blood flow, particularly the medullary blood flow, and attenuate the vasoconstrictor influence of angiotensin II. These actions as well as the increase of GFR are dependent on the status of the renin-angiotensin system. Thus, prostaglandins enhance renal blood flow and GFR only in states of high plasma concentrations of angiotensin II (volume depletion, dietary sodium restriction...). Prostaglandins increase urine sodium excretion and several intrarenal mechanisms are involved. Again the action of prostaglandins on salt excretion depends on the state of hydration. Prostaglandins impair the urine concentrating mechanisms. Prostaglandins lower the medullary osmolality by increasing the medullary blood flow and reducing sodium reabsorption. Prostaglandins inhibit the action of vasopressin in the collecting tubule. It has been shown that prostaglandins inhibit vasopressin-induced cyclic AMP production. The cellular mechanisms of the interaction are not clear but prostaglandins do not simply inhibit vasopressin-adenylate cyclase interaction. *Beck* et al. [1] have shown that high doses of prostaglandins have an opposite effect and increase cyclic AMP synthesis.

Urinary cyclic AMP was measured in 11 children before and after Indomethacin. The variations observed were not correlated with the reduction of diuresis.

Thus, it seems that the antidiuretic effect of Indomethacin cannot be explained by the restoration of a state of relative sensitivity of the collecting duct to vasopressin as we discussed previously.

Several other actions of Indomethacin may explain the reduction of urine output in nephrogenic diabetes insipidus: increase in the water permeability of the collecting tubule, phosphodiesterase inhibition, increase of sodium reabsorption, increase of the corticopapillary osmotic gradient and/or decrease of GFR. The results of our study indicate a good correlation between the reduction of diuresis and the reduction of the sodium delivery to the distal tubule. We did not find significant variations of GFR or of the distal sodium reabsorption. Thus, we believe, like others [2, 9, 10], that the action of Indomethacin is in part due to an increase of proximal reabsorption.

Indomethacin may be used in infants. In our cases, it did not impair GFR which normally increased with age. The combined therapy with Indomethacin and thiazides considerably improve the clinical condition of children with nephrogenic diabetes insipidus.

References

1 Beck, N.P.; Kaneto, T.; Zor, V.; Field, J.; Davis, B.B.: Effect of vasopressin and prostaglandin E_2 on adenyl-cyclase cyclic 3',5'-adenosine monophosphate system on the renal medulla of the rat. J. clin. Invest. *50:* 2461–2465 (1971).
2 Blachar, Y.; Zadik, Z.; Shemesh, M.; Kaplan, B.S.; Levin, S.: The effect of inhibition of prostaglandin synthesis on free water and osmolar clearance in patients with hereditary nephrogenic diabetes insipidus. Int. J. Ped. Nephrol. *1:* 48–52 (1980).
3 Dousa, T.P.; Valtin, H.: Cellular actions of vasopressin in mammalian kidney. Kidney int. *10:* 46–63 (1976).
4 Fichman, M.P.; Brooker, G.: Deficient renal cyclic adenosine 3'–5'-monophosphate production in nephrogenic diabetes insipidus. J. clin. Endocr. Metab. *35:* 35–47 (1972).
5 Jackson, B.A.; Edwards, R.M.; Dousa, T.P.: Vasopressin-prostaglandin interactions in isolated tubules from rat outer medulla. J. Lab. clin. Med. *96:* 119–128 (1980).
6 Levenson, D.J.; Simmons, C.E.; Brenner, B.M.: Arachidonic acid metabolism, prostaglandins and the kidney. Am. J. med. *2:* 354–374 (1982).
7 Monn, E.: Prostaglandin synthetase inhibitors in the treatment of nephrogenic diabetes insipidus. Acta paediat. scand. *70:* 39–42 (1981).

8 Stokes, J.B.: Integrated actions of renal medullary prostaglandins in the control of water excretion. Am. J. Physiol. *240:* F471–F480 (1981).
9 Turi, S.; Merth, I.; Sztriha, L.: Indomethacin treatment of children suffering from nephrogenic diabetes insipidus or secondary tubulopathy associated severe polyuria. Int. J. Ped. Nephrol. *2:* 263–268 (1981).
10 Usberti, M.; Dechaux, M.; Guillot, M.; Seligmann, R.; Pavlovitch, H.; Loirat, C.; Sachs, C.; Broyer, M.: Renal prostaglandin E_2 in nephrogenic diabetes insipidus. Effects of inhibition of prostaglandin synthesis by indomethacin. J. Pediat. *97:* 476–480 (1980).
11 Weber, P.C.; Scherer, B.; Siess, W.: Prostaglandins and the regulation of renin release: prostaglandins and the regulation of extra cellular fluid; in Berti, Velo, The prostaglandin system. Endoperoxides, prostacyclin ad thromboxanes. Nato Advanced Study Institutes Series, pp. 247–270 (Plenum Publishing, New York 1981).
12 Zimmermann, D.; Green, O.C.: Nephrogenic diabetes insipidus type II: defect distal to adenyl-cyclase step. Pediat. Res. *9:* 381 (1975).

Patrick Niaudet, Service de Néphrologie Pédiatrique, Hôpital Necker Enfants-Malades, 149, rue de Sèvres, F-75730 Paris Cedex 15 (France)

Role of Autoimmunity in Central Diabetes insipidus

W.A. Scherbaum[a,1], *G.F. Bottazzo*[b], *P. Czernichow*[c], *J.A.H. Wass*[d], *D. Doniach*[b]

[a] Zentrum für Innere Medizin I, Univeristy of Ulm, Ulm, FRG; [b] Department of Immunology, Middlesex Hospital, Medical School, London, England;
[c] Département de Pédiatrie, Service d'Endocrinologie Pédiatrique et Diabète, Hôpital des Enfants Malades, Paris, France; [d] Endocrine Unit, St. Bartholomew's Hospital, London, England

Autoimmune variants have been recognized in most endocrine disorders previously thought to be idiopathic in nature, and in the majority of the cases, a deficiency state of the respective hormone is the predominant clinical feature. Classical examples are Hashimoto's thyroiditis, primary myxoedema, and autoimmune Addison's disease. These autoimmune conditions are characterized by the presence in the serum of autoantibodies directed to the hormone-producing cells [6]. The indirect immunofluorescence (IFL) test has provided a potent tool to detect new antibody specificities in complex organs composed of different endocrine cells [10]. New discoveries by this method comprise the detection of antibodies to anterior pituitary cells in partial pituitary deficiency [2, 3] and in the preclinical state of type I diabetes [7], and most recently the demonstration of autoantibodies to hypothalamic vasopressin cells in patients with central diabetes insipidus (DI) [9].

Where diagnosis of central DI is established by clinical, biochemical and metabolic criteria, the symptoms are easily 'cured' by vasopressin hormone or its analogues. However, every effort should be made to determine the cause of the syndrome since an underlying disease may require specific therapy. Such a disease may sometimes only be detected through repeated evaluation of the patient, but still, excluding postoperative DI, the aetiology of central diabetes insipidus remains ob-

---

[1] *W.A. Scherbaum* was a research associate at the Middlesex Hospital, London, in 1981–1982, and is now supported by the 'Deutsche Forschungsgemeinschaft', Sche 225/1–2.

Table I. Incidence of vasopressin cell antibodies on different tissues of hypothalamus in 10 positive cases

Source of hypothalamus	Age of donor	Time freezing h	Lipo-fuscin contents	Vasopressin cell antibodies: number positive
Post delivery				
Human fetal	21 weeks	2	–	10
	21 weeks	1.5	–	8
	18 weeks	1	–	6
Postmortem				
Baboon adult	?	0.25	(+)	8
Human adult	50 years	3	+	4
	38 years	6.5	(+)	1
	91 years	2.5	+++	1

scure in about 30–40% of the cases [1]. This is the major group where the detection of vasopressin-cell antibodies helps to understand the pathogenesis of DI.

Demonstration of Vasopressin Cell Antibodies

Arginine vasopressin (AVP) cell antibodies are detected by indirect IFL on cryostat sections of fresh hypothalamus. Human fetal tissue obtained from therapeutic hysterotomies provides the best substrate (table I). Adult postmortem tissue obtained several hours after death gives poor results since the cytoplasmic autoantigens accounting for the specific AVP cell staining are rapidly inactivated. Older adult donors are also unsuitable because the natural accumulation of lipofuscin granules in secretory hypothalamic cells interferes with the reading of the specific cytoplasmic IFL. Fresh primate (baboon) hypothalamus obtained from young animals gives a good cross-reactivity with the human antibodies (table I) and may therefore be used for clinical tests.

Antibodies to large secretory cells of the supraoptic and paraventricular nucleus are visualized by IFL when positive sera are applied to cryostat sections of hypothalamus and counterstained with fluoresceinated antihuman immunoglobulin. Reactivity to vasopressin cells can be proven by a four-layer double fluorochrome IFL test where antivasopressin serum (raised in rabbits) and rhodaminated antirabbit Ig serum are applied in the second sandwich (fig. 1a, b). Sera containing AVP cell antibodies stain the same cells as does the antivasopressin serum, and this can be shown when the same field under the fluorescence microscope is first viewed with a red, and then with a green filter which gives a yellow colour when superimposed in double-exposure photography.

Immunological Characteristics of AVP Cell Antibodies

About half the sera reacting with AVP cells also stain oxytocin cells of the hypothalamus. This may be due to the close similarity in the chemical structure of the two neurohormones. Sera containing oxytocin-cell antibodies alone have not been detected so far, but this specificity remains to be precisely characterized. AVP cell antibodies are of IgG and/or IgA class and about half the positive sera also fix complement (table II). Antibody titres have so far ranged from 1:1 to 1:64 so that undiluted sera have to be applied for screening tests.

It has been argued that the AVP cell antibodies detected by IFL could reflect antibodies to the hormone itself which may be induced by treatment with vasopressin or its analogues. However, there is clear evidence that this is not the case. In 2 patients with positive AVP cell antibodies, blood had been taken before hormone therapy was initiated. The

---

Fig. 1. Unfixed cryostat section of human supraoptic nucleus stained by the double fluorochrome technique. *a* Immunofluorescent complement fixation test with serum from a patient with autoimmune central DI which contained complement-fixing antibodies to the cytoplasm of vasopressin cells shown in green. The nucleus was unstained. *b* The positive cells were identified in the same field by applying as the second sandwich a rabbit vasopressin antiserum counterstained with goat antirabbit immunoglobulin. The positive cells are now stained red. No blocking occurred as the patient's antibodies are against the cytoplasmic membranes and the rabbit reacts with the hormone itself.

Fig. 2. Cryostat section of baboon hypothalamus stained by immunofluorescent complement fixation test. This patient with autoimmune central DI was positive on vasopressin cells giving the same staining pattern as the case shown in figure 1. In addition, his serum also stained small cells shown here in green, which reacted neither with antivasopressin nor with anti-oxytocin. The granular cytoplasm suggests an endocrine cell and its hormone remains to be identified. The use of monoclonal antibodies to the other known hypothalamic hormones will help to characterize these cells.

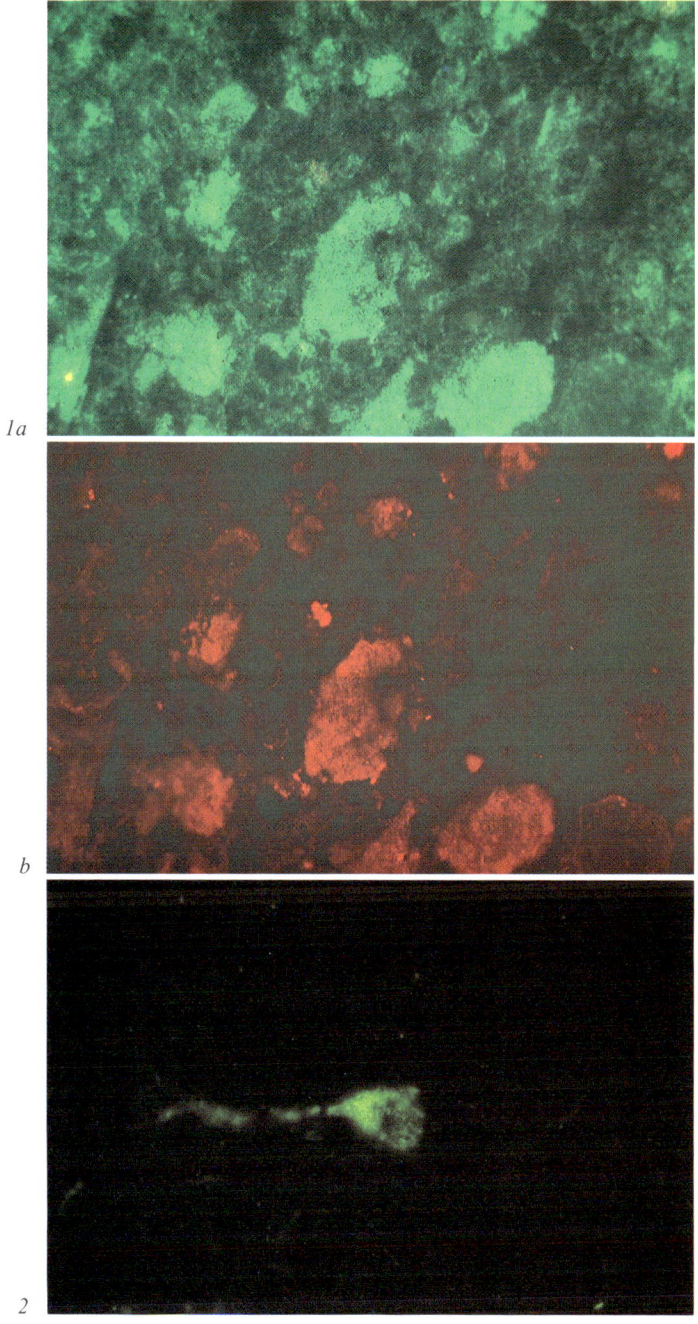

Table II. Immunoglobulin classes and complement-fixing ability of AVP-cell antibodies in 24 positive sera

Immunoglobulin class of AVP cell antibodies	Number positive
IgG alone	6
IgA alone	9
IgG + IgA	9
IgM	0
Complement-fixing	10

antibody reactivity is not diminished by pre-absorption of positive sera with an excess of synthetic vasopressin, oxytocin, or their corresponding neurophysins in vitro. This is in accordance with our knowledge on other organ-specific autoantibodies such as thyroid microsomal antibodies, adrenocortical, and pancreatic islet cell antibodies which react with membrane-bound autoantigens distinct from the hormone [6].

Prevalence of AVP Cell Antibodies in DI and Controls

One third of the sera from adult patients with so-called idiopathic central DI contain autoantibodies to AVP cells (table III). This proportion still remains the same when we look at patients with onset in childhood, indicating that in these cases an autoimmune process to hypothalamic vasopressin cells is involved in the pathogenesis of the disease. AVP cell antibodies have not been detected in control sera from patients without DI including hypothalamic and pituitary disorders, endocrine autoimmune diseases and other conditions. Cases of familial central DI, and DI diabetes mellitus, optic atrophy and deafness (DIDMOAD) syndrome are also constantly negative (table IV). Only 2 cases of nephrogenic DI were tested to date, both with negative results.

More than 50% of the sera from patients with DI due to histiocytosis X contain autoantibodies to AVP cells whereas other forms of symptomatic DI are only positive in occasional cases. This may be explained by the fact that histiocytosis X cells bear class II major histocompatibility antigens on their surface so that specific infiltration of the hypothalamus

Table III. Prevalence of vasopressin cell antibodies and associated organ-specific autoimmune conditions in 108 patients with different forms of central diabetes insipidus (DI)

Form of diabetes insipidus	Number of cases	Vasopressin cell antibodies	Other organ-specific autoantibodies or diseases	
			adults	children
Idiopathic central DI	47	18 (38)	12 (36)	1 (7)
Histiocytosis-X with DI	8	5 (63)	0	0
Other forms of central DI	53	1 (2)	5 (16)	0

Percentages are given in parentheses.

Table IV. Symptomatic and familial central DI cases tested for AVP cell antibodies

	Number tested	Number positive AVP cell antibodies
Histiocytosis-X	8	5
Craniopharyngioma	11	1
Pituitary adenoma with postoperative DI	13	0
Germinoma	7	0
DIDMOAD	7	0
Familial DI	6	0
Malformation of hypothalamus	2	0
Metastatic tumours	3	0
Traumatic	1	0
Postradiation	1	0
Sarcoidosis	1	0
Acute myeloid leukaemia	1	0

DI = Diabetes insipidus; DIDMOAD = diabetes insipidus, diabetes mellitus, optic atrophy, and deafness.

may trigger T-helper cells to induce an autoimmune reaction to hypothalamic antigens. Such a response may be increased in the presence of a T-suppressor cell defect which has been demonstrated in histiocytosis X. Also cytotoxic lymphocytes have been detected in this disease which can now be separated into a tumorous and an autoimmune subgroup [8]. AVP cell antibodies may be considered an early marker of hypothalamic invasion in cases of histiocytosis X.

Deficiencies of hypothalamic pituitary hormone releasing factors, i.e. corticotropin releasing factor, thyrotropin releasing hormone, luteinizing hormone releasing hormone, somatotropin release inhibiting factor, growth hormone releasing factor, probably exist in some cases of histiocytosis X. When the respective endocrine cells can be reliably located and identified by immunofluorescence, it will be of great interest to look for autoantibodies to them in patients with the appropriate pituitary functional defects.

Postmortem Studies

Postmortem examination of the hypothalamus has been reported in only a few cases of familial [4] and non-familial [1] idiopathic central DI. Each showed a heavy loss of secretory cells and severe gliosis of the supraoptic and similar but moderate changes in the paraventricular nuclei. All these patients had the disease for many years, and no lymphoid infiltration was observed.

Association of 'Idiopathic' Central DI with Endocrine Autoimmune Diseases

Autoimmune diseases may sometimes affect more than one endocrine gland in the same individual and such polyendocrine autoimmune conditions [5] were also noted in patients with 'idiopathic' central DI. We have collected 13 cases of idiopathic central DI who had one or more associated autoimmune disorders. 7 of these had AVP cell antibodies in their sera. There were 2 cases of thyrotoxicosis, 4 of primary myxoedema, 3 Hashimoto goitres, 3 cases of 'autoimmune' diabetes mellitus, 2 pernicious anaemia and 1 each with the following: idiopathic hypoparathyroidism with chronic mucocutaneous candidiasis (C-E syndrome), alopecia totalis, unexplained growth retardation, myasthenia gravis and Sjögren's syndrome. 7 of these patients had 'polyendocrine' disease with more than one associated disorder. The appropriate organ-specific autoantibodies were detected in most of the sera [11]. In idiopathic central DI with onset in childhood, associated autoimmune diseases are rare, and this may reflect a later involvement of other endocrine cells or an aetiologically different subgroup of the disease.

Taking both together, the demonstration of specific autoantibodies directed to hypothalamic AVP cells and the association with well-recognized autoimmune disorders, there is strong evidence for the existence of an autoimmune form of central DI which can now be separated from the unexplained idiopathic cases.

Future Prospects

The recent discoveries in DI here described pose a panel of new questions and open a wide field for future research. Autoantibodies to cytoplasmic components of hypothalamic AVP cells (as detected on cryostat sections) may be regarded as markers for an autoimmune process. However, an attack to the cells by the immune system in vivo has to involve the cell surface. It will, therefore, be a major task to look for hypothalamic AVP cell surface antibodies and find out if such antibodies exert a cytotoxic effect by the addition of complement in vitro. It will also be interesting to look for a blocking effect of AVP cell antibodies on vasopressin production in whole explant cultures of the hypothalamus.

The hypothalamus is still poorly investigated in postmortem studies and, in fact, very few autopsied cases of idiopathic and familial central DI have been described so far [1, 4]. Gliosis and atrophy of the supraoptic and paraventricular nuclei in these long-standing cases may be compared with the final fibrosis of the respective glands in primary myxoedema and Addison's disease. It will be of great interest to see if there is a lymphocytic infiltration of the hypothalamus at the onset of the inflammatory process as seen in the lesions of experimental myasthenia gravis. 'Autoimmune hypothalamitis' could be induced in animals by injections of cell extracts and followed from its onset. The association with autoimmune diseases in the absence of AVP cell antibody suggests that specific autoantibodies might have disappeared over the years as in primary myxoedema and Addison's disease [12]. In these cases it seems worthwhile to look for cell-mediated autoimmunity which may help to reveal autoimmune sensitization to hypothalamic tissue. HLA tissue typing may be useful in future studies to see the relationship of idiopathic DI with the other autoimmune endocrinopathies.

The sera of some patients with DI or other endocrine diseases contain autoantibodies to small hypothalamic cells which do not produce oxytocin or vasopressin (fig. 2). It will be a challenge to characterize the peptide hormone of the reacting cells which may allow us to draw conclusions as to the significance of these antibodies. The detection of vasopressin cell antibodies in central DI suggests that autoimmunity may also play a role in other hypothalamic endocrine defects which have hitherto been regarded as idiopathic.

References

1 Blotner, H.: Primary or idiopathic diabetes insipidus: a system disease. Metabolism. *7:* 191–200 (1958).
2 Bottazzo, G.F.; Pouplard, A.; Florin-Christensen, A.; Doniach, D.: Autoantibodies to prolactin-secreting cells of human pituitary. Lancet *ii:* 97–101 (1975).
3 Bottazzo, G.F.; McIntosh, C.; Stanford, W.; Preece, M.: Growth-hormone-cell antibodies and partial growth-hormone deficiency in a girl with Turner's syndrome. Clin. Endocrinol. *12:* 1–9 (1980).
4 Braverman, L.E.; Mancini, J.P.; McGoldrick, D.M.: Hereditary idiopathic diabetes insipidus. A case report with autopsy findings. Ann. intern. Med. *63:* 503–508 (1965).
5 Doniach, D.; Bottazzo, G.F.: Polyendocrine autoimmunity; in Franklin, Clinical immunology update, pp. 95–121 (Elsevier, Amsterdam 1981).
6 Doniach, D.; Bottazzo, G.F.; Drexhage, H.A.: The autoimmune endocrinopathies; in Lachman, Peters, Clinical aspects of immunology; 4th ed., vol. II, pp. 903–937 (Blackwell, Oxford 1983).
7 Mirakian, R.; Cudworth, A.G.; Bottazzo, G.F.; Richardson, C.A.; Doniach, D.: Autoimmunity to anterior pituitary cells and the pathogenesis of type I (insulin-dependent) diabetes mellitus. Lancet *i:* 755–759 (1982).
8 Nesbit, M.E.; O'Leary, M.; Dehner, L.P.; Ramsay, N.K.C.: The immune system and the histiocytosis syndromes. Am. J. Pediat. Hematol. Oncol. *3:* 141–149 (1981).
9 Scherbaum, W.A.; Bottazzo, G.F.: Autoantibodies to vasopressin cells in idiopathic diabetes insipidus: evidence for an autoimmune variant. Lancet *i:* 897–901 (1983).
10 Scherbaum, W.A.; Mirakian, R.; Pujol-Borrell, R.; Dean, B.M.; Bottazzo, G.F.: Indirect immunofluorescence in the study and diagnosis of organ-specific autoimmune diseases; in Polak, van Noorden, Immunocytochemistry. Practical applications in pathology and biology, pp. 346–361 (Wright, Bristol 1983).
11 Scherbaum, W.A.; Wass, J.A.H.; Besser, G.M.; Mirakian, R.; Doniach, D.; Bottazzo, G.F.: Autoimmune central diabetes insipidus: its association with endocrine disorders and with histiocytosis X (submitted).
12 Sotsiou, F.; Bottazzo, G.F.; Doniach, D.: Immunofluorescence studies on antibodies to steroid-producing cells, and to germline cells in endocrine disease and infertility. Clin. exp. Immunol. *39:* 97–111 (1980).

G.F. Bottazzo, Department of Immunology, Arthur Stanley House, Middlesex Hospital Medical School, 40–50 Tottenham Street, GB-London W1P 9PG (England)

Intracranial Germinoma in Children and Diabetes insipidus: Clinical Description and Search for Tumor Markers

R. Pomarede, P. Czernichow, R. Brauner, R. Rappaport

Unité d'Endocrinologie Pédiatrique et Diabète, Clinique Robert Debré,
Hôpital Necker Enfants-Malades, Paris, France

Intracranial germinomas are frequently associated with diabetes insipidus. The latter is often isolated, without any neurological symptoms, nor any neuroradiological signs. In this case, diagnosis of the tumor and of its origin is difficult. Presence of tumor markers takes on an important diagnostic significance because the sole treatment of these tumors is radiotherapy. Our preliminary observations have led us to undertake a prospective study of tumor markers at the onset of polyuria and regularly during follow-up.

Germinomas and Diabetes insipidus

14 patients with intracranial germinoma were studied – 4 girls and 10 boys, 8–19 years of age at the time of diagnosis. The diagnosis of germinoma was made after histologic examination of the tumor by surgical biopsy in 9 cases and by stereotaxic biopsy in 2. In the 3 remaining cases, the diagnosis was made on indirect evidence: tumor cells in the spinal fluid associated with the presence of human chorionic gonadotropin (hCG) and/or α-fetoprotein in plasma, and pineal tumor.

Diabetes insipidus

Diabetes insipidus was present in all patients. In the literature, this symptom was found in germinomas with a frequency of 70–100% [1–5]. In 5 patients, the intensity of polyuria varied. When 11 out of 14 patients were studied by a short dehydration test [*Czernichow* et al., this volume; 7], diabetes insipidus was partial in 5 cases and complete in the 6 remaining cases. These findings were not different from diabetes insipidus with other tumors [8]. Adipsia, however, was especially frequent: 8 patients out of 11 did not experience thirst when plasma osmolality was greater than 300 and up to 340 mosm/kg. When a patient had partial diabetes insipidus and adipsia, polyuria was generally less severe and might be missed. This was the case of 1 girl in our series, presenting with abdominal pain, fever (40°C) and vomiting. Appendectomy was performed. Following surgery, she developed thrombosis of the saphenous vein. She was found to be dehydrated (Na^+ = 160 mEq/l). Rehydration therapy led to polyuria. Central diabetes insipidus was then diagnosed and related to an intracranial germinoma. In 2 other cases, partial diabetes insipidus associated with adipsia led to variable polyuria which was considered to be of psychogenic origin. *Sklar* et al. [3] have also reported frequent adipsia in germinoma.

Associated Clinical Symptoms

At the onset, diabetes insipidus is not always associated with other neurological symptoms. In our series, only 2 cases presented with diabetes insipidus and intracranial hypertension leading to an early diagnosis of a pineal tumor. In the other cases, the clinical signs were progressive and the diagnosis was made 3 months to 3 years after the onset of diabetes insipidus when other symptoms developed. Intracranial hypertension appeared in 4 other cases of pineal germinomas. Parinaud's syndrome (n = 3) and visual abnormalities (n = 2) were present in suprasellar lesions. In 4 cases, short stature was noticed and we observed 1 case of precocious puberty. In our experience when diabetes insipidus became associated with other symptoms, the tumor was visible on neuroradiological examination.

Endocrine Functions

All patients, except 1, were tested for growth hormone (GH) and adrenocorticotropic hormone (ACTH) secretion after arginine-insulin [9]; 10 were tested for thyroid stimulating hormone (TSH) secretion after thyrotropin releasing hormone (TRH). 7 had the short metyrapone test [10]. The anterior pituitary functions were frequently deficient. In 11 patients including the 4 with short stature, GH secretion was abnormal. Low cortisol levels in 9 patients after insulin-induced hypoglycemia indicated insufficient ACTH secretion. The diagnosis was confirmed in 4

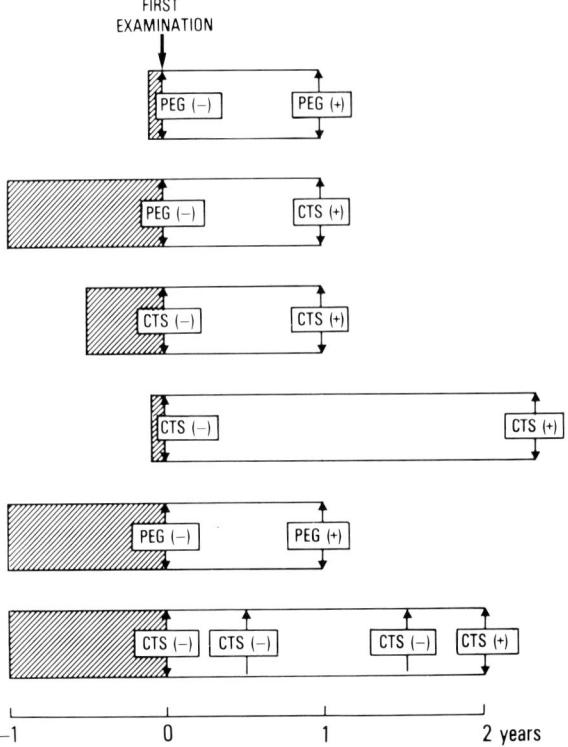

Fig. 1. Neuroradiological survey of the 6 patients who had a first normal neuroradiological examination. Hatched bars represent duration of diabetes insipidus before the first examination. Tumor diagnosis was made between 1 and 3 years after onset of diabetes insipidus.

patients by the metyrapone test. TSH response to TRH was abnormal in 9 cases out of 10. In 4 cases we noted a TSH deficiency and in 5 cases a prolonged response of hypothalamic type [11]. These results are comparable to those of *Sklar* et al. [3] who reported a frequency of 100% of GH deficiency, 75% of TSH deficiency, and 50% of ACTH deficiency.

Neuroradiological Aspects

All patients underwent neuroradiological examinations either by pneumoencephalogram (5 patients) or by computed tomographic scan (9 patients). Calcification of the pineal gland was observed in 7 patients. In 4 cases, the tumor was confined to the pineal gland. 10 patients had

Table I. Results of tumoral markers in 6 patients with a normal first neuroradiological examination. n.d. = Not done

Cases	Plasma α-fetoprotein ng/ml	Plasma β-hCG mU/ml	Spinal fluid β-hCG	Immuno-histochemical analysis for hCG
1	1	2.5	n.d.	+
2	5	<2.5–<2.5–<2.5	<2.5–<2.5–5	+
3	3	<2.5	n.d.	–
4	14	<2.5–<2.5–14–36	n.d.	+
5	n.d.	LH = 90 LH after LHRH = 90	n.d.	–
6	3	2.5–17	n.d.	n.d.

N<10 ng/ml; N<3 mU/ml.

suprasellar tumors, 3 of whom had pineal lesions as well. In 8 cases, the tumors were detected during the first radiological examination after the onset of diabetes insipidus (in 1 case, the first examination was not performed until 4 years after the onset of diabetes insipidus). In 6 patients, the first radiological examination was normal and the tumor was not diagnosed until the examination was repeated 15 months to 3 years later (fig. 1). *Sklar* et al. [3] described a similar case in which the diagnosis was made 2 years later on the third neuroradiological examination. This silent period seems to be characteristic of germinoma.

Tumoral Markers and Germinomas

It is well known that precocious puberty can be a symptom of intracranial germinoma in the boy. Puberty may be due to an abnormal secretion of hCG by the trophoblastic cells of the tumor, which are often present in germinoma [12–16]. Excluding these cases where the tumor marker was evident by the clinical symptomatology, we studied systematically the presence of hCG and other tumor markers during the survey of diabetes insipidus.

Three tumor markers were studied. (1) hCG was assayed [17] in plasma and in the spinal fluid when possible. It is the tumor marker of trophoblastic cells. (2) α-fetoprotein was measured in plasma. It is the tumor marker of endodermal sinus cells. (3) Finally, spinal fluid cytology was performed to detect tumoral cells (Dr. *Pfister*) [18].

When the diagnosis was ascertained and the surgical biopsy was performed, immunohistochemical staining for hCG was done in 5 cases out of 6 (Dr. *Pfister*). The results are shown in table I.

Plasma hCG and β-hCG

Plasma hCG and β-hCG was elevated in 2 cases (Nos. 4, 6). In case 5, very high plasma levels of luteinizing hormone (LH) unresponsive to luteinizing hormone releasing hormone (LHRH) were found and thought to be due to hCG. Plasma concentration of these tumor markers increased in patients 4 and 6 when the tumor became visible on CT scan, after a lag time of 1 year.

Immunohistochemistry (fig. 2)

Immunohistochemical analysis of the tumor showed hCG in 3 cases. In 1 male patient (No. 4) plasma hCG levels increased gradually and he developed precocious puberty. In another case, repeated determinations of hCG in plasma and in spinal fluid were always negative until discovery of the tumor. Biopsy showed large amounts of hCG staining tumoral cells. The tumor may therefore synthesize hCG even when its plasma levels are undetectable. Finally, immunohistochemical analysis was not performed on the last patient (No. 6) with high hCG plasma levels.

α-Fetoprotein (fig. 3)

This second marker was assayed in these patients at the time of diagnosis. It was moderately elevated in 1 case associated with high plasma levels of hCG. Similar observations have been reported in the literature [17, 18].

Spinal Fluid Cytology

Spinal fluid cytology was systematically studied. Two different cells were observed. (a) Large neoplastic cells with a hyperchromatic nucleus containing several small nucleoli and an abundant hyperbasophilic cytoplasm. They are specific of germinoma. (b) Large reactive lymphocytes, the existence of which contrasted with the absence of other signs of subacute inflammation. The coexistence of these two types of cells in the cerebrospinal fluid (CSF) is very specific of germinoma. On the contrary,

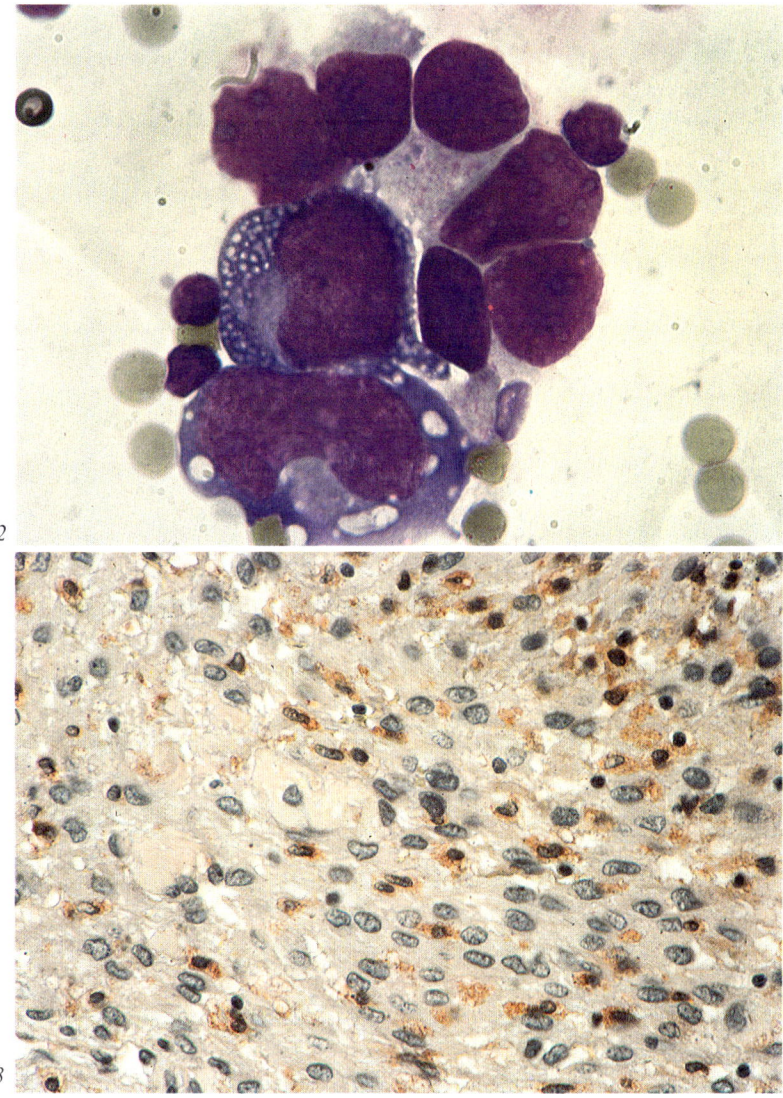

Fig. 2. Hypothalamic germinoma in an 11-year-old girl. Tumoral cells are brown stained by horseradish peroxydase-labeled anti-hCG antibody. ×1,200.

Fig. 3. Spinal fluid cells. On the upper part, lymphocytes are small round cells with large nucleus. Typical germinomal cells in the middle of the preparation are very large cells with hyperbasophilic, microvacuolated cytoplasm and a large, irregular, hyperchromatic multinucleated nucleus.

the presence of isolated large reactive lymphocytes without neoplastic cells can lead to false positivity of spinal fluid cytology. In our 14 cases, cytology was performed on CSF in 12 cases and coexistence of these two types of cells (neoplastic cells and reactive lymphocytes) was found in 5 cases. In 1 case, tumor cells were found before the tumor was visible on neuroradiological investigation. In this case, a second cytological study was performed 1 year later; it was also positive and the tumor was then visible. Therefore, CSF cytology should be studied systematically along with other tumor markers. In patients who have germinoma-specific tumor cells in the CSF, we are currently performing immunohistochemical staining for hCG. Presence of tumor cells staining for hCG in a patient with central diabetes insipidus might establish the diagnosis of germinoma without any biopsy even if the tumor were not visible by neuroradiological investigation. These patients could then be treated earlier by radiotherapy before a larger tumor developed. Although there is no experience with such cases, one would have to consider whether positive cells in the CSF in the presence of diabetes insipidus was sufficient evidence to initiate radiotherapy before a larger (and apparent) tumor developed.

References

1 Sung, D.; Harisiadis, L.; Chang, C.H.: Midline pineal tumors and suprasellar germinomas; highly curable by irradiation. Radiology *128:* 745–751 (1978).
2 Simson, L.R.; Lampe, I.; Abell, M.R.: Suprasellar germinomas. Cancer *22:* 533–544 (1968).
3 Sklar, C.A.; Grumbach, M.M.; Kaplan, S.L.; Conte, F.A.: Hormonal and metabolic abnormalities associate with central nervous system germinoma in children and adolescents and the effect of therapy. Report of 10 patients. J. clin. Endocr. Metab. *52:* 9–16 (1981).
4 Camins, M.B.; Mount, L.A.: Primary suprasellar atypical teratoma. Brain *97:* 447–451 (1974).
5 Pomarede, R.; Czernichow, P.; Finidori, J.; Pfister, A.; Roger, M.; Kalifa, C.; Zucker, J.M.; Pierre-Kahn, A.; Rappaport, R.: Endocrine aspects and tumoral markers in intracranial germinoma: an attempt to delineate the diagnosis procedure in 14 patients. J. Pediat. *101:* 374–378 (1982).
6 Takeuchi, J.; Handa, H.; Nagata, I.: Suprasellar germinoma. J. Neurosurg. *49:* 41–48 (1978).
7 Czernichow, P.; Pomarede, R.; Basmaciogullari, A.; Rappaport, R.: Diabetes insipidus in children. I. Arginine-vasopressin determination in plasma during short dehydration test. Acta paediat. scand. suppl. 277, pp. 64–68 (1979).

8   Pomarede, R.; Czernichow, P.; Rappaport, R.; Royer, P.: Le diabète insipide pipresso sensible chez l'enfant. II. Etude de 93 cas observés entre 1955 et 1978. Archs. fr. Pédiat. *37:* 37–44 (1980).
9   Penny, R.; Blizzard, R.M.; Davis, W.T.: Sequential arginine and insulin tests. J. clin. Endocr. Metab. *29:* 1499–1501 (1969).
10 Limal, J.M.; Basmaciogullari, A.; Rappaport, R.: The evaluation of single dose metyrapone tests in children with hypopituitarism. Acta paediat. scand. *66:* 177–183 (1976).
11 Illig, R.; Krawczynska, H.; Torresani, T.; Prader, A.: Elevated plasma TSH and hypothyroidism in children with hypothalamic hypopituitarism. J. clin. Endocr. Metab. *41:* 722–724 (1975).
12 Romshe, C.A.; Sotos, J.F.: Intracranial human chorionic gonadotropin secreting tumor with precocious puberty. J. Pediat. *86:* 250–252 (1975).
13 Rosenberg, D.: Puberté précoce par tumeur pinéale secrétant des gonadotrophines chorioniques. Annls pediat. *27:* 179–184 (1980).
14 Sklar, C.A.; Conte, F.A.; Kaplan, S.L.; Grumbach, M.M.: Human chorionic gonadotropin-secreting pineal tumor: relation to pathogenesis and sex limitation of sexual precocity. J. clin. Endocr. Metab. *53:* 656–660 (1981).
15 Ahmed, S.R.; Shalet, S.M.; Price, D.A.; Pearson, D.: Human chorionic gonadotrophin secreting pineal germinoma and precocious puberty. Archs. Dis. Childh. *58:* 743–744 (1983).
16 Pomarède, R.; Finidori, J.; Czernichow, P.; Pfister, A.; Hirsh, J.F.; Rappaport, R.: Germinoma in a boy with precocious puberty: evidence of hCG secretion by the tumoral cells. Child's Brain *11:* 298–303 (1984).
17 Chartier, M.; Roger, M.; Barrat, J.; Michelon, B.: Measure of plasma human chorionic gonadotropin (hCG) and β-hCG activities in the late luteal phase: evidence of the occurrence of spontaneous menstrual abortions in infertile women. Fert. Steril. *31:* 134–137 (1979).
18 Pfister, A.; Delezoide, A.; Vendrely, E.; Da Lage, G.: Détection et diagnostic des tumeurs du système nerveux central par l'étude cytologique du LCR et des liquides de kystes tumoraux. Archs. Anat. Cytol. path. *29:* 295 (1981).
19 Norgaard Pedersen, J.; Lindholm, J.; Labrechtsen, R.; Arends, J.; Diemer, N.H.; Riishede, J.: Alpha-fetoprotein and human chorionic gonadotropin in a patient with a primary intracranial germ cell tumor. Cancer *41:* 2315–2320 (1978).
20 Lee, S.H.; Sundaresan, N.; Jere, B.; Galicich, J.H.: Endodermal sinus tumor of the pineal region. Neurosurgery *3:* 407–411 (1978).

Dr. R. Pomarede, Unité d'Endocrinologie Pédiatrique et Diabète, Clinique Robert Debré, Hôpital Necker Enfants-Malades, 149, rue de Sèvres, F-75743 Paris Cedex 15 (France)

Postoperative and Post-Traumatic Diabetes insipidus

Joseph G. Verbalis[a], *Alan G. Robinson*[a], *Arnold M. Moses*[b]

[a] Department of Medicine, University of Pittsburgh, Pittsburgh, Pa.;
[b] State University Hospital, Syracuse, N.Y., USA

Neurosurgical procedures and head trauma have long been recognized as causes of diabetes insipidus (DI) [1, 2]. Early studies demonstrated a wide range of clinical incidences for postoperative DI depending upon the extent of neurosurgical resection required for mass lesions (as high as 100% transient DI and 60–80% permanent DI following palliative hypophysectomy) [3, 4]. Closed head trauma has been less frequently associated with DI, with incidences of <1% reported in large series [2]. In recent years several factors have contributed to an increased frequency of both postoperative and post-traumatic DI, and these are now the most common causes of DI in developed nations. Specifically, advances in the diagnosis of sellar and suprasellar masses, both via improved radiographic techniques and by more sensitive hormone markers of pituitary tumors, have greatly increased the volume of pituitary surgery. Although pituitary trans-sphenoidal microsurgery has reduced the frequency of DI following resection of smaller lesions [5], the net result is still a greater overall number of cases of postoperative DI. Secondly, both DI and anterior pituitary necrosis are more likely to be associated with severe degrees of head trauma involving loss of consciousness, skull fracture and more generalized neurological deficits [6, 7]. In the past, survival after such extensive injuries was rare, but with better techniques

---

[a] Supported by NIH Grant MO1 RR00056, NIH Grant AM 16166, NIH Grant NS 17138, The Veterans Administration Research Career Development Program.
[b] Supported by General Clinical Research Center Program NIH Grant RR 229, and The Veterans Administration.

of managing severe head trauma, more patients survive long enough to manifest deficits in posterior pituitary function.

Table I presents data on 28 patients from our experience with DI following head trauma. Virtually all patients had severe enough injury to be rendered unconscious and 86% had skull fracture. The male preponderance and young age of the patients may reflect the major cause of injury, automobile accidents. Some anterior pituitary deficiency and permanent neurologic deficit occurred in about 40% of the cases. DI was usually manifest by 12–24 h after injury. Unlike the therapy of stable chronic DI described by *Robinson and Verbalis* [this volume], where convenience to the patient is usually the prime consideration influencing therapy, patients with DI following neurosurgery or head trauma may for periods of time be unable to regulate their own fluid intake because of loss of consciousness, altered mental status, and/or disordered thirst regulation. In these patients appropriate therapeutic management is crucial to optimize their chances for health and survival. Familiarity with the diagnosis, time-course and management of postoperative and posttraumatic DI is therefore an important aspect of posterior pituitary dysfunction for practicing clinicians.

Pathophysiology

Experimental studies have long confirmed that damage to the supraoptic nucleus (SON) and paraventricular nucleus (PVN), or the hypothalamo-neurohypophyseal tracts, is necessary to produce significant DI. In both dogs [8] and monkeys [9] pituitary stalk section alone did not cause permanent DI unless more proximal lesions in the median eminence were produced. Early studies by *Lipsett* et al. [3] of palliative hypophysectomy in humans confirmed these experimental results: virtually all of 34 patients studied had transient DI following hypophysectomy; none of the 8 in whom the hypophysectomy was performed trans-sphenoidally without violation of the diaphragm sella developed permanent DI, but 22 of 26 patients in whom the pituitary stalk was completely cauterized via a transfrontal approach did have permanent DI [3]. Furthermore, as in experimental animals, the level of the stalk section is also a determinant of the development of DI in man. In a series of 24 patients receiving a low pituitary stalk section at the level of the diaphragm sella only 62% developed permanent DI [4], compared to an

Table I. Clinical data on 28 patients with post-traumatic diabetes insipidus

Parameter	Cases	Percentage
Age, years: 5–48 (21 median)		
Sex		
Male	21	75
Female	7	25
Trauma		
Automobile	20	73
Motorcycle	3	9
Bicycle	2	7
Other	3	11
Skull fracture	24	86
Unconscious (minutes to months)	27	96
Anterior pituitary dysfunction >6 months		
TSH deficiency	11	39
ACTH deficiency	10	36
Gonadotropin deficiency[1]	9	32
hGH deficiency[1]	4	14
Prolactin increased[1]	4	14
Permanent neurologic deficit	19	46
Cranial Nerves	11	39
First CN	4	14
Second CN	6	21
Other	5	18
Diffuse brain damage	4	14

[1] Not tested in all patients.

incidence of 80–100% with higher stalk cauterization [3] or clipping [10]. The pathophysiological basis for these observations is related to the number of vasopressin-secreting magnocellular neurons which remain viable following the lesion or injury. Figure 1 shows the relationship between the development of polyuria and the percent of residual normal cells in the SON of dogs with experimental stalk lesions. If >15% of magnocellular neurons remained viable significant DI did not occur [11]. Similar neuroanatomical studies in humans following stalk sections have not demonstrated so clear a relationship between the percent of remaining magnocellular neurons and the development of DI, but have shown decreased mean residual SON ($\sim$15%), and to a lesser extent PVN

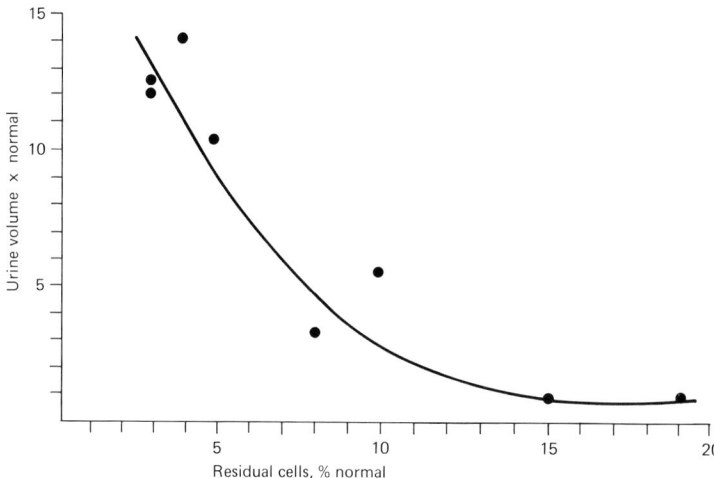

Fig. 1. Relationship between the degree of polyuria and the percentage of residual magnocellular cells in the SON following pituitary stalk section in dogs. Reproduced from ref. [11] by permission of the publishers.

(~50%), cell counts following pituitary stalk section [12]. Both the human and animal studies demonstrated retrograde degeneration of magnocellular neurons following stalk lesions, with cell swelling, dissolution of Nissl substance, vacuolization, glial reaction and eventual cell loss which was complete in the SON by 6–8 months following stalk section, but continued in the PVN for up to 22 months [9, 12]. Thus, the major determinant of whether a given lesion causes DI is the extent of retrograde axonal and neuronal degeneration. If <10–20% of magnocellular neurons remain viable, permanent DI will occur, and the greater the cell loss beyond this threshold the more severe will be the DI. The relationship between the level of stalk section and the occurrence of DI is understandable since lesions more proximal to the cell bodies in the SON and PVN are more likely to cause neuronal death. To some extent this may also account for the greater degrees of cell loss in the SON than the PVN with lesions of the posterior pituitary and the pituitary stalk, although a more likely explanation is that virtually all magnocellular neurons of the SON project to the posterior pituitary while many of the magnocellular neurons of the PVN project to other areas of the brain such as the median eminence, hippocampus, brainstem and spinal cord [13]. It should also be remembered that factors other than the level of axonal damage may influence the extent of axonal degeneration following neurosurgical or traumatic lesions, including: altered blood flow to the pituitary stalk secondary to edema and/or traction, decrease in blood flow to the SON and PVN secondary to hypotension or hypovolemia, and possibly increased metabolic activity of magnocellular neurons caused by stimulated secretion of vasopressin. Finally, as in most human disease, considerable individual variation exists with regard to cellular response to injury and noxious stimuli. An occasional patient may develop permanent DI following a pure posterior pituitary lesion, and another may not develop DI even after extensive trauma, and even section, of the pituitary stalk.

Patterns of DI Following Neurosurgery or Trauma

Three distinct patterns of DI may follow neurosurgical or traumatic damage to the neurohypophysis as shown in figure 2. The most common pattern is transient DI with fairly rapid resolution over several days, accounting for approximately 50–60% of all cases of postoperative and post-traumatic DI. The onset of polyuria is abrupt, almost always within 24 h of the initial insult. Both the intensity and duration are quite variable. Most cases resolve by 3–5 days but DI occasionally persists for several weeks. This is the usual pattern when DI occurs following resection of pituitary adenomas confined to the sella, with an incidence of 10–20% [3, 4]. Local instillation of toxic agents, such as absolute alcohol applied to the margins of a resected pituitary adenoma, or trauma to the posterior pituitary may be responsible for the DI. Retrograde axonal degeneration must be minimal, since most such patients show no evidence of decreased vasopressin secretory capacity after resolution of the initial DI. The next most common pattern is permanent DI, accounting for 30–40% of postoperative and post-traumatic DI. As noted previously, production of this pattern requires more proximal damage to the pituitary stalk, median eminence or hypothalamus, and is therefore more frequent following transfrontal resection of larger masses with suprasellar components. Permanent DI may be complete or partial depending upon the number of viable vasopressin-secreting neurons. In addition, 'permanent' is somewhat of a misnomer and might be replaced by 'prolonged' DI, since some of these patients may show eventual resolution of DI. The third pattern, the 'triphasic response', was first described following bilateral lesions of the supraopticohypophyseal tracts in cats (in what

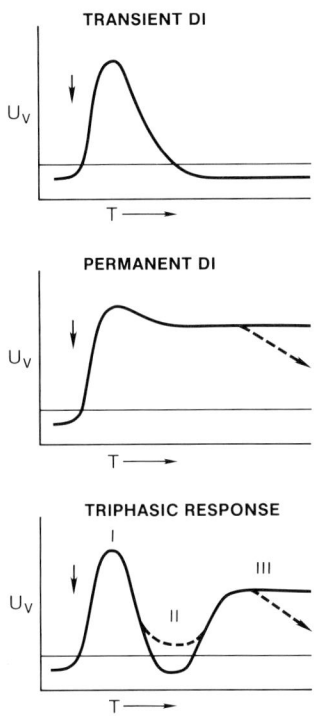

Fig. 2. Diagrammatic summary of the major patterns of postoperative and post-traumatic DI. The abcissa represents time following the initial injury (arrow) and the ordinate urinary volume relative to a hypothetical 'normal' urine output of 2–3 l/day (solid line).

was actually the first demonstration that hypothalamic lesions could cause DI) [14]. It is the least frequent, but the most interesting and potentially dangerous pattern. The occurrence of this pattern has been reported in as many of 70–80% of patients following complete stalk section [10] but most report a 10–30% incidence in such patients [3, 4]. We find this pattern in no more than 5–10% of all cases of postoperative and post-traumatic DI. The triphasic response is characterized by abrupt onset of DI as with the previous two patterns, but DI is then followed by a period of antidiuresis typically beginning several days after the initial insult and lasting 2–14 days before eventual return of DI completing the third phase. Earlier studies interpreted the second phase as a 'normal interphase' with normal regulation of fluid balance, but *O'Connor* [15] clearly demonstrated that this phase was actually a period of antidiuresis associated with inappropriate release of antidiuretic hormone. His and other studies in animals have found that the triphasic response depended on the presence of residual neurohypophyseal tissue following pituitary stalk section. Complete removal of the posterior pituitary, pituitary stalk and median eminence resulted in permanent DI without a phase of antidiuresis [8, 9, 11]. Furthermore, reimplantation of the posterior pituitary into the sella turcica [16] or beneath the kidney capsule [17] reproduced to varying degrees the antidiuresis seen in the second phase. Thus, in animal studies degeneration of residual denervated neurohypophyseal tissue with release of vasopressin is a valid explanation for the antidiuretic phase, and when all such tissue is destroyed DI recurs. Data from humans, however, has not been so consistent. Numerous cases of triphasic response have occurred even after complete hypophysectomy. In the series of *Ikkos* et al. [10] 18 of 24 patients had some form of a triphasic response after clipping and sectioning the pituitary stalk and evacuating all pituitary contents from the sella. Such cases have sometimes been cited as evidence that the cause of the triphasic response in humans is different than in animals since residual posterior pituitary tissue does not appear to be as crucial in humans. In fact, such cases are not that surprising since in animal studies removal of virtually all neurohypophyseal tissue, including the stalk and median eminence, was necessary to completely abolish any triphasic response. Thus, even after hypophysectomy in humans, sufficient neurohypophyseal tissue may remain to cause nonspecific release of stored neurohypophyseal peptides and a period of antidiuresis. There seems little doubt that the occurrence of a triphasic response is absolutely dependent upon damaged neurohypophyseal tissue which degenerates in situ, and the magnitude and duration of the second, or antidiuretic, phase is roughly proportional to the amount of such residual tissue. It follows that the cases most likely to show this pattern involve pituitary stalk trauma without hypophysectomy such as resection of purely suprasellar masses, and additionally traumatic DI caused by pituitary stalk trauma, or even traumatic section, with *all* neurohypophyseal tissue left in place. Figure 3 shows one example of a patient with closed head trauma and a basilar skull fracture after an automobile accident. A typical triphasic pattern was observed with severe DI initially, followed by inappropriate antidiuresis resulting in hyponatremia, $[Na^+] = 125-130$ mEq/l, requiring fluid restriction and therapy with a hypertonic saline-lasix regimen, and finally return of DI. This patient survived with permanent DI and anterior pituitary dysfunction. Many patients with post-traumatic DI may have some degree of triphasic re-

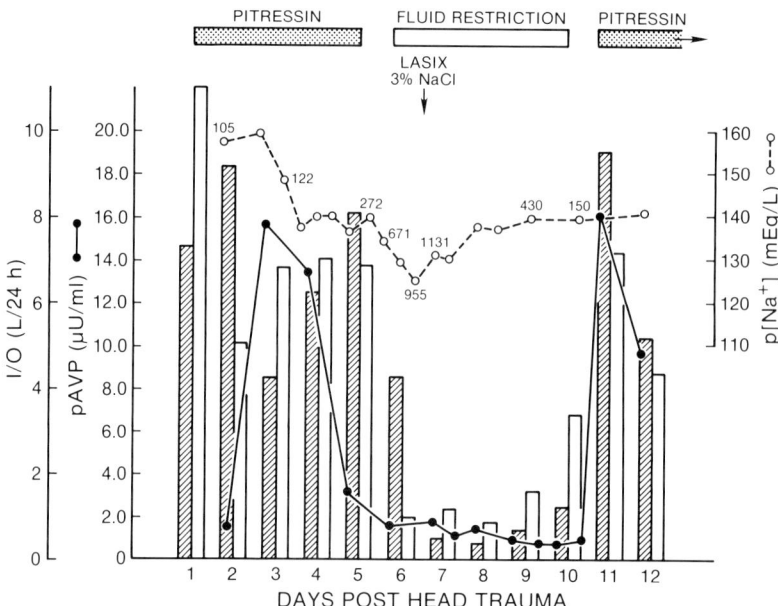

Fig. 3. Clinical summary of a 28-year-old male with a triphasic response following head trauma. Total fluid intake (solid bars) and urine output (open bars) is shown for 12 days following head trauma. Plasma sodium concentration (dotted line) falls to hypoosmolar levels coincident with decreasing urine output and inappropriate urinary concentration (urine osmolarity is shown above corresponding plasma sodium concentrations). Therapy is summarized above the figure showing that inappropriate antidiuresis and hyponatremia developed 24 h following the last dose of Pitressin and persisted for 4 days in the absence of any antidiuretic therapy before eventual return of DI on days 10–11. Plasma AVP levels (solid line) are elevated during the periods of Pitressin therapy, but fall to low levels (<2 μU/ml) during the second antidiuretic phase.

sponse. The time course of the antidiuretic phase is somewhat more predictable than its occurrence. In 4 patients followed at the University of Pittsburgh, antidiuresis began 4–8 days following neurosurgery or trauma, and lasted 5–6 days (table II). This is in agreement with earlier human studies which showed a 4- to 5-day onset and 5- to 6-day duration of antidiuresis after stalk section [10], and is in contrast to animal studies which have shown much greater variability in both the onset and duration of the second phase [8, 9, 14, 16]. Even though the presence of residual neurohypophyseal tissue seems critical for production of a triphasic response, it has been difficult to measure inappropriate release

Table II. Triphasic response in humans

Clinical characteristics				Phase II		
Case	p[Na+], mEq/l			onset (days)	duration (days)	Uosm (max.)
	I	II	III			
1	146±2	124±2	141±1	4.5	5.6	490±57
2	145±1	120±3	139±1	8.4	6.2	855±45
3	147±1	134±1	141±1	6.9	5.8	462±11
4	145±4	130±1	141±1	4.8	4.7	759±88
	146±1	127±3	140±1	6.2±0.9	5.6±0.3	642±98

Neurohypophyseal hormone levels

Phase	[Na+]	pAVP, µU/ml	pOT, µU/ml	hAVP-Np, ng/ml
I	145.7±0.5	1.60±0.06	0.48±0.05	0.22±0.12
II	127.4±3.1	1.54±0.11	0.49±0.03	0.16±0.07
III	140.4±0.4	1.38±0.10	0.45±0.11	0.11±0.06

of neurohypophyseal peptides during the second phase. Plasma vasopressin levels in the patient shown in figure 3 were high during treatment with arginine vasopressin (AVP), but during the second phase plasma AVP levels were at the limit of detection at a time when urine osmolality was 800–1,000 mosm/kg. Table II summarizes similar serial hormonal measurements in the 4 patients with a triphasic pattern and shows no elevation of any neurohypophyseal peptide (AVP, oxytocin or neurophysin) during the period of antidiuresis. This is not what would be expected with degeneration of residual neurohypophyseal tissue and subsequent release of all neurohypophyseal peptides stored in neurosecretory granules. Therefore, several questions remain about the cause of antidiuresis in humans, and specifically whether biologically active but immunologically unrecognized forms of neurohypophyseal hormones are released by degenerating neurohypophyseal tissue. Alternatively, there may be enhanced sensitivity of the renal collecting tubules to low levels of vasopressin, but it is not clear what might cause such an effect.

Diagnosis

The diagnosis of DI in a postoperative or post-traumatic setting is generally straightforward, although two important points must always be considered in the evaluation of all such cases. The first is to be certain that polyuria represents DI and is not secondary to either fluid overload or an osmotic diuresis. Fluid replacement in excess of body losses, both intraoperatively and postoperatively, is normally retained to varying degrees depending upon the magnitude of sympathetic stimulation, secretion of aldosterone and vasopressin, and alterations in renal blood flow. Eventually excretion of excess retained fluid occurs, usually abruptly within a several-hour period. This physiologic diuresis may be confused with the onset of DI and antidiuretic therapy begun prematurely, thwarting appropriate excretion of the retained fluid. Thereafter, each time the antidiuretic effects of a given dose diminishes the ensuing diuresis may be interpreted as evidence of recurrence of DI and another dose of an antidiuretic agent administered. This pattern of 'chasing one's tail' results in needless therapy and complicates management of postoperative and head trauma patients. It can be avoided quite easily by requiring that the usual criteria for the diagnosis of DI be rigorously applied in the postoperative setting, i.e. excretion of inappropriately dilute urine in the presence of true hyperosmolality of plasma. In practice one should therefore avoid making a diagnosis of DI even in the postoperative setting until plasma [Na$^+$] is >143–145 mEq/l with continuing diuresis of hypotonic urine. The second situation potentially capable of causing an erroneous diagnosis of DI is the presence of an osmotic diuresis. This can be seen in the postoperative setting either from the use of mannitol as an osmotic diuretic, or more commonly from glucosuria secondary to the hyperglycemia sometimes induced by high doses of corticosteroids administered to patients undergoing pituitary surgery or following head trauma. Less commonly, diuresis following obstructive uropathy or nonoliguric acute renal failure may cause a similar picture. The above possibilities may be eliminated by a review of medications and by checking for glucosuria – and when appropriate for the absence of a rising blood urea nitrogen (BUN) and serum creatinine level.

Two other situations may mask the presence of DI, and both are more likely to be a problem with post-traumatic than postoperative DI. The first is the occurrence of a triphasic response. A patient presenting in the second, or antidiuretic, phase may superficially appear to have normal regulation of vasopressin secretion unless excess fluid intakes cause hyponatremia. The eventual return of DI in the third phase is then unexpected and often poorly diagnosed. Since the antidiuretic phase of the triphasic response is virtually always preceded by DI (in 10 years the authors observed only 1 case where a patient went directly into the second phase postoperatively without preceding DI), initial antidiuresis is unlikely to be unexpected in a postoperative setting where the patient is followed carefully from the day of surgery. However, patients with head trauma may not be seen until after a delay of several days, either because of a later transfer from another hospital or alternatively the absence of any initial hospitalization because of an asymptomatic interval when head trauma is less severe. Furthermore, the large fluid volumes used in the management of trauma patients with hemorrhage and hypotension sometimes make a diagnosis of DI difficult immediately following the injury. Thus, in all patients with recent severe head trauma the possibility of neurohypophyseal damage should be considered despite normal plasma and urine osmolality. Any patient with head trauma having hyponatremia with inappropriately concentrated urine should be assumed to be in the second phase of a triphasic response, treated accordingly, and a third phase anticipated.

The second situation which may mask underlying DI is the presence of anterior pituitary deficiency, specifically hypocortisolemia. The inability of patients with cortisol insufficiency to excrete a maximally dilute urine is well known [20, 21], and amelioration of DI following the development of anterior pituitary insufficiency is well described. A graphic example of a patient following palliative pituitary stalk section without anterior pituitary replacement therapy [10] is shown in figure 4. There was apparent normal water balance for several weeks, but DI became obvious immediately after initiation of corticosteroid therapy. This effect is not likely to be misinterpreted in the postoperative setting since most patients with sellar and suprasellar surgery are empirically given high doses of corticosteroids. However, patients with extensive head trauma may not be so treated, and in such patients pituitary damage may go unsuspected because hypocortisolemia secondary to adrenocorticotropic hormone (ACTH) deficiency may mask the presence of DI. Note that in our series, table I, more than one third of the cases were eventually proven to have adrenal insufficiency. If there is any suspicion of cortisol deficiency, patients with severe head trauma should have a

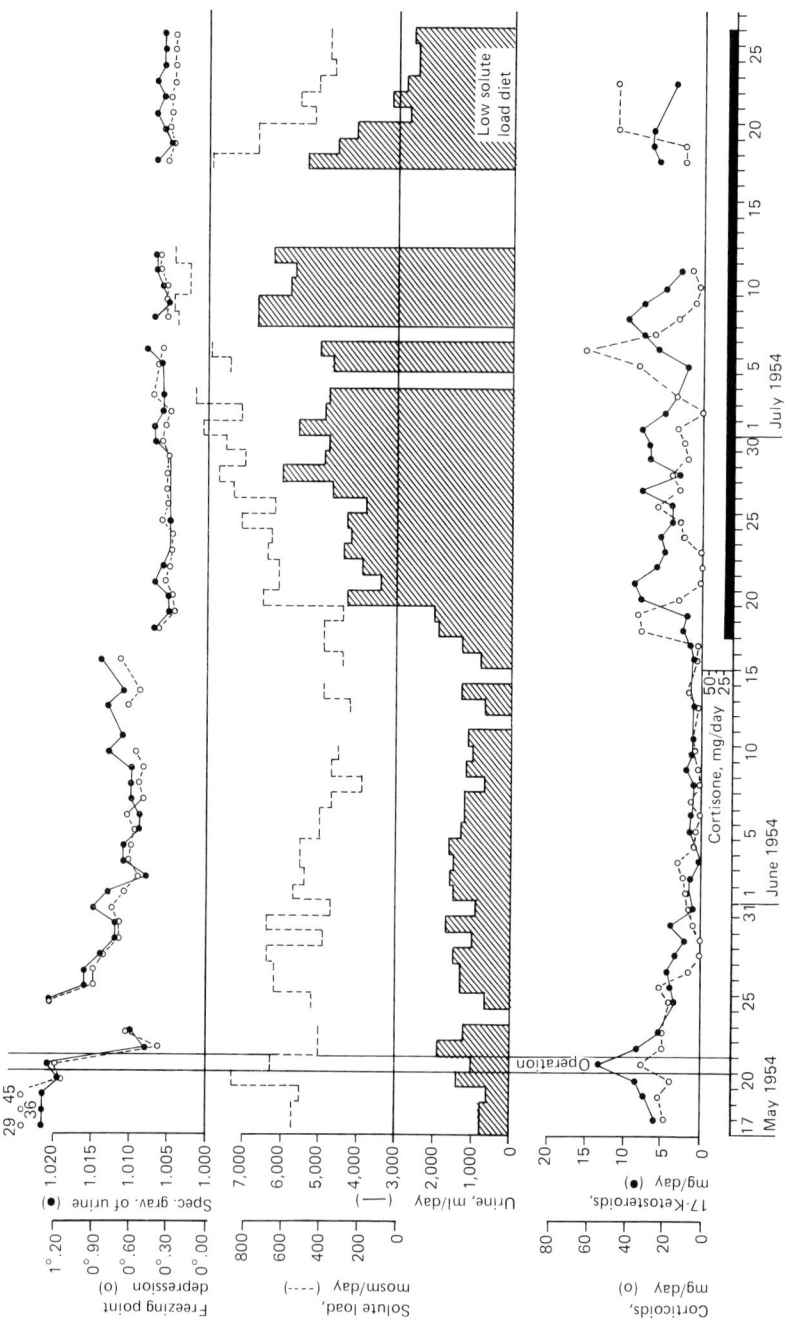

serum cortisol level drawn and then be empirically placed on 'stress' doses of corticosteroids (hydrocortisone 100 mg i.v. infusion every 8 h or equivalent amounts of nonsodium retaining glucocorticoids).

Treatment

As with all types of DI, the mainstay of therapy is ingestion or administration of sufficient fluid to replace urinary losses. Fluid replacement should be in the form of low Na^+ fluids, such as 5% dextrose in water, since the deficits with DI are primarily free water rather than salt. However, treatment with fluid replacement alone is often difficult in the postoperative and post-traumatic setting because: (1) such patients frequently are unable to regulate oral fluid intake because of changes in level of consciousness or abnormalities of thirst regulation, and fluid replacement must be given intravenously; (2) the DI in this setting is frequently of greater severity than in chronic DI, and (3) the tenuous state of fluid balance makes inadequate replacement potentially dangerous since failure to keep pace with such losses can quickly result in volume depletion and hyperosmolality. Therefore, treatment with antidiuretics is indicated in all cases of postoperative and post-traumatic DI. The timing and choice of antidiuretic agents is largely a matter of preference as long as the goal of maintaining adequate overall fluid balance is achieved. In general, the risk of overtreatment producing free water retention and hyponatremia, which can cause increased cerebral edema and seizures, is greater in the postoperative or post-traumatic patient with DI than is the risk of undertreatment. Consequently, treatment with antidiuretic agents should await a definite diagnosis as described previously. Even after the diagnosis is confirmed many surgeons choose to treat only with intravenous fluids while awaiting possible recovery although, as already mentioned, this is often difficult. The authors begin antidiuretic therapy as soon as DI is confirmed and adjust the patients' oral and intravenous fluid intake to maintain fluid balance. In the past,

---

Fig. 4. Effect of hypocortisolism to mask DI following pituitary stalk section and hypophysectomy in a 59-year-old woman with metastatic breast carcinoma. For the first 28 days postoperatively no evidence of DI was seen, but following initiation of cortisone therapy on day 29 marked polyuria with decreased urinary concentration ensued. Reproduced from ref. [10] by permission of the publishers.

one had to choose between a short-acting antidiuretic agent, such as aqueous Pitressin, or a long-acting one, such as Pitressin tannate in oil. Use of aqueous Pitressin requires frequent injections (5 U subcutaneously every 3–6 h) and may cause vasoconstriction, with some rare cases of angina reported in patients with ischemic heart disease. Pitressin tannate in oil (5–10 U intramuscularly) produces a steadier control of polyuria lasting 48–72 h and has less vasoconstrictive effect because of the delayed absorption and lower levels of vasopressin, but has greater risk of water intoxication because of the prolonged period of antidiuresis. In addition, use of a long-acting agent may result in uncertainty as to whether sustained antidiuresis is secondary to the onset of a second phase or to prolonged Pitressin effect. Consequently, 1-desamino-8-D-arginine-vasopressin (DDAVP) is preferred in the postoperative and post-traumatic setting. This antidiuretic analog produces antidiuresis from 12–24 h and has virtually no pressor effects (allowing safety in patients with heart disease). In patients undergoing a transfrontal neurosurgical approach, DDAVP may be administered intranasally at a dose of 0.1 ml (10 µg) every 8–24 h; it can be administered by a nurse or physician if the patient is unable to administer his own dose. In patients undergoing transsphenoidal surgery with nasal packing postoperatively, or head trauma involving facial injury, parenteral DDAVP is the treatment of choice. Experience with this drug in Western Europe has confirmed its safety and efficacy, and we have recently evaluated parenteral DDAVP in patients with postoperative DI. A dose of 2 µg subcutaneously maintains antidiuresis for 12–24 h without side effects.

Attention must always be given to fluid replacement when antidiuretic agents are used. If the patient is awake and able to respond to thirst, the patient's own thirst is the best guide to water replacement. One exception to this is the first few days following trans-sphenoidal hypophysectomy. Because of nasal packing and congestion, such patients are forced to mouth breath and develop dry mouth and excessive thirst. They must be reminded to drink only in response to thirst and to use ice chips or other mouth-lubricating agents (hard candy) for comfort. If the patient is unable to respond to thirst, either because of decreased consciousness or because of hypothalamic damage to thirst centers, fluid balance must be maintained by intravenous infusion. Urine osmolality and serum sodium levels must be followed every several hours during the initial therapy and at least daily until stabilization or resolution of diabetes insipidus. In general, the serum sodium concentration is the best

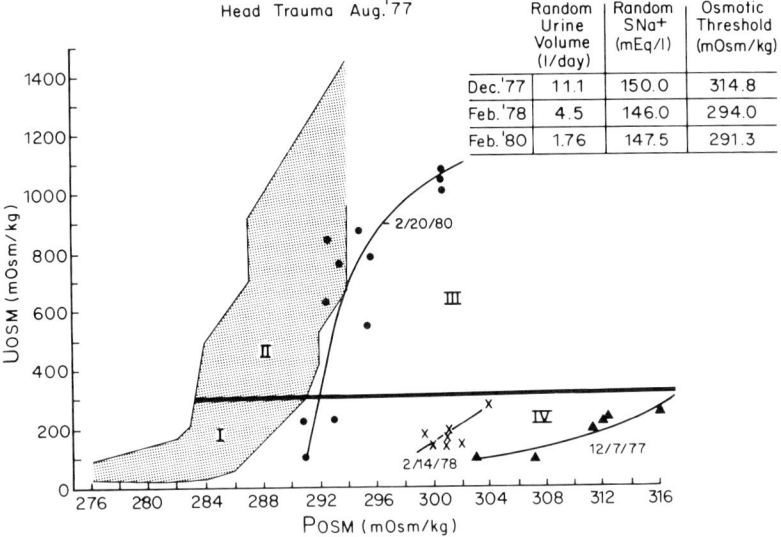

Fig. 5. Plasma vs urine osmolality relationships over a 26-month period in a 16-year-old male patient with traumatic DI. During this time the plasma vs urine osmolality relationship progressively improved. Urine volumes and osmotic threshold values also improved, while plasma sodium improved slightly but remained elevated. Shaded area represents data obtained from normal subjects. Horizontal line represents equal plasma and urine osmolality.

single guide to therapy with antidiuretic agents and water. To avoid fluid retention and hyponatremia, each dose of DDAVP should be given only after recurrence of polyuria, confirming continued DI. Once the diagnosis of DI has been made, it is neither necessary nor desirable that the patient actually become hyperosmolar before each subsequent dose of antidiuretic therapy, but simply to show continued diuresis of hypotonic urine. Excretion of 200–250 ml/h with urine osmolality <200 (or specific gravity <1.005) demonstrates the need for retreatment with DDAVP. With this regimen the risk of cumulative fluid retention and hyponatremia is minimized, and resolution of the DI, or development of the antidiuretic phase of a triphasic pattern, can be promptly recognized. Monitoring of each day's intake and output is also useful to detect early development of a triphasic pattern. Positive daily fluid balance (≥ 2 l) suggests the possibility of inappropriate antidiuresis, and antidiuretic therapy should be witheld to see if polyuria indicates continued DI. It cannot be overly stressed that patients undergoing a triphasic pattern usually get into difficulty only when inappropriate fluid therapy is given without monitoring of daily fluid balances.

A final practical therapeutic consideration is that any patient with postoperative or post-traumatic DI should be assumed to have anterior pituitary insufficiency. Accordingly, all such patients should be covered with appropriate doses of corticosteroids (hydrocortisone 100 mg every 8 h during stress) until anterior pituitary function can be thoroughly evaluated. Replacement of other pituitary hormones is not critical acutely, although thyroxine may be added if thyroid function tests and thyroid-stimulating hormone (TSH) become depressed after neurosurgery or head trauma.

Long-Term Resolution

Patients with persistent DI >2 weeks following head trauma or neurosurgery have prolonged DI and should be treated accordingly for partial or complete DI as described by *Robinson and Verbalis* [this volume]. Amelioration of DI, if it occurs, usually takes place within the first year. However, we have had patients who improved to the point of not requiring therapy after intervals of 2, 3 and 10 years. The data on 1 of our cases who continued to improve for at least 2.5 years are depicted in figure 5 [22]. During that period of time his Uosm improved in relation to his Posm, his urine volume decreased from 11.1 to 1.76 l/day, while his saline-induced osmotic threshold (vide infra) decreased from 314.8 to 291.3 mosm/kg. Therapy was discontinued between the second and third dehydration tests. The patient's serum sodium concentration remained slightly elevated. When patients like this have partial AVP deficiency, as did this patient between the second and third dehydration

Fig. 6. Reorganization of the human pituitary stalk following hypophysectomy. The top panel represents the pituitary stalk of a normal subject stained for CAH-positive material and the bottom panel a similarly stained normal posterior pituitary showing significantly greater amounts of neurosecretory material. The middle panel shows a sagittal section through the pituitary stalk of a patient 18 months following hypophysectomy. This pituitary stalk now has an increased amount of CAH-positive material clustered around blood vessels such that it bears a closer resemblance to normal posterior pituitary tissue than normal pituitary stalk tissue. Reproduced from ref. [25] by permission of the publishers.

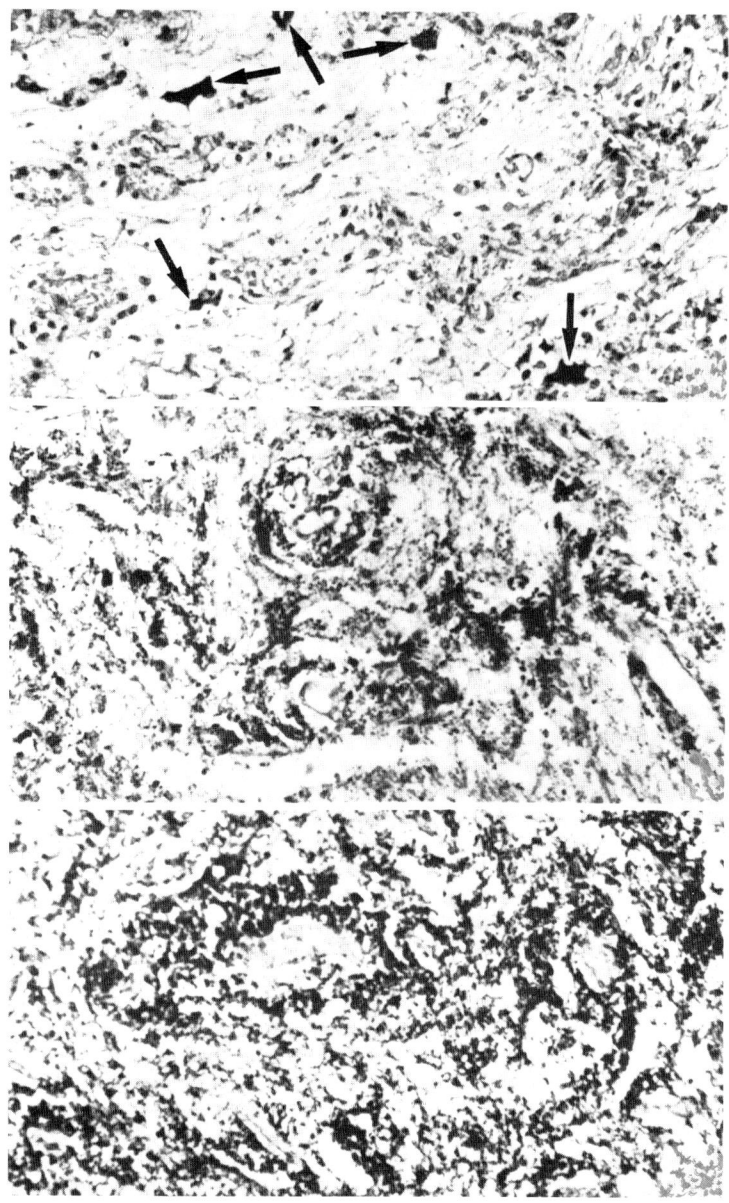

tests, the sensitivity of the thirst mechanism determines the degree of polydipsia and polyuria. They may have severe polydipsia and polyuria if the thirst mechanism is sensitive, or essential hypernatremia if the thirst sensation is blunted. The reason for amelioration and resolution is apparent from pathological and histological examination of neurohypophyseal tissue following pituitary stalk section in both animals [23, 24] and humans [25]. Neurohypophyseal neurons which have intact perikarya are able to regenerate axons and form new nerve terminal endings capable of releasing vasopressin into nearby capillaries. In animals this may be accompanied by a bulbous growth at the end of the severed stalk which represents a new, albeit small, neural lobe. In humans the regeneration process appears to proceed more slowly, and formation of a new neural lobe has not been noted. Nonetheless, histological examination of the severed human stalk shown in figure 6 [25] demonstrates the reorganization of neurohypophyseal fibers in the stalk of a patient 18 months after hypophysectomy (middle panel) and shows neurosecretory granules in close proximity to nearby blood vessels, more closely resembling the histology of a normal posterior pituitary (bottom panel) than a normal pituitary stalk (top panel).

References

1 Randall, R.V.; Clark, E.C.; Dodge, H.W.; Jr.; Love, J.G.: Polyuria after operation for tumors in the region of the hypophysis and hypothalamus. J. clin. Endocr. Metab. *20:* 1614–1621 (1960).
2 Porter, R.J.; Miller, R.A.: Diabetes insipidus following closed head injury. J. Neurosurg. Psychol. *2:* 258–262 (1948).
3 Lipsett, M.B.; MacLean, I.P.; West, C.D.; Li, M.C.; Pearson, O.H.: An analysis of the polyuria induced by hypophysectomy in man. J. clin. Endocr. Metab. *16:* 183–195 (1956).
4 Sharkey, P.C.; Perry, J.H.; Ehni, G.: Diabetes insipidus following section of hypophyseal stalk. J. Neurosurg. *18:* 445–460 (1961).
5 Hardy, J.: Transsphenoidal microsurgical treatment of pituitary tumors; in Linfoot, Recent advances in the diagnosis and treatment of pituitary tumors (Raven Press, New York 1979).
6 Goldman, K.P.; Jacobs, A.: Anterior and posterior pituitary failure after head injury. Br. med. J. *5217:* 1924–1926 (1960).
7 Daniel, P.M.; Treip, C.: The pathology of the pituitary gland in head injury; in Gardiner-Hill, Modern trends in endocrinology, pp. 55–69 (Hoeber, New York 1961).
8 Pickford, M.; Ritchie, A.E.: Experiments on the hypothalamic-pituitary control of water execretion in dogs. J. Physiol., Lond. *104:* 105–128 (1945).

9 Magoun, H.W.; Fisher, C.; Ranson, S.W.: The neurohypophysis and water exchange in the monkey, Endocrinology 25: 161–174 (1939).
10 Ikkos, D.; Luft, R.; Olivecrona, H.: Hypophysectomy in man: effect on water excretion during the first two postoperative months. J. clin. Endocr. Metab. 15: 553–567 (1955).
11 Heinbecker, P.; White, H.L.: Hypothalamico-hypophyseal system and its relation to water balance in the dog. Am. J. Physiol. 133: 582–593 (1941).
12 MacCubbin, D.A.; VanBuren, J.M.: A quantitative evaluation of hypothalamic degeneration and its relation to diabetes insipidus following interruption of the human hypophyseal stalk. Brain 86: 443–464 (1963).
13 Swanson, L.W.; Sawchenko, P.E.: Hypothalamic integration: organization of the paraventricular and supraoptic nuclei; in Cowan, Shatter, Stevens, Champson, Annual review of neuroscience, vol. 6, pp. 269–324 (Annual Reviews, Palo Alto 1983).
14 Fisher, C.; Ingram, W.R.; Ranson, S.W.: Diabetes insipidus and the neurohumoral control of water balance. A contribution to the structure and function of the hypothalamico-hypophyseal system (Edward Brothers, Ann Arbor 1938).
15 O'Connor, W.J.: The normal interphase in the polyuria which follows section of the supraoptico-hypophyseal tracts in the dog. Q. Jl exp. Physiol. 37: 1–10 (1952).
16 Hollinshead, W.H.: The interphase of diabetes insipidus. Mayo Clin. Proc. 39: 92–100 (1964).
17 Laszlo, F.A.; De Wied, D.: Antidiuretic hormone content of the hypothalamo-neurohypophyseal system and urinary excretion of antidiuretic hormone in rats during the development of diabetes insipidus after lesions in the pituitary stalk. J. Endocr. 36: 125–137 (1966).
18 Miller, M.; Moses, A.M.: Potentiation of vasopressin action by chlorpropamide in vivo. Endocrinology 86: 1024–1027 (1970).
19 Pokracki, F.J.; Robinson, A.G.; Seif, S.M.: Chlorpropamide effect: measurement of neurophysin and vasopressin in humans and rats. Metabolism 30: 72–78 (1981).
20 Mandell, I.N.; DeFronzo, R.; Robertson, G.L.; Forrest, J.N.: Role of plasma arginine vasopressin in the impaired water diuresis of isolated glucocorticoid deficiency in the rat. Kidney int. 17: 186–196 (1980).
21 Linas, S.L.; Berl, T.; Robertson, G.L.; Aisenbrey, G.A.; Schrier, R.W.; Anderson, R.J.: Role of vasopressin in the impaired water excretion of glucocorticoid deficiency. Kidney int. 18: 58–67 (1980).
22 Moses, A.M.: Long-standing posttraumatic diabetes insipidus; in Medical grand rounds, vol. 2, pp. 117–128 (Plenum Publishing, New York 1983).
23 Adams, J.H.; Daniel, P.M.; Prichard, M.M.L.: Transaction of the pituitary stalk in man: anatomical changes in the pituitary glands of 21 patients. J. Neurol. Neurosurg. Psychol 29: 545–555 (1966).
24 Daniel, P.M.; Prichard, M.M.L: Regeneration of hypothalamic nerve fibers after hypophysectomy in the goat. Acta endocr., Copenh. 64: 696–704 (1970).
25 Daniel, P.M.; Prichard, M.M.L.: The human hypothalamus and pituitary stalk after hypophysectomy or pituitary stalk section. Brain 95: 813–824 (1972).

Joseph G. Verbalis, MD, Department of Medicine, University of Pittsburgh, Pittsburgh, PA 15261 (USA)

Diabetes insipidus and Pregnancy

Janet A. Amico

Department of Medicine, University of Pittsburgh School of Medicine, Pittsburgh, Pa., USA

Introduction

A deficiency of vasopressin or resistance to its effect on the kidney may occur in pregnancy and will produce polyuria and polydipsia. The coexistence of diabetes insipidus and pregnancy is rare [5, 17, 21–23], probably because the former is uncommon [41]; but an alternate possibility is that the diabetes insipidus may have an adverse effect on the ability of an individual either to become pregnant or to sustain a normal pregnancy. The possibility that diabetes insipidus will affect fertility, gestation, parturition, or lactation is reasonable because the synthesis and/or release of oxytocin is closely linked anatomically to vasopressin. Further, some of the biochemical and physiological changes which occur during pregnancy might alter the course of diabetes insipidus.

Osmoregulation in Normal Pregnancy

Physiologic changes which occur in normal gestation include decreases in the osmotic thresholds for both secretion of vasopressin and for thirst [11, 12]. Pregnant women have decreased plasma osmolality of 8–10 mosm/kg below that of non-pregnant women [11]. The change in osmolality occurs in the first trimester and continues throughout gestation. Despite the lower osmolality, both basal and stimulated levels of vasopressin are similar in pregnant and postpartum women [12]. In response to dehydration or hypertonic saline, the increase in levels of vasopressin in plasma and in osmolality of urine is comparable to that observed postpartum [12]. The correlation which exists between plasma vasopressin and plasma osmolality in non-pregnant women is maintained in pregnant women, but with lower osmotic thresholds for both secretion of vasopressin and thirst [12], and pregnant women maintain a lower osmolality within a narrow range.

Pathology

Central and nephrogenic diabetes insipidus both occur in pregnant women, but the central form has received the most attention. The lesion producing central diabetes insipidus is usually in the supraopticohypophyseal tract [15, 19, 20, 25], and must be above the median eminence or in the supraoptic nucleus for the disorder to be permanent [15]. As described in *Verbalis* et al. [this volume], severing the supraoptico-hypophyseal tract below the median eminence, or removing the posterior lobe of the pituitary gland, produces only transient diabetes insipidus because sufficient vasopressin is produced from regeneration of nerve terminals in the median eminence or stalk [19, 20, 25]. Progressive loss of neurons in the supraoptic nucleus results in increasingly severe vasopressin deficiency [19, 20, 25]. The paraventricular nucleus, which lies in close proximity to the supraoptic nucleus, synthesizes less vasopressin but is a major site of synthesis of oxytocin [41]. The neurons of the paraventricular nucleus are less frequently involved in disorders which affect the supraoptic nucleus [42]. The reason is in part because the axons from the paraventricular nucleus have a longer tract to the posterior pituitary and are less susceptible to retrograde degeneration when the stalk is cut; also, the paraventricular nucleus may be less vulnerable to blood-borne injury or hemorrhage because its capillary blood supply is less rich than that of the supraoptic nucleus [34, 42]. Production and secretion of oxytocin and vasopressin are not always diminished in parallel. The heterogeneity in structural lesions of the neurohypophysis results in variable degrees of deficiency of vasopressin and oxytocin. Renal refractoriness to vasopressin also occurs in pregnancy but will not be discussed in detail as it has been reviewed recently [3].

Effect of Diabetes insipidus on Pregnancy

Effect of Diabetes insipidus on Fertility

An isolated deficiency of vasopressin without concomitant loss of hormones of the anterior pituitary does not result in altered fertility. *Blotner and Kunkel* [5] reviewed all published cases of diabetes insipidus complicating pregnancy prior to 1942. Several women with diabetes insipidus had repeated pregnancies [5], and the authors concluded that patients with diabetes insipidus are usually normally fertile. Subsequent publications also cited examples of women with diabetes insipidus who have had multiple pregnancies [21, 22]. The possibility of altered fertility in patients with central diabetes insipidus is raised because of a possible deficiency in oxytocin. There is a peak secretion of oxytocin at mid-cycle in humans which might indicate that the hormone has a role in ovulation [2]. However, synthesis of oxytocin is preserved in cases of central diabetes insipidus in which the neurons of the paraventricular nucleus are spared; as a definitive role for oxytocin in ovulation has not been established, the absence of the hormone as a possible cause of infertility is speculative.

Effect of Diabetes insipidus on Gestation

With the exception of polyuria and polydipsia, gestation is uncomplicated [5, 21, 22]. Urinary tract obstruction has been reported in a few individuals, probably due to the large volumes of urine [5, 26, 27, 39] and the obvious abdominal mass. No increase in the number of spontaneous abortions has been noted [8]. Reports of the association of preeclampsia or toxemia with diabetes insipidus are rare [4, 7, 14]. Interestingly, diabetes insipidus improved when the preeclampsia became manifest even when the diabetes insipidus was deteriorating prior to the onset of the preeclampsia [4]. A decrease rather than an increase in the requirement for vasopressin replacement has been noted when preeclampsia develops in patients with preexisting diabetes insipidus [4], perhaps due to decreased renal perfusion. The improvement persists for several weeks postpartum making it unlikely that factors from the placenta contribute to the clinical improvement [4].

Effect of Diabetes insipidus on Labor and Delivery

Physiologic concentrations of vasopressin do not stimulate uterine muscle, but oxytocin has a potent stimulatory effect upon the parturient uterus [6]. Uterine receptors for oxytocin increase as pregnancy advances [16, 36], but the role that oxytocin plays in the onset of labor and parturition is not known. During labor, a quantitative rise in oxytocin in human plasma is not regularly found [24], but this does not disprove that the qualitative effects of the hormone may be important for normal labor [16, 36]. Other factors, especially prostaglandins, are important in producing uterine contractions, and oxytocin may act in concert with the other factors [16]. The presence of abnormal labor does not necessarily mean abnormal secretion of oxytocin, but the presence of normal labor may not mean normal secretion of oxytocin. Failure of labor is the exception rather than the rule in women with diabetes insipidus [5, 17, 21–23]. *Blotner and Kunkel* [5] reported 2 cases and reviewed the literature prior to 1942. One of their patients had spontaneous onset of labor and normal vaginal delivery, and the other patient had elective cesarean section. They concluded that usually the onset of labor was spontaneous and delivery was normal.

Hendricks [21] studied 2 patients. One, a primiparous woman with diabetes insipidus, had spontaneous onset of labor and normal delivery. The other, a multiparous woman, also had spontaneous onset of labor in each of three pregnancies. *Hendricks* [21] cited 4 cases of uterine atony after delivery which required infusion of oxytocin. He agreed that most women with diabetes insipidus have normal labor and delivery and concluded that only the women with very extensive lesions in the neurohypophysis will have abnormal labor and delivery. *Carfagno* et al. [8] studied 2 women with the disorder, reviewed the literature, and concluded that most women have spontaneous onset of labor and normal delivery. *Hime and Richardson* [22] published an extensive review in 1978 and studied 1 woman who did not have spontaneous labor but required infusion of oxytocin to initiate uterine contractions. The authors concluded that the patient was not representative of most individuals with the disorder.

Studies in which levels of oxytocin have been measured in the plasma of pregnant women with diabetes insipidus during labor and delivery are few because measurement of oxytocin in plasma is difficult [24]. *Hawker* et al. [18] reported 2 individuals with diabetes insipidus in whom levels of oxytocin were measured by bioassay. They found measurable levels of the hormone in the women with diabetes insipidus, and the concentrations were equivalent to those in women without the disorder who had normal labor. *Shangold* et al. [32] used radioimmunoassay to measure levels of oxytocin during gestation and labor. Oxytocin was measurable, but the levels were variable ranging from less than 1–76 µU/ml. *Sende* et al. [31] studied 1 pregnant woman with diabetes insipidus who had the onset of spontaneous labor but eventually required cesarean section because of cephalopelvic disproportion. Oxytocin was measured by radioimmunoassay and was secreted in a pulsatile fashion. The authors concluded that in some cases oxytocin synthesis is not decreased and that labor proceeds normally. Study of a larger number of patients will be necessary to determine whether in individuals with extensive lesions of the supraoptic and paraventricular nuclei, oxytocin synthesis is impaired.

Effect of Diabetes insipidus on Lactation

Although vasopressin does not have a physiologic effect upon the myoepithelial cells of the breasts, oxytocin promotes contraction of the myoepithelial cells and evokes milk ejection [41]. In cases in which labor and delivery proceed normally, lactation usually occurs normally also. The case reviewed by *Hime and Richardson* [22] was that of a woman who did not have spontaneous labor, and she lactated for only 2 weeks after delivery. Several reviews suggest that milk production and lactation occur normally, but the details of 'successful lactation' are not well described [5, 17, 21–23]. *Chau* et al. [9] assessed oxytocin activity during breast-feeding by obtaining intramammary pressure recordings from one breast of a woman with diabetes insipidus while the infant was suckling the contralateral breast. A rise in intramammary pressure was noted during suckling, consistent with the release of oxytocin.

As with labor and delivery, the degree of damage of the paraventricular nucleus can give rise to a varied clinical picture. Although lactation is normal in most cases of diabetes insipidus, it must be acknowledged that the role of the neurohypophysis in successful lactation is not fully understood. Very little is known about secretion and necessity of secretion of oxytocin during breast-feeding in normal women. Observations from this laboratory [unpublished data] and other laboratories have documented that oxytocin but not vasopressin is secreted in response to infant suckling [40]. What is not known is whether lactation can occur in the absence of secretion of oxytocin. Studies in animals suggest that oxytocin may be necessary [41]. Determination by radioimmunoassay of oxytocin in plasma obtained from women with diabetes insipidus who are breast-feeding their infants is necessary. Infant versus maternal factors as causes for lactational failure must be differentiated before concluding that lactational failure with diabetes insipidus is due solely to deficits of the maternal neurohypophysis.

Effect of Pregnancy upon Diabetes insipidus

Blotner and Kunkel [5] noted a variable effect of pregnancy upon diabetes insipidus and their original observations have been verified in subsequent reviews. The effects of pregnancy upon diabetes insipidus were categorized by *Blotner and Kunkel* [5] as follows: (1) Aggravation of diabetes insipidus during pregnancy. Polyuria and polydipsia may increase by varing degrees and during various stages of gestation. Several examples of worsening of symptoms during pregnancy are cited [5, 22]. The exacerbation might be due to the effect of oxytocinase on vasopressin, although the effect of this enzyme to change the half-life of vasopressin may not be great [38]. Some increase in polydipsia may be due to the reset thirst center as described earlier in this chapter. (2) Improvement of the diabetes insipidus during pregnancy. *Blotner and Kunkel* [5] cited an example of a woman who had improvement of polyuria and polydipsia in the seventh month of pregnancy, had normal urine output in the weeks preceding delivery, but had return of diabetes insipidus after delivery [5]. Some work in rats has suggested that estrogen may increase the synthesis of vasopressin [35], but this has not been shown in humans. (3) No effect of pregnancy upon the disease. Several cases are referenced in which the disorder is unchanged during pregnancy [5, 22]. (4) Transient diabetes insipidus appearing during pregnancy. Transient diabetes insipidus may occur in normal women during any stage of pregnancy but, in most cases, is apparent after the fifth or sixth month of gestation [5, 22]. In several individuals the diabetes insipidus ceases a few days after delivery. One might wonder whether these patients have limited vasopressin reserve before pregnancy, but this has not been documented. A report of transient nephrogenic diabetes insipidus occurring in 3 pregnant women was presented by *Barron* et al. [3]. Nephrogenic diabetes insipidus was documented by failure to respond to exogenous vasopressin, and in 1 woman by polyuria despite high levels of vasopressin in plasma. Some women had other pregnancies in which diabetes insipidus was not present, although the phenomenon of transient diabetes insipidus occurring in a woman during each of several consecutive pregnancies and occurring at the same time during each pregnancy has been reported [5], which would support some underlying pathology. (5) Onset of diabetes insipidus during pregnancy and persistence after delivery. The onset of diabetes insipidus during pregnancy which persists postpartum suggests the possibility of a tumor of the pituitary which enlarged during pregnancy, necessitating thorough evaluation postpartum. (6) Transient appearance of diabetes insipidus after delivery which resolves. Individuals in this category may represent women with Sheehan's syndrome [33]. Although diabetes insipidus rarely occurs with Sheehan's syndrome, because of the difference in blood supply to the anterior and posterior lobes of the pituitary [34, 42], an occasional individual with anterior pituitary necrosis also has hypothalamic damage and atrophy of the supraoptic nucleus [34, 42]. Diabetes insipidus, which was masked by the presence of cortisol deficiency, would resolve if Sheehan's syndrome were not recognized and treated, but would reappear when the cortisol deficiency was treated [1]. An interesting case of diabetes insipidus complicating Sheehan's syndrome was presented by *Schwartz and Leddy* [30].

The author has had the opportunity to study an individual with permanent central diabetes insipidus during each of two consecutive pregnancies and postpartum.

Effect of Pregnancy on Treatment of Diabetes insipidus

Because the major problem in diabetes insipidus is either an absolute or relative deficiency in vasopressin, the treatment is replacement of the deficient hormone. Vasopressin does not cross the placenta [37], and absence of the hormone from the maternal circulation or the administration of exogenous hormone should not affect fetal development. The development of the vasopressin axis in the fetus is discussed in the chapter by *Leake and Fisher.* Standard treatment with lysine vasopressin may produce effective antidiuresis, but large doses of vasopressin may stimulate the uterus [28]. An analogue with maximum antidiuretic and minimum vasopressor properties, 1-desamino-8-*D*-arginine vasopressin (DDAVP), is available for clinical use in the treatment of diabetes insipidus [29]. The elimination of the free amino group in the hemicystine in position one increases the antidiuretic activity, and the substitution of the '*D*' for the '*L*' arginine in position 8 decreases the pressor activity [29].

The oxytocin-like activity of the compound also is decreased, making DDAVP an ideal agent for treatment of diabetes insipidus in pregnancy [29]. It was reported that pregnancy plasma does not destroy DDAVP in vitro [13]. Studies of the use of DDAVP in pregnant women are limited and no controlled studies in pregnant women or nursing mothers have been done. Isolated reports of the use of DDAVP in pregnant women suggest no harm to the fetus [10, 28]. *Oravec and Lichardus* [28] reported 1 woman whom he treated with the compound. The patient did not respond to vasopressin tannate and had uterine contractions when treated with lysine vasopressin. DDAVP produced no uterine contraction and produced adequate antidiuresis. *Cort* et al. [10] reported the safe and efficacious use of DDAVP in a pregnant woman. The author has used DDAVP in 2 patients with diabetes insipidus, in 1 during two pregnancies, with no toxicity to the mother or fetus. Studies of 1 patient's response to the agent are outlined below. Whether DDAVP crosses the human placenta is unknown. A single study in a postpartum woman demonstrates a change in plasma levels of DDAVP, but little assayable DDAVP in breast milk after an intranasal dose [unpublished observations].

Illustrative Case

A 20-year-old woman with traumatic central diabetes insipidus required treatment with 0.1 ml (100 µg/ml) of DDAVP intranasally every 8 h since the onset of the diabetes insipidus. During the initial 4 months of her first pregnancy, she required the same dose. At 5 months gestation, she complained of increasing thirst and urination and stated that 0.1 ml of DDAVP produced no antidiuretic effects. Electrolytes were unchanged, and attempts to quantitate the urine output were not successful. A water deprivation test was performed during the 36th week of pregnancy (fig. 1). Four weeks later, she had the spontaneous onset of uterine contractions with rupture of amniotic membranes. A continuous infusion of oxytocin was started because of hypocontractile labor. She had cesarean delivery of a healthy infant. In the 24-hour period during and immediately post delivery, she required only one dose of DDAVP, 0.1 ml intranasally, and subsequently 0.1 ml every 8 h. She chose not to breast-feed her infant; however, she had breast engorgement and milk production. The patient was restudied 3 months postpartum (fig. 1). The results

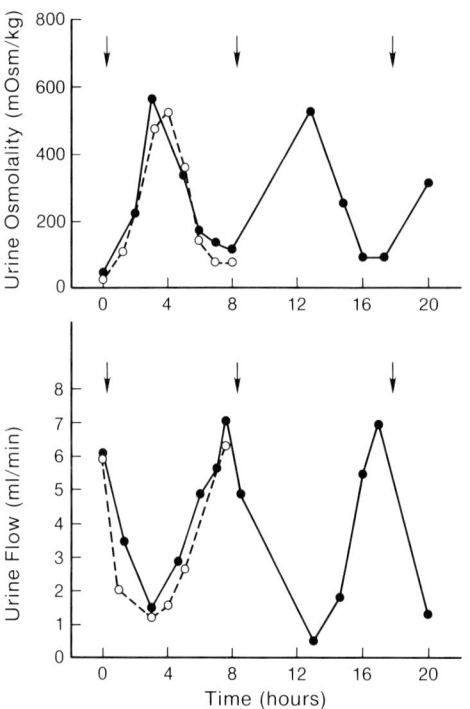

Fig. 1. The upper graph shows urine osmolality and the lower graph shows urine flow in the same patient at 36 weeks of gestation (○) and 3 months postpartum (●). The arrows indicate the time of administration of 10 µg of DDAVP intranasally.

of this test confirmed the need for 0.1 ml of DDAVP once every 8 h. Two years later, she had a second pregnancy. During the fifth month of pregnancy, she again complained of increased thirst and urination and that 0.1 ml of DDAVP no longer produced antidiuresis. At 33 weeks gestation, she underwent water deprivation test which established the need for 0.1 ml of DDAVP once every 8 h. She had a cesarean delivery of a healthy child 6 weeks later. She required one dose of DDAVP in the 24 h following delivery. Her requirement at present is 0.1 ml DDAVP every 8 h.

Although the patient described worsening of the polyuria and polydipsia and the need for increased doses of DDAVP in two pregnancies, no difference in the duration of action of DDAVP or its effectiveness was noted when pregnant and when not pregnant. Therefore, the clinical response was consistent with the data quoted above that DDAVP is not destroyed by pregnancy plasma. The increased thirst may be due to the reset of the thirst osmostat. In a normal woman the threshold for release of vasopressin is correspondingly decreased and the patient readjusts to a lower serum sodium. In a patient with diabetes insipidus vasopressin is administered and (unless overtreated) does not induce a lowering of serum sodium. The patient may then have persistent thirst. Careful evaluation of pregnant patients with diabetes insipidus is necessary to avoid overdosage and dilutional hyponatremia. Discontinuation of DDAVP because of 'ineffectiveness' would be unfortunate because the agent may be effective even if the patient complains of thirst. Stopping treatment would likely lead to severe thirst and hypernatremia.

References

1 Aguilo, F.; Vega, L.A.; Haddock, L.; Rodriguez, O.: Diabetes insipidus syndrome in hypopituitarism of pregnancy. Case report and a critical review of the literature. Acta endocr., Copenh. *60:* suppl. 137, p. 7 (1969).
2 Amico, J.F.; Seif, S.M.; Robinson, A.G.: Elevation of oxytocin and the oxytocin-associated neurophysin in the plasma of normal women during mid-cycle. J. clin. Endocr. Metab. *53:* 1229 (1981).
3 Barron, W.M.; Cohen, L.H.; Ulland, L.A.; Lassiter, W.E.; Fulgham, E.M.; Emmanouel, D.; Robertson, G.; Lindheimer, M.D.: Transient vasopressin-resistant diabetes insipidus of pregnancy. New Engl. J. Med. *310:* 442 (1984).
4 Bloemers, D.: Diabetes insipidus and toxaemia of pregnancy. J. Obstet. Gynaec. Br. Commonw. *68:* 322 (1961).
5 Blotner, H.; Kunkel, P.: Diabetes insipidus and pregnancy: report of two cases. New Engl. J. Med. *227:* 287 (1942).
6 Caldeyro-Barcia, R.; Poseiro, J.J.: Oxytocin and contractility of the pregnant human uterus. Ann. N.Y. Acad. Sci. *75:* 813 (1958).
7 Campbell, J.W.: Diabetes insipidus and complicated pregnancy. J. Am. med. Ass. *243:* 1744 (1980).
8 Carfagno, S.C.; Durant, T.M.; Shuman, C.R.: Diabetes insipidus in pregnancy. Archs intern. Med. *92:* 542 (1953).
9 Chau, S.S.; Fitzpatrick, R.J.; Jamieson, B.: Diabetes insipidus and parturition. J. Obstet. Gynaec. Br. Commonw. *76:* 444 (1969).
10 Cort, J.H.; Schuck, O.; Stribrna, J.; Skopkova, J.; Jost, K.; Mulder, J.L.: Role of the disulfide bridge and the C-terminal tripeptide in the antidiuretic action of vasopressin in man and the rat. Kidney int. *8:* 292 (1975).
11 Davison, J.M.; Vallotton, M.B.; Lindheimer, M.D.: Plasma osmolality and urinary concentration and dilution during and after pregnancy: evidence that lateral recumbency inhibits maximal urinary concentration ability. Br. J. Obstet. Gynaec. *88:* 472 (1981).
12 Davison, J.M.; Gilmore E.A.; Durr, J.; Robertson, G.L.; Lindheimer, M.D.: Altered

osmotic thresholds for vasopressin secretion and thirst in human pregnancy. Am. J. Physiol. *246:* F105 (1984).

13 Edwards, C.R.W.; Kitau, M.J.; Chard, T.; Besser, G.M.: Vasopressin analogue DDAVP in diabetes insipidus: clinical and laboratory studies. Br. med. J. *iii:* 375 (1973).

14 Ferrara, J.M.; Malatesta, R.; Kemmann, E.: Transient nephrogenic diabetes insipidus during toxemia in pregnancy. Diagn. Gynecol. Obstet. *2:* 227 (1980).

15 Fisher, C.; Ingram, W.R.; Ranson, S.W.: Diabetes insipidus and the neurohormonal control of water balance: a contribution to the structure and function of the hypothalamico-hypophyseal system (Edwards, Ann Arbor 1938).

16 Fuchs, A.R.; Fuchs, F.; Husslein, P.; Soloff, M.S.; Fernstrom, M.J.: Oxytocin receptors and human parturition: a dual role for oxytocin in the initiation of labor. Science *215:* 1396 (1981).

17 Gulotta, C.E.; Beacham, W.D.; Webster, H.D.: Diabetes insipidus with superimposed pregnancy. J. La. St. med. Soc. *115:* 383 (1963).

18 Hawker, R.W.; North, W.G.; Colbert, I.C.; Lang, J.P.: Oxytocin blood levels in two cases of diabetes insipidus. J. Obstet. Gynaec. Br. Commonw. *74:* 430 (1967).

19 Heinbecker, P.; White, H.L.: Hypothalamico-hypophyseal system and its relation to water balance in the dog. Am. J. Physiol. *133:* 582 (1941).

20 Heinbecker, P.; White, H.L.; Rolf, D.: The essential lesion in experimental diabetes insipidus. Endocrinology *40:* 104 (1947).

21 Hendricks, C.H.: The neurohypophysis in pregnancy. Obstetl gynec. Surv. *9:* 323 (1954).

22 Hime, M.C.; Richardson, J.A.: Diabetes insipidus and pregnancy: case report, incidence and review of the literature. Obstetl gynec. Surv. *33:* 375 (1978).

23 Hirota, T.; Kassai, A.: A case of diabetes insipidus associated with pregnancy. Mie med. J. *14:* 205 (1965).

24 Leake, R.D.; Weitzman, R.E.; Glatz, T.H.; Fisher, D.A.: Plasma oxytocin concentrations in men, non-pregnant women, and pregnant women before and during spontaneous labor. J. clin. Endocr. Metab. *53:* 730 (1981).

25 Magoun, H.W.; Fisher, C.; Ranson, S.W.: The neurohypophysis and water exchange in the monkey. Endocrinology *25:* 161 (1939).

26 Mahanon, G.: Diabetes insipidus and uterine atony: a case observed over a period of 26 years. Br. med. J. *ii:* 769 (1947).

27 McLaren, H.C.; McLeod, M.: Diabetes in pregnancy. J. Obstet. Gynaec. Br. Emp. *49:* 51 (1942).

28 Oravec, D.; Lichardus, B.: Management of diabetes insipidus in pregnancy. Br. med. J. *iv:* 114 (1972).

29 Robinson, A.G.: DDAVP in the treatment of central diabetes insipidus. New Engl. J. Med. *294:* 507 (1976).

30 Schwartz, A.R.; Leddy, A.L.: Recognition of diabetes insipidus in postpartum hypopituitarism. Obstet. Gynec., N.Y. *59:* 394 (1982).

31 Sende, P.; Pantelakis, N.; Sazuki, K.; Bashore, R.: Plasma oxytocin determinations in pregnancy with diabetes insipidus. Obstet. Gynec., N.Y. *48:* 385 (1976).

32 Shangold, M.M.; Freeman, R.; Kumaresan, P.; Feder, A.S.; Vasicka, A.: Plasma oxytocin concentrations in a pregnant woman with total vasopressin deficiency. Obstet. Gynec., N.Y. *61:* 662 (1983).

33 Sheehan, H.L.; Murdock, M.B.: Postpartum necrosis of the anterior pituitary: pathological and clinical aspects. J. Obstet. Gynaec. Br. Emp. *45:* 456 (1938).
34 Sheehan, H.I.; Whitehead, R.: The neurohypophysis in postpartum hypopituitarism. J. Path. Bact. *85:* 145 (1963).
35 Skowsky, W.R.; Swann, L.; Smith, P.: Effects of sex steroid hormones on arginine vasopressin in intact and castrated male and female rats. Endocrinology *104:* 105 (1979).
36 Soloff, M.S.; Alexandrova, M.; Fernstrom, M.J.: Oxytocin receptors: triggers for parturition and lactation? Science *204:* 1313 (1979).
37 Stegner, H.; Leake, R.D.; Palmer, S.M.; Fisher, D.A.: Permeability of the sheep placenta to ^{125}I-arginine vasopressin. Dev. Pharmacol. Ther. (in press).
38 Tuppy, H.: The influence of enzymes on neurohypophysial hormones and similar peptides; in Berde, Handbook of experimental pharmacology, vol. XXXIII, p. 67 (Springer, Berlin 1968).
39 Vickers, D.M.: Diabetes insipidus with acute urinary retention in pregnancy: with report of a case. Surgery Gynec. Obstet. *38:* 223 (1924).
40 Weitzman, R.E.; Leake, R.D.; Rubin, R.T.; Fisher, D.A.: The effect of nursing on neurohypophyseal hormones and prolactin in human subjects. J. clin. Endocr. Metab. *51:* 836 (1980).
41 Weitzman, R.; Vorherr, H.; Kleeman, C.R.: Water metabolism and the neurohypophysial hormones: vasopressin and oxytocin; in Bondy, Rosenberg, Metabolic control and disease, pp 1241–1324 (Saunders, Philadelphia 1980).
42 Whitehead, R.: The hypothalamus in postpartum hypopituitarism. J. Path. Bact. *86:* 55 (1963).

Janet A. Amico, MD, Department of Medicine, University of Pittsburgh, School of Medicine, Pittsburgh, PA 15261 (USA)

Pharmacology of Deamino-*D*-Arginine Vasopressin[1]

Vladimir Pliška

Institut für Tierproduktion, Eidgenössische Technische Hochschule, Zürich, Switzerland

Applicability and usefulness of any substance in substitution therapy of a hormone deficiency syndrome are bound primarily to the following conditions: (1) the agent is not toxic in a long-term administration schedule; (2) the agent is not antigenic in humans; (3) the side effects are minimal; (4) the therapeutically desired potency is high, and finally (5) the duration of the elicited effect is long.

It has been recognized for a long time that in the case of diabetes insipidus, the naturally occurring hormones (arginine or lysine vasopressin) are not ideal as therapeutic agents, mainly because they do not fit the third and fifth conditions mentioned above. First, their vasopressor activity is not negligible, and second (except for depot parenteral administration of an oil suspension) the persistence of vasopressins in the body is low and their tissue concentrations drop quickly down to values below an effective level.

A vasopressin-like peptide that would elicit a long-lasting antidiuretic response and, at the same time, would possess a pronounced antidiuretic potency, became therefore a typical target of the drug design.

Prolongation of Biological Responses to Peptides of Vasopressin Series

Enhancement of drug persistence in a target tissue can be achieved in at least two ways: (1) by a change of the structure such that the distri-

---

[1] Dedicated to *Frederik Paulsen* on the occasion of his 75th birthday.

bution coefficient expressed in the simplest way as a ratio 'concentration in target tissue/concentration in plasma', is high; (2) by enhancing the resistance of the molecule against inactivating enzymes. Little attention was given to the former possibility because of its uncertain experimental background. However, a high persistence for some hydrophobic antagonists and agonists of oxytocin in the rat uterus suggests that this approach may have some potential. On the other hand, the second alternative received much more attention. Attempts to synthesize enzyme-resistant analogues can be summarized as follows [1]: (1) stabilization of the molecule against aminopeptidase splitting by omitting or by substituting the N^α-amino group [2]; (2) substitution of the disulfide bridge by a nonreducible bond, as in the so-called carba-analogues [3, 4]; (3) stabilization of the penultimate peptide bond against trypsin-like enzymes or carboxamidopeptidases, among others by substitution of 8-arginine with its *D*-enantiomer.

Although some of the analogues stabilized on a single molecular site elicit a prolonged response in certain biological systems, this prolongation has not been systematic. Multiple protection against various enzymes, on the other hand, increases the chance of prolongation of all or most of the effects. The subject of this communication, deamino-*D*-arginine vasopressin (DDAVP) can serve here as a good example.

Relationships between Antidiuretic Response and Primary Structure

In contrast to the design of analogues with prolonged action, structure-activity rules for biological activities of neurohypophyseal peptides were derived purely inductively from the empiric knowledge collected over decades. In many instances, it is not clear which phase of action (i.e. affinity to the receptor, transport to the receptor, inactivation, etc., cf. [5]) was influenced by a given change of molecular structure. Some of these rules, therefore, may have only limited validity, in particular with respect to molecules concomitantly changed on several positions.

The most important general structural features which exercise a clear-cut effect on antidiuretic potency and its ratio to vasopressor potency are as follows (table I):

(1) Deletion of the N-Terminal α-Amino Group
Increase of activities following this structural change has long been known [6]. It results in elevated antidiuretic potency and slightly diminished vasopressor potency [7, 8]

Table I. Antidiuretic activities (AD) and ratios antidiuretic/vasopressor activities (AD/BP) of vasopressin (VP) and deaminovasopressin (dVP) analogues substituted in position 8 with L- and D-enantiomers of basic amino acids. Potency data by various authors

8-Amino acid	Peptide chain	L-Enantiomers			D-Enantiomers		
		AD[1]	AD/BP	impact index[2]	AD[1]	AD/BP	impact index[2]
Arginine	VP	1[3]	1[3]	1[3]	0.26	27.8	2.69
	dVP	2.99	3.52	3.23	4.60[4]	4,000[4]	125.6[4]
Lysine	VP	0.60[5]	0.91[5]	0.74[5]	0.046	20.0	0.96
	dVP	0.69	2.38	1.28	0.009	3.62	0.18
Ornithine	VP	0.20	0.24	0.23	0.16	291.7	6.83
	dVP	0.46	0.57	0.51	1.36	287.8	19.6
Homoarginine	VP	0.97	1.71	1.29	0.19	101.2	4.38
	dVP	1.05	4.79	2.24	0.45	3,267	38.3
Homolysine	VP	0.36	0.64	0.48	0.48	113.1	7.37
	dVP	1.46	4.14	2.46	0.39	340.0	11.5
2,4-Diamino-butyric acid	VP	0.28	0.81	0.48	0.28	33.3	3.05
	dVP	3.15	3.04	3.09	0.83	175.6	12.1
2,3-Diamino-propionic acid	VP	0.14	0.55	0.28	0.39	8.02	1.77
	dVP	2.48	77.1	13.8	2.29	43.9	10.0

[1] Potencies related to arginine vasopressin (absolute values for both antidiuretic and vasopressor potencies are 435 IU/μmol).
[2] Impact index defined as geometric mean of AD and AD/BP.
[3] Arginine vasopressin; naturally occurring hormone in the majority of mammalian species.
[4] DDAVP.
[5] Lysine vasopressin; naturally occurring hormone of the suborder Suiformae.

(table I). For analogues having an L-enantiomer in the 8-position, however, the ratio is not changed dramatically. Despite some circumstantial evidence for increased affinity to the receptors particularly in the oxytocin series (pA_2 of deaminated inhibitors is, as a rule, higher than that of aminated), the main contribution to the potency enhancement is likely to be enhanced metabolic stability, as the substance cannot be inactivated by aminopeptidases. Also, the omission of a polar NH_2 group results in an increase of overall hydrophobicity [9] and, consequently, in better transport into the target tissue.

(2) Physicochemical Features of the Side Chain in Position 8

There is no doubt that these features, in particular the change and hydrophobicity of the side chain, play a crucial role in the antidiuretic activity. The potency values of individual analogues, of both vasopressin and deamino-vasopressin, correlate well with the hy-

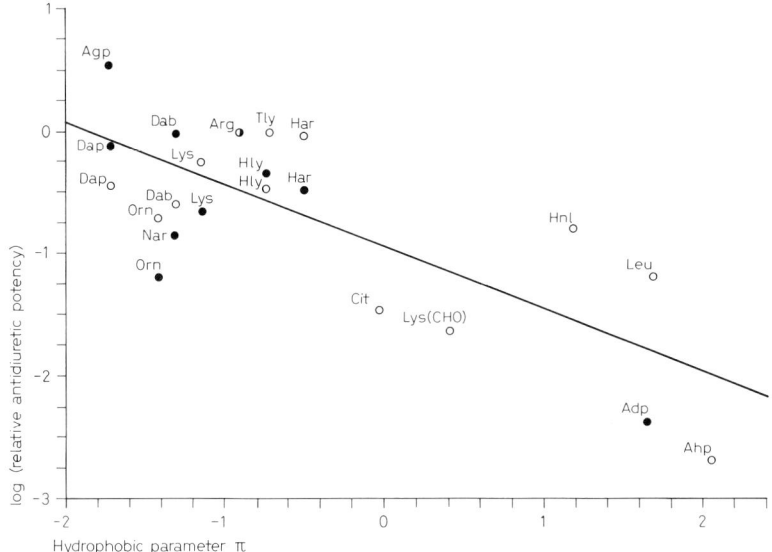

Fig. 1. Relationship between antidiuretic potency and hydrophobicity of 8-side chain for *L*-enantiomers. Potency data by various authors. Values of hydrophobicity constants π from *Pliška* et al. [46] and from *Fauchère and Pliška* [47]. ○ = Vasopressin analogues (potency relative to arginine vasopressin); ● = deaminovasopressin analogues (potency relative to [8-arginine]deaminovasopressin). *L*-Amino acid in position 8 is indicated at each point (in addition to standard IUPAC-IUB nomenclature, the following abbreviations are used: Adb = 2-amino-3,3-dimethylbutyric acid; Agp = 2-amino-3-guanidino-propionic acid; Ahp = 2-aminoheptanoic acid; Cit = citrulline; Dab = diaminobutyric acid; Har = homoarginine; Hly = homolysine; Dap = diaminopropionic acid; Lys(CHO) = N^ε-formyllysine; Hnl = ε-hydroxynorleucine; Nar = norarginine; Orn = ornithine; Tly = thialysine).

drophobicity constants π of the side chain [10, 11] (fig. 1): the potency is high for negative π-values (low hydrophobicity) and decreases when π-values become more positive. However, hydrophobicity may not be the only critical feature of the 8-side chain: π-values themselves correlate with the polarizability (expressed, for example, as molar refraction), with the electrostatic charge, with the ability to establish hydrogen bonds, and also with other properties of the substituent group.

(3) D-Enantiomers in Position 8

With few exceptions, introduction of *D*-enantiomers somewhat decreases the antidiuretic potency but dramatically increases the antidiuretic-to-pressor ratio (AD/BP ratio, see table I). DDAVP is the most important exception: its antidiuretic potency is roughly 50% *higher* than in the corresponding *L*-analogue; the ratio is increased by a factor of 1,000 [12–14]. Apparently, besides enhancement of resistance against C-terminal splitting enzymes, the *D*-configuration stabilizes in some way the conformation of the peptide molecule, perhaps by stabilization of a β-turn (for review cf. [15]). It may also improve the receptor binding of the peptide, for instance by facilitating an induced fit, due to the conformational change mentioned above.

(4) Physicochemical Properties of the Side Chain in Position 4

Certain changes in position 4 may bring about an enhancement of antidiuretic activity. By analogy with the inhibitors of uterotonic response where equal or similar changes increase the pA_2 values, the potency increase can be again accounted for by an increase in the hormone-receptor association constant. Threonine and valine are particularly efficient in this respect. Much attention was paid to the possible physicochemical background of this phenomenon; both hydrophobicity [16] and substituent volume of the 4-side chain [17] have been thought to play a decisive role here. However, recent studies seem to indicate a combined effect of several molecular features, including the two mentioned above [18].

Design of DDAVP and of Related Peptides

Deamino-*D*-arginine vasopressin (in IUPAC-IUB nomenclature [1-mercaptopropionyl, 8-*D*-arginine]-vasopressin), shown schematically in figure 2, was first synthesized by *Zaoral* et al. [13] in the late 1960s, with the aim to amplify the resistance against inactivation enzymes by protection of two critical sites of the molecule. However, the expected prolongation of the antidiuretic response was not unambiguously achieved in the common animal models (hydrated rat) but the antidiuretic activity of the 8-*D*-arginine analogue was found to increase by a factor of 10–100. The pressor activity, on the other hand, dropped down to negligible values. Due to the strong shift in the activity spectrum (see AD/BP ratio in table I), the peptide became interesting for clinical use.

Similar changes, however, have *not* been observed for all analogues so designed. For instance, deamino-*D*-lysine vasopressin [13] displays exceptionally strongly reduced antidiuretic activity and, as one of very few peptides, decreased AD/BP ratio when compared to the corresponding peptide with an N-terminal amino group. Also, no activity increase was observed in similar analogues containing 8-homolysine [19, 20]. It can therefore be concluded that the effects of N-terminal deamination and of $D \rightarrow L$ substitution in position 8 are superimposable only in certain instances and to a limited extent.

Correlations between antidiuretic and pressor activities of 8-substituted analogues are insignificant, both overall and within individual groups seen in table I (VP, dVP, *L*- and *D*-enantiomers). Objectives for selection of the 'optimal' analogue for substitution therapy may there-

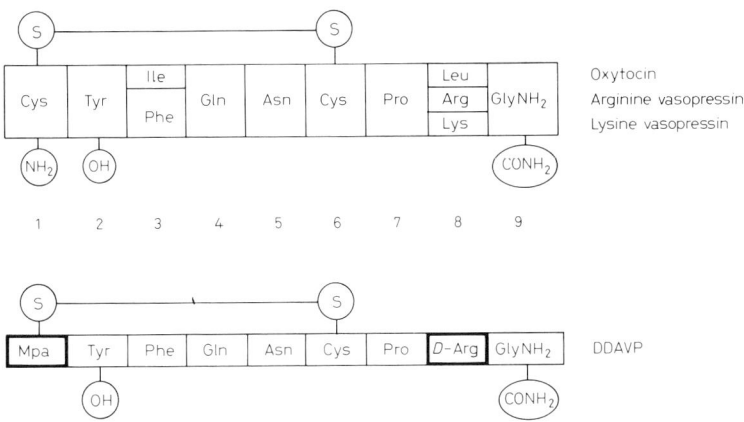

Fig. 2. Amino acid sequence of neurohypophyseal hormones (upper panel) and [1-β-mercaptopropionic acid, 8-*D*-arginine]-vasopressin (DDAVP, lower panel). Important functional groups are indicated in circles. Heavy borders indicate differences between DDAVP and AVP. Standard abbreviations are used for common amino acids. Mpa = β-Mercaptopropionic acid; *D*-Arg = *D*-arginine; GlyNH$_2$ = glycine amide. From ref. [21], with permission.

fore become difficult, since two noncorrelated properties should be maximized: the antidiuretic potency and the AD/BP ratio. A simple, though not theoretically founded, solution has been suggested in our earlier article [21]: the selection may be carried out according to an 'impact index of therapeutic value', represented, for instance, by the geometric mean of the maximized parameters related to a standard (a slightly different definition was used by us earlier). These indicators are shown in table I: the value of 1 characterizes the selected standard (arginine vasopressin), values smaller than 1 are for substances less suitable than the standard, values greater than 1 indicate an increase of the therapeutic value. The highest impact index belongs to DDAVP, but also *D*-homoarginine [22], *D*-ornithine [23], *L*-2,3-diaminopropionic acid [24] and *D*-2,4-diaminobutyric acid analogues [25] display high impact index values.

As mentioned above, 4-substitution by certain amino acids may render substances with still higher impact indices. Probably the most attractive substance at present is [4-valine, 8-*D*-arginine]-deamino-vasopressin [26] with a relative antidiuretic potency of 5.92, AD/VP ratio 123,000 and an impact index of 853, followed by 4-threonine analogue

[27] showing values of 2.46, 39,000 and 310 (impact index of DDAVP is 126, cf. table I). However, a prediction of an optimal side chain in position 4 is still uncertain (see above), and until now the design was merely intuitive. Further development of new substances on this basis with even higher therapeutic value is possible.

Basic Pharmacology of DDAVP

Antidiuretic activity is certainly the dominant biological property of this substance. The *relative* potency based on equipotent doses of DDAVP and arginine vasopressin (in moles) is dependent on the level of antidiuretic response for which it is expressed. This phenomenon is well understood, as the dose-response curves for the two substances are parallel only in a low dose region [14]. Potency values between 3 and 100 (hydrated rat) have been reported from various laboratories. When keeping an unwritten, not fully justified convention of taking equipotent doses obtained by extrapolation to the zero response [28], a value between 4 and 5 can be obtained (cf. table I). The nonparallelity of the dose-response curves for an integrated (total) response is in itself an indicator of different elimination rates of the two substances from their target tissue (s) [28]. In the case of DDAVP, this rate seems to be particularly strongly concentration-dependent [14, 29], as can be deduced from a detailed analysis of the dose-response curves. Measurements in humans, on the other hand, indicate prolonged antidiuretic effect, regardless of the route of administration which was used [30–34].

> Nonparallelism of the two curves brings about problems in estimating therapeutic doses. Due to different slopes of dose-response curves, the response obtained by, for example, doubling of the DDAVP dose is different than that for the equipotent dose of arginine vasopressin. Expression of antidiuretic potency in international units is therefore useless in such cases; the dose can be only estimated empirically and its conversion by means of factors correcting for age, body weight, etc., should, sensu stricto, not be used. This difficulty may – and hopefully will soon – be removed by introduction of separate DDAVP standards.

The cellular mechanism of action of DDAVP on the kidney has been found similar to that of vasopressins [*Jard*, this volume]. The response is elicited on the V_2-receptors in the medullopapillary section and associated with an activation of adenylate cyclase: enhanced production of adenosine 3':5'-cyclic phosphate (cAMP) causes an increase of cell membrane permeability for water. On the physiological level, the elevated permeability is reflected in a higher rate of tubular water reabsorption in the collecting duct. Reabsorption of salts is virtually unchanged. Therefore, clearance of osmotically free water is a more relevant measure for the antidiuretic response of vasopressin-like substances than overall urine excretion, urine osmolality and related effects. The sensitivity of the adenylate cyclase system seems to be strongly species-dependent: in pig kidney membrane preparation [35], the equipotent-DDAVP-concentration for cAMP production is over three orders of magnitude higher than the one of lysine vasopressin, whereas DDAVP was found slightly more potent than arginine vasopressin in rat kidney membranes [36]. It seems that a strongly enhanced affinity of kidney receptors for DDAVP cannot be the sole and general explanation of its high in vivo potency.

The pronounced decrease of vasopressor activity, measured on rats premedicated with alpha-adrenergic inhibitors, of 8-*D*-enantiomers of basic amino acids, has already been mentioned above. Also the uterotonic activity in vitro (isolated rat uterus in magnesium-free medium) is somewhat decreased: uterotonic-to-antidiuretic potency for arginine vasopressin is 0.03, whereas for DDAVP it is 0.003.

Hemostatic Effects of DDAVP

Some rather unexpected effects of DDAVP on blood clotting processes were reported by *Mannucci* et al. [38] in 1975. Similar to drugs increasing vascular tonus, like vasopressin and catecholamines, DDAVP enhances plasma levels of both inactive factor VIII-complex (F VIII R:AG) and its hemostatically active component, F VIII:C. The elevation of F VIII:C indicates not only accelerated release from endothelial cells but also an effect of DDAVP upon activation of the factor VIII complex. At the same time, DDAVP exerts an effect upon fibrinolysis by elevating plasma levels of plasminogen activator [38]. Originally, it was assumed that a vascular 'squeezing' effect might explain the hemostatic response to vasoconstrictor agents; however, the ability to release factor VIII is high for DDAVP which causes only a minor vasoconstriction in the central vascular system. Rather, these effects can be accounted for by existence of a separate receptor on factor VIII-releasing endothelial cells [39].

Since hemostatic response to DDAVP is lacking in patients with severe hemophilia A, no DDAVP effect upon de novo synthesis of a factor VIII component can be expected. However, for mild and medium severe forms of hemophilia A, DDAVP has become a widespread agent for control of blood clotting, in particular during surgery.

All these effects have been demonstrated in humans. Recent findings from our laboratory, however, indicate that experimental animals may behave differently in F VIII:C response to DDAVP. We did not find any regular response to both DDAVP and lysine vasopressin in sheep and in rats. Also, the increase of plasminogen activator was very irregular in our experiments. As a preliminary explanation, we suggest that the spontaneous release of factor VIII may have reached an upper limit and cannot be further increased by any hemostatic agent; plasma levels in these animals are 6–10 times higher than in humans. Other reasons are, of course, conceivable.

Central Nervous Effects of DDAVP Associated with Memory Processes

Behavioral and antiamnesic effects of neurohypophyseal hormones have been known for the last two decades (for a recent review cf. [40]). Structural requirements for some of these effects – abolishment of puromycin-induced amnesia in mice was used as the first model – have been demonstrated in the mid-1970s. No general structure-activity rules could be deduced but DDAVP was found to be quite active in these experiments [41]. Prolonged action of DDAVP compared to lysine vasopressin has been reported for retention of passive avoidance behavior in rats. It was hoped that these may be useful in therapy of memory disorders, and several clinical observations were already published. Recently, the critical study by *Jolles* [42] showed that many of these hopes were false. Central effects of peptides in a clinical situation are difficult to prove unambiguously, since the methodology is not sufficiently developed and the diseases themselves not sufficiently characterized biochemically, pathologically, anatomically, etc.

Pharmacokinetic of DDAVP

Generally, prolonged action assumes long survival of the substance in the organism. Indeed, blood levels of DDAVP and of lysine vasopressin in humans (administered intranasally) parallel with the antidiuretic response in their time course [43]. The survival is usually expressed in terms of metabolic clearance which can be measured in individual organs or as a total clearance (Cl_{tot}) with respect to the total body disappearance rate. As for the other neurohypophyseal hormones, both total and organ clearances have been studied [44], the latter by various methods (organ perfusion in vivo or in vitro, or surgical blockage of the organ in question). One of the design targets which led to the synthesis of DDAVP was associated with total metabolic clearance of the substance, and it is surprising that until now this was not properly investigated. The only relevant data come from experiments using the technique of single injection in humans [30, 45], which is definitely an inferior method to the commonly used continous infusion. We have recently tried to evaluate them [21]. The disappearance rate follows approximately a double exponential. Irreversible disappearance of the substance, by enzymic degradation, detoxication, excretion to urine, bile, feces, etc., can be estimated by the single injection method first when concentrations of all participating compartments are in a steady state. Then, usually a single exponential function reflects the irreversible elimination and the corresponding exponential constant can be taken for computation of total metabolic clearance in a usual way [21]. When evaluating the values of Cl_{tot} from literature data, one obtains values of around 3 ml/min/kg body wt for normal persons; values for patients with various forms of cranial diabetes insipidus range between 0.4 and 2.7. Values for healthy and polyuric subjects are apparently not different (a statistical proof was problematic, due to a very restricted group of subjects). The arithmetic mean of the whole group was 2.1 ± 1.0 ml/min/kg body wt. The average value of Cl_{tot} for lysine vasopressin in humans was reported in the literature [41] to be 14.8 ml/min/kg body wt. The total disappearance rate of DDAVP in humans is therefore about 12% of that of lysine vasopressin.

Since the inactivation processes on peptide bonds 1–2 and 8–9 seem to be decisive for inactivation of neurohypophyseal peptides, the question remains as to the main step of DDAVP inactivation. Very likely, the major part of this inactivation can be accounted for by reduction of the disulfide bridge and subsequent splitting by less specific endopeptidases. Other possible mechanisms, like oxidation of the tyrosine ring, are probably less common.

Concluding Remarks

Introduction of DDAVP into clinical practice is of immense value in the therapy of diabetes insipidus, and more recently also of hemophilia A. However, investigation of *all* features of this substance has not yet been completed.

Firstly, although no patient has developed hypersensitivity or resistance by raising antibodies against DDAVP, should this occur, an alternative substance would be useful for continued treatment. Secondly, although no chronic changes of patients blood clotting status have been reported so far, effects on clotting must be considered as an *undesirable side effect* in treatment of diabetes insipidus. Dissociation of the two activities is a future task of design. Thirdly, new pharmaceutical forms for more simple administration would be welcome both by patients and physicians; efficient oral administration remains a target. Therefore, development of new, potent antidiuretic peptides with long response duration and high specificity still remains a desirable goal.

Acknowledgement

Unpublished experiments from the author's laboratory have been kindly supported by Ferring AB, Malmö, and by Swiss National Science Foundation, Grant No. 3.705–80.

References

1 Pliška, V.; Rudinger, J.: Modes of inactivation of neurohypophysial hormones: significance of plasma disappearance rate for their physiological responses. Clin. Endocrinol. *5:* suppl., pp. 73s–84s (1976).
2 Jošt, K.; Rudinger, J.; Šorm, F.: Amino acids and peptides. XXXIX. Analogues of oxytocin exerting protracted biological effects. Coll. Czech. Chem. Commun. *28:* 2021–2030 (1963).
3 Rudinger, J.; Jošt, K.: A biologically active analogue of oxytocin not containing a disulphide group. Experientia *20:* 570–571 (1964).
4 Pliška, V.; Rudinger, J.; Douša, T.; Cort, J.H.: Oxytocin activity and the integrity of the disulphide bridge. Am. J. Physiol. *215:* 916–920 (1968).
5 Rudinger, J.; Pliška, V.; Krejčí, I.: Oxytocin analogs in the analysis of some phases of hormone action. Recent Prog. Horm. Res. *28:* 131–166 (1972).
6 Vigneaud, V. du; Winestock, G.; Murti, V.V.S.; Hope, D.B.; Kimbrough, R.D., Jr.: Synthesis of 1-β-mercapto-propionic acid oxytocin (desamino-oxytocin), a highly potent analogue of oxytocin. J. biol. Chem. *235:* PC64–PC66 (1960).
7 Kimbrough, R.D., Jr.; Cash, W.D.; Branda, L.A.; Chan, W.Y.; Vigneaud, V. du: Synthesis and biological properties of 1-desamino-8-lysine-vasopressin. J. biol. Chem. *238:* 1411–1414 (1963).
8 Huguenin, R.L.; Stürmer, E.; Boissonnas, R.A.; Berde, B.: Desamino-arginine-vasopressin, an analogue of arginine vasopressin with high antidiuretic activity. Experientia *21:* 68–69 (1965).
9 Lindberg, G.; Vilhardt, H.; Larsson, L.-E.; Melin, P.; Pliška, V.: Effect of O-alkylated analogues of lysine vasopressin on adenylate cyclase of pig kidney membranes. J. Receptor Res. *1:* 389–402 (1980).
10 Pliška, V.: Semiempirical structure-activity relationships in peptide pharmacology; in Eberle, Geiger, Wieland, Perspectives in peptide chemistry, pp. 221–235 (Karger, Basel 1981).
11 Pliška, V.: Phylogeny of neurohypophyseal hormones: parsimonial phylogenetic trees and evolution of some biological activities; in Baertschi, Dreifuss, Neuroendo-

crinology of vasopressin, corticoliberin and opiomelanocortins, pp. 177–189 (Academic Press, London 1982).

12 Zaoral, M.; Kolc, J.; Šorm, F.: Synthesis of D-Arg[8]-and D-Lys[8]-vasopressin. Coll. Czech. Chem. Commun. *31:* 382–383 (1966).

13 Zaoral, M.; Kolc, J.; Šorm, F.: Amino acids and peptides. LXXI. Synthesis of 1-deamino-8-D-amino-butyrine vasopressin, 1-deamino-8-D-lysine vasopressin and 1-deamino-8-D-arginine vasopressin. Coll. Czech. Chem. Commun. *32:* 1250–1257 (1967).

14 Vávra, I.; Machová, A.; Holeček, V.; Cort, J.H.; Zaoral, M.; Šorm, F.: Effects of a synthetic analogue of vasopressin in animals and in patients with diabetes insipidus. Lancet *i:* 948–952 (1968).

15 Hruby, V.: Structure and conformation related to the activity of peptide hormones; in Eberle, Geiger, Wieland, Perspectives in peptide chemistry, pp. 207–220 (Karger, Basel 1981).

16 Manning, M.; Balaspiri, L.; Judd, J.; Acosta, M.; Sawyer, W.H.: Probing the molecular basis of antidiuretic specificity and duration of action with specific peptides. FEBS Lett. *44:* 229–232 (1974).

17 Zaoral, M.; Brtník, F.; Barth, T.; Machová, A.: Specific antidiuretic effect of [Leu[4],D-Arg[8]] vasotocin and [Mpr[1],Leu[4],D-Arg[8]]vasotocin. Comment on the idea of lipophilic properties and position 4. Endocrinol. exp. *10:* 183–191 (1976).

18 Pliška, V.; Melin, P.; Carlsson, L.: 4-substituted analogs of deamino-D-argininevasopressin: biological potencies and structure-activity relationships related to position 4; in Rich, Gross, Peptides: synthesis-structure-function, pp. 117–120 (Pierce Chemical Company, Rockford 1982).

19 Lindeberg, E.G.; Bodanszky, M.; Acosta, M.; Sawyer, W.H.: Synthesis and some pharmacological properties of 8-L-homoarginine-vasopressin and of 1-deamino-8-L-homoarginine-vasopressin. J. medicinal Chem. *17:* 781–783 (1974).

20 Lindeberg, E.G.: Solid phase synthesis and some pharmacological properties of 8-D-homolysine-vasopressin and 1-deamino-8-D-homolysine-vasopressin. Int. J. Pept. Prot. Res. *7:* 395–401 (1975).

21 Pliška, V.; Vilhardt, H.: Design of DDAVP: the path between ideas and pharmacologic reality; in Sutor, Minirin, DDAVP-Anwendung bei Blutern, pp. 12–29 (Schattauer, Stuttgart 1980).

22 Lindeberg, E.G.G.; Melin, P.; Larsson, L.-E.: Solid phase synthesis and some pharmacological properties of 8-D-homoarginine-vasopressin and 1-deamino-8-D-homoarginine-vasopressin. Int. J. Pept. Prot. Res. *8:* 193–198 (1976).

23 Zaoral, M.; Brtník, F.; F. Barth, T.; Machová, A.: [1-β-Mercaptopropionic acid, 8-α,γ-diaminobutyric acid]vasopressin and [1-β-mercaptopropionic acid, 8-D-ornithine]vasopressin. Synthesis and biological effects. Coll. Czech. Chem. Commun. *41:* 2088–2095 (1966).

24 Krchňák, V.; Zaoral, M.; Machová, A.: [1-β-Mercaptopropionic acid, 8-α,β-diaminopropionic acid]vasopressin and [1-β-mercaptopropionic acid, 8-D-α,β-diaminopropionic acid]vasopressin. Two lysine-vasopressin analogs with considerable antidiuretic effect. Coll. Czech. Chem. Commun. *44:* 2161–2164 (1979).

25 Zaoral, M.; Kolc, J.; Šorm, F.: Amino acids and peptides. LXXI. Synthesis of 1-deamino-8-D-γ-aminobutyrine-vasopressin, 1-deamino-8-D-lysine-vasopressin and

1-deamino-8-*D*-arginine-vasopressin. Coll. Czech. Chem. Commun. *32:* 1250–1257 (1967).

26 Manning, M.; Balaspiri, L.; Acosta, M.; Sawyer, W.H.: Solid phase synthesis of [1-deamino, 4-valine]-8-*D*-arginine-vasopressin (DVDAVP), a highly potent and specific antidiuretic agent possessing protracted effects. J. medicinal Chem. *16:* 975–978 (1973).

27 Manning, M.; Balaspiri, L.; Moehring, J.; Haldar, J.; Sawyer, W.H.: Synthesis and some pharmacological properties of deamino[4-threonine, 8-*D*-arginine]vasopressin and deamino[8-*D*-arginine]vasopressin, highly potent and specific antidiuretic peptides, and [8-*D*-arginine]vasopressin and deamino-arginine-vasopressin. J. medicinal Chem. *19:* 842–845 (1976).

28 Pliška, V.: On the in vivo kinetics of drug action. Arzneimittel-Forsch. *16:* 886–893 (1966).

29 Vávra, I.; Machová, A.; Krejčí, I.: Antidiuretic action of 1-deamino-[8-*D*-arginine]-vasopressin in unanesthetized rats. J. Pharmac. exp. Ther. *188:* 241–247 (1974).

30 Edwards, C.R.W.; Kitau, M.J.; Chard, T.; Besser, G.M.: Vasopressin analogue DDAVP in diabetes insipidus: clinical and laboratory studies. Br. med. J. *1973:* 375–378.

31 Aronson, A.S.; Andersson, K.-E.; Bergstrand, C.G.; Mulder, J.L.: Treatment of diabetes insipidus in children with DDAVP, a synthetic analogue of vasopressin. Acta paediat. scand. *62:* 133–140 (1973).

32 Ward, M.K.; Fraser, T.R.: DDAVP in treatments of vasopressin-sensitive diabetes insipidus. Br. med. J. *1974:* 86–89.

33 Irmscher, K.; Sennejunker, K.; Wiegelmann, W.; Solbach, H.G.: Behandlung des Diabetes insipidus mit 1-Desamino-8-*D*-Arginin-Vasopressin. Dt. med. Wsch. *99:* 2431–2437 (1974).

34 Radó, J.P.; Marosi, J.; Szende, L.; Borbely, L.; Takó, J.; Fischer, J.: The antidiuretic action of 1-deamino-8-*D*-arginine vasopressin (DDAVP) in man. Int. J. clin. Pharmacol. *13:* 199–209 (1976).

35 Jard, S.; Bockaert, J.; Butlen, D.; Rajerison, R.; Roy, C.: Vasopressin-sensitive adenylate cyclase from mammalian kidney; in Klinge, Proc. 6th Int. Congr. of Pharmacology, vol. 1, pp. 121–130 (Pergamon Press, New York 1976).

36 Seif, S.M.; Zenser, T.V.; Ciarochi, F.F.; Davis, B.B.; Robinson, A.F.: DDAVP (1-desamino-8-*D*-arginine-vasopressin) treatment of central diabetes insipidus – mechanism of prolonged antidiuresis. J. clin. Endocr. Metab. *46:* 381–388 (1978).

37 Mannucci, P.M.; Åberg, M.; Nilsson, I.M.; Robertson, B.: Mechanism of plasminogen activator and factor VIII increase after vasoactive drugs. Br. J. Haematol. *30:* 81–93 (1975).

38 Gader, A.M.A.; Costa, J. da; Cash, J.D.: A new vasopressin analogue and fibrinolysis. Lancet *ii:* 1417–1418 (1973).

39 Sutor, A.H.: Gegenwärtiger Stand der DDAVP-Anwendung bei Blutern; in Sutor, Minirin, DDAVP-Anwendung bei Blutern, pp. 2–9, 178–179 (Schattauer, Stuttgart 1980).

40 De Wied, D.: Central actions of neurohypophysial hormones; in Cross, Leng, The neurohypophysis: structure, function and control, pp. 155–167 (Elsevier, Amsterdam 1983).

41 Walter, R.; Hoffman, P.L.; Flexner, J.B.; Flexner, L.B.: Neurohypophyseal hor-

mones, analogs, and fragments: their effect on puromycin-induced amnesia. Proc. natn. Acad. Sci USA 72: 4180–4184 (1975).

42 Jolles, J.: Vasopressin-like peptides and the treatment of memory disorders in man; in Cross, Leng, The neurohypophysis: structure, function and control, pp. 169–182 (Elsevier, Amsterdam 1983).

43 Limal, J.-M.; Mugner, E.; Czernichow, P.: Traitment du diabète insipide pitressosensible de l'enfant par le DDAVP. Archs. fr. Pédiat. 34: 965–972 (1977).

44 Lawson, H.D.: Metabolism of the neurohypophysial hormones; in Knobil, Sawyer, Handbook of physiology, vol. IV, sect. 7, part I, pp. 287–393 (Am. Physiological Society, Washington 1974).

45 Pullow, P.T.; Burger, H.G.; Johnston, C.I.: Pharmacokinetics of 1-desamino-8-D-arginine vasopressin (DDAVP) in patients with central diabetes insipidus. Clin. Endocrinol. 9: 273–278 (1978).

46 Pliška, V.; Schmidt, M.; Fauchère, J.-L.: Partition coefficients of amino acids and hydrophobic parameters π of their side chains as measured by thin-layer chromatography. J. Chromatogr. 216: 79–92 (1981).

47 Fauchère, J.-L.; Pliška, V.: Hydrophobic parameters π of amino-acid side chains from the partitioning of N-acetyl-amino-acid amides. Eur. J. Med. Chem. 18: 369–375 (1983).

PD Dr. V. Pliška, Institut für Tierproduktion, ETH-Zürich, TAN F1, CH-8092 Zürich (Switzerland)

Treatment of Central Diabetes insipidus[1]

Alan G. Robinson, Joseph G. Verbalis

Department of Medicine, School of Medicine, University of Pittsburgh, Pittsburgh, Pa. USA

Introduction

Diabetes insipidus does not cause progressive morbidity due to the lack of vasopressin nor due to secondary complications in other vital systems. Therefore, in many cases the major morbidity is the excessive thirst and frequent urination. Most patients with diabetes insipidus have a normal thirst, and in most situations are able to drink sufficient water to maintain normal metabolic balance. Therapy then should be convenient for the patient's life-style, easy to administer and monitor, and effective in preventing polyuria and nocturia. The patient should understand the therapy so that individual judgement can be used for flexibility of therapy depending upon daily activities. The safety of any therapy must be an overriding consideration because of the benign course of the disease. Overtreatment should be avoided because this might produce a degree of hyponatremia which would be more detrimental than undertreatment of the disease.

The description of diabetes insipidus as benign is based upon the ability of the patient to respond appropriately to thirst. If the patient is either unable to sense thirst or unable to respond by drinking water, the disease is potentially life-threatening. After head trauma, a surgical procedure, or any time that the patient is unconscious, the patient may have normal serum and urine composition when first seen by a physician because of the persistent effect of therapy. As the therapeutic agent

[1] Supported by NIH Grant MO1 RR00056, NIH Grant AM 16166, NIH Grant NS 17138, The Veterans Administration Research Career Development Program.

reaches the end of its normal duration of action, excessive urine output may occur abruptly and in a few hours produce severe dehydration and cardiovascular collapse. To avoid this every patient with diabetes insipidus should carry in their wallet or purse a medical card indicating that diabetes insipidus is present and should wear a Medic-Alert tag indicating the presence of this disorder.

Treatment of Diabetes insipidus

Water is emphasized as therapy for diabetes insipidus because water alone taken in sufficient quantity will correct any metabolic abnormality secondary to excessively dilute urine. Therapy is designed to reduce the required water intake to tolerable levels.

Arginine vasopressin is the natural vasopressin of all mammals (with the exception of the pig). 'Aqueous vasopressin' is a buffered solution of L-arginine vasopressin which can be given parenterally. Vasopressin may be purified from bovine pituitary or chemically synthesized. It is usually packaged in 1-ml snap-top vials at a concentration of 20 U/ml. Given subcutaneously it has an onset of action within 1–2 h and a duration of action of 4–8 h. Intravenous bolus administration should be avoided because of a shorter duration of action, and because of potentially hazardous pressor effects (hypertension, angina) [5]. Natural arginine vasopressin was also available as a 'snuff', but this crude preparation when administered intranasally often produced allergic reactions [3].

'Pitressin tannate' in oil is a relatively crude extract of bovine posterior pituitary containing arginine vasopressin in a suspension of peanut oil. The snap-top vials contain vasopressin at a concentration of 5 U/ml. This is a suspension and the vasopressin will sediment during storage, sometimes in the hollow snap-top. It is imperative that prior to administration the vial be examined to locate this brown pellet and warmed and shaken vigorously until the pellet is completely suspended. A single dose of 5–10 U intramuscularly will usually provide a 24- to 72-hour duration of action. The oily suspension causes slow absorption, but because of the oil base sterile abscesses have occurred. At times these become so large as to require surgical removal [personal observation].

'Lysine vasopressin' is the vasopressin of the pig and is available commercially as a purified synthetic vasopressin in an aqueous buffer at

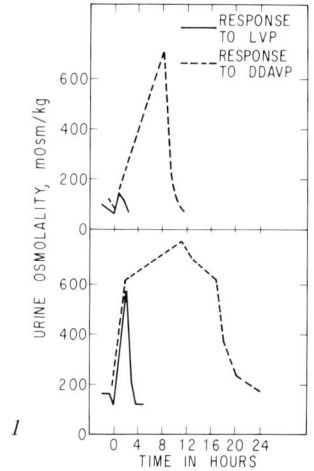
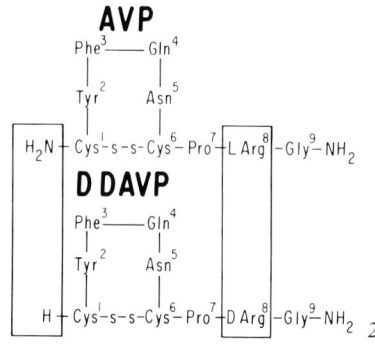

Fig. 1. Effect of lysine vasopressin, 18.5 µg (solid lines), on urine osmolality as compared with DDAVP, 5 µg (broken lines), in 2 patients.

Fig. 2. Comparison of the structure of 1-desamino-8-D-arginine-vasopressin (DDAVP) with the natural arginine vasopressin (AVP). The boxes indicate the differences between the two molecules.

a concentration of 50 U/ml [9, 19]. It can only be used as a nasal spray. Absorption of the vasopressin from the nasal mucosa is rapid and the duration of action is quite variable, but has a maximum of 4–6 h (fig. 1). Because it has a lesser antidiuretic effect but as much pressor effect as arginine vasopressin, angina and flushes may occur [9].

'DDAVP', 1-desamino-8-D-arginine-vasopressin, is a synthetic analogue of natural vasopressin in which the amino group of cystine has been removed and D-arginine has been substituted for L-arginine in position 8 (fig. 2). Removal of the terminal NH_2 on cysteine prolongs the half-life [6, 13, 25, 26, 29] and substitution of D- for L-arginine reduces the pressor activity of the molecule [13, 25, 26, 29, 33]. The agent is approximately 2000 times more specific for antidiuresis than is L-arginine vasopressin [1, 3, 25, 26, 32]. DDAVP is available in a buffered aqueous solution of 100 µg/ml. A soft plastic tube is used to administer 50–200 µl by blowing into the nose. The onset of action is within 1 h and the duration of effect is from 6 to 24 h [7, 12, 24, 30, 34]. Despite the variation of duration of effect between individuals, in a given patient the effect is quite reproducible (fig. 3) [7, 12, 24, 30, 34]. DDAVP is also available in

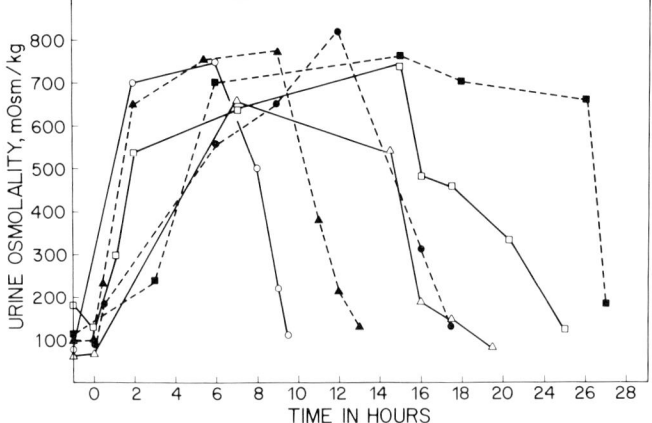

Fig. 3. Urine osmolality of individual patients with diabetes insipidus in response to 20 µg of DDAVP administered at zero time.

2-ml vials of 4 µg/ml for parenteral injection. When administered parenterally about ⅕th to ⅒th the dose of DDAVP will produce an effect similar to DDAVP administered intranasally (fig. 4) [10, 23].

Chlorpropamide was discovered by serendipity to cause decreased diuresis in patients with diabetes insipidus [2]. The drug acts to enhance the effect of vasopressin on the renal tubule and to increase the hydroosmotic action of vasopressin [15, 17]. For chlorpropamide to exert its antidiuretic effect some vasopressin must be present [18]. The drug is of no use in patients with complete absence of vasopressin, but will cause a decreased diuresis in patients with partial diabetes insipidus and some ability to secrete vasopressin [2, 15, 17, 22]. The usual dose is 100–500 mg orally per day. Maximum antidiuresis is observed after 4 days of treatment. One must always be cautious about hypoglycemia when treating diabetes insipidus with chlorpropamide, especially in cases of hypopituitarism [31].

Carbamazepine causes release of vasopressin in patients with partial diabetes insipidus [14]. As with chlorpropamide, the agent is of no use in complete diabetes insipidus. The dosage is 200–600 mg/day, but the physician should be thoroughly familar with the potential toxicity of this agent before using it to treat diabetes insipidus [28].

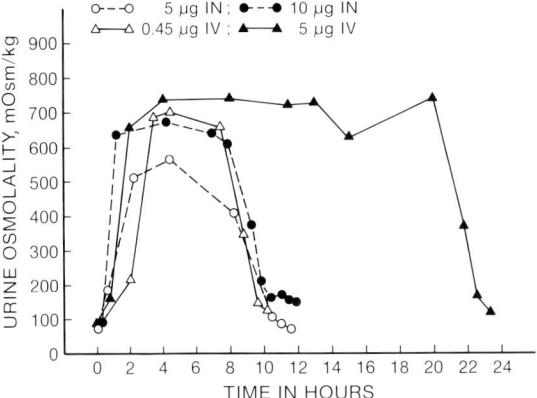

Fig. 4. Urine osmolality in 1 patient with diabetes insipidus. Response to 5 and 20 µg of DDAVP administered intranasally (i.n.) and 0.45 and 5 µg administered intravenously (i.v.) – each at time zero.

Clofibrate also stimulates release of endogenous vasopressin [20]. Patients with partial diabetes insipidus may respond to 500 mg every 6 h, but because of possible increased incidence of gall bladder disease and carcinoma of the gall bladder in patients taking clofibrate [35], the agent is not recommended for routine treatment of diabetes insipidus.

Thiazide diuretics will decrease urine volumes in patients with both hypothalamic and nephrogenic diabetes insipidus [8, 11]. Since this effect is seen even with complete central diabetes insipidus, the action does not require vasopressin and is probably secondary to volume contraction, decreased ultrafiltrate, and increased proximal tubular absorption of Na^+ and water. For hydrochlorthiazide, usually 50–100 mg/day is sufficient. Potassium replacement may be necessary to prevent hypokalemia.

Initiating Therapy – Complete Diabetes insipidus

In complete diabetes insipidus there is no measurable vasopressin and no ability to concentrate the urine during dehydration [16]. These patients require replacement with some form of vasopressin – preferably DDAVP. To use DDAVP the patient must be taught to use the rhinyl catheter and to measure the prescribed dose. Patients may practice with a solution of saline to fill the tube. When the measured dose is in the catheter it is held in a U-shape with both ends up to allow the solution to run to the bottom of the loop of the tube. The catheter is maintained in a U-shape and raised to the level of the mouth while about 0.5 inch is inserted into the nose. Without sucking on the tube a moderate inspiration is taken and held while the other end of the catheter is placed in the mouth. A swift puff through the catheter as one might use to blow a ball out of a straw will blow the hormone high into the nose. A physician or an assistant must be trained in the use of the rhinyl catheter to assist in the training of the patient. Poor technique will cause wasted hormone by dripping hormone out the end of the catheter, swallowing the hormone, or blowing it too low in the nose to allow absorption.

Initially the patient should try 50, 100 and 200 µl of DDAVP in a controlled environment where urine output can be measured for volume and osmolality (or specific gravity). If the patient is on other therapy, it is discontinued and urine output measured until polyuria of about 4 ml/min is established. The first dose of 50 µl of DDAVP is administered. Urine output will decrease 1–2 h later and the duration of action will be 6–24 h. After each dose the patient is allowed to return to baseline polyuria before the next dose. Establishing the duration of action of 50, 100 and 200 µl in each patient is helpful in planning therapy and in assuring the patient (and the physician) that administration of a second dose of drug earlier than prescribed will not lead to any adverse effect.

From the response to the three doses of drug a dose and time of day to administer the drug is prescribed. Some patients will require only a single dose per day, but most patients will require two doses. Two small doses (e.g. 50 µl) is more cost-effective than one large dose (e.g. 200 µl). When given intranasally the biologic half-life of DDAVP is about 4 h [10, 27]. Therefore, doubling the dose will only extend the effect by about 4 h [27]. When a single daily dose of drug is used it may be given at bedtime to allow a full night of sleep, but many patients prefer to take the drug in the morning to allow a full day of work or other activities. A single episode of nocturia may be preferable to frequent voiding during the work day. If two doses of DDAVP are used the time of administration should be based upon the daily activity of the patient. Almost always the first dose should be given early in the morning but the time of the second dose should be individualized. If the patient is home for the evening the second dose is best given late in the evening to allow a full night's sleep. If the patient is going out for the evening, it is best to administer the second dose in the late afternoon or early evening to allow an evening free of polyuria. If this dose is not sufficient for the night a third dose of medication may be taken on that particular day just prior to bedtime. Three doses of 50 µl will be readily accepted by a patient who has already taken 200 µl in a single test dose as described above.

The danger of a flexible program is that a patient may take enough DDAVP to always maintain urine at a low output. They then may become volume-expanded, natriuretic, and hyponatremic, i.e. develop the syndrome of inappropriate antidiuretic hormone. To avoid this, once or twice a week the patient should withold DDAVP until pronounced polyuria and thirst recurs. This protects the patient from hyponatremia due to excessive DDAVP and provides ongoing documentation of the persistence of diabetes insipidus. This is especially useful in post-surgical or post-traumatic diabetes insipidus where the ability to secrete vasopressin may recover (even years) after the initial insult.

Pitressin tannate in oil provided satisfactory therapy for many patients for many years, but the injection must be intramuscular and most patients object to therapy with this agent.

Because DDAVP is expensive some physicians also prescribe an oral agent to prolong the action of DDAVP. Both chlorpropramide and indomethacin [21] will prolong the effect of DDAVP, but the potential complications from drugs which are not otherwise indicated make this attempt to decrease the cost undesirable. It is important to recognize that if the oral agents described above are given for other reasons they may prolong the effect of DDAVP and predispose the patient to water intoxication. The dose of DDAVP may have to be decreased during treatment with any of the oral agents described above.

Initiating Therapy – Partial Diabetes insipidus

In 'partial' diabetes insipidus the patients have some endogenous vasopressin but not sufficient to maintain maximum urinary concentration at a normal serum Na^+. Most patients will not become hypernatremic because of their excessive thirst [16]. Probably the best therapy for these patients is also DDAVP. Treatment is as described above except that these patients may tolerate a less frequent administration of DDAVP because the polyuria and thirst is less severe. In some of these patients the oral pharmacologic agents may be indicated. If a patient with partial diabetes insipidus is unable to utilize DDAVP because of poor vision or for economic reasons, chlorpropamide may be used. Therapy is initiated with 100 mg/day and increased every 4 days until 500 mg/day or until appropriate antidiuresis is obtained. For an oral agent the appropriate measure of effectiveness is the 24-hour urine volume and the patient's symptomatic response. In an occasional patient with diabetes insipidus and either diabetes mellitus or congestive failure, it may be appropriate to treat both diseases with either chlorpropamide or thiazide diuretics, respectively. The dose should be prescribed to treat the primary disease, e.g. diabetes mellitus or congestive heart failure, and the effect upon urine volume observed. Further therapy with vasopressin will be based on the response to these agents.

In patients with hypopituitarism and lack of ACTH and growth hormone, the threat of hypoglycemia during treatment with chlorpropamide must be recognized [31] and frequent feedings given and blood sugars obtained. In some cases of mild partial diabetes insipidus thiazide diuretics may give sufficient reduction of urine output and decrease of thirst to allow satisfactory control.

Chlofibrate, carbamazepine or indomethacin are not recommended for the treatment of partial diabetes insipidus if other agents are available. Most patients with partial diabetes insipidus who will respond to any oral agents will respond to either chlorpropamide or thiazides.

Therapy of Diabetes insipidus with Inadequate Thirst

Diabetes insipidus with inadequate thirst consitutes a most difficult management problem. Some patients may have partial diabetes insipidus but do not respond normally with thirst or with secretion of vasopressin to increased serum osmolality. In this variant of partial diabetes insipidus there is the potential for response to oral agents [4]. A trial of chlorpropamide up to 500 mg/day is indicated because some cases will not only decrease the urine volume but increase the thirst. If chlorpropamide is not effective, DDAVP should be used. The duration of action of DDAVP can be determined after a water load given orally or intravenously. When the duration of effect of DDAVP is established, the patient must be prescribed a rigid regimen of DDAVP and water. Even with careful monitoring of DDAVP and water, regular measurement of serum sodium is necessary because these patients are prone to develop water intoxication with hyponatremia or alternatively dehydration with hypernatremia.

Therapy of Postoperative Diabetes insipidus

Diabetes insipidus occurs frequently following surgery of the hypothalamic pituitary area. If the patient is receiving high doses of glucocorticoids which cause hyperglycemia and glycosuria the diagnosis of diabetes insipidus may be in doubt [*Pliška*, this volume]. The osmotic diuresis must be corrected to establish the diagnosis of diabetes insipidus and for vasopressin to exert its appropriate effect. Therapy for this acute condition is vasopressin, but many neurosurgeons fear water overload and brain edema and the patient may be treated with fluid intravenously for a considerable time prior to the use of vasopressin. If the patient is able to respond to thirst one can treat with vasopressin and allow thirst to be the guide for water replacement. If the patient is not able to sense thirst normally, they may be thirsty even though water has been adequately replaced or may not sense thirst even with severe dehydration. In either case fluid balance will need to be maintained by intravenous fluid. It is imperative that urine osmolality and serum sodium be checked every several hours during the initial therapy and at least daily thereafter.

Serum sodium is an excellent guide to adequate replacement with fluid and vasopressin. Aqueous vasopressin or parenteral DDAVP may be the initial therapy. Usually the patient is hypernatremic (thus establishing the diagnosis of diabetes insipidus in the face of hypotonic urine). The serum sodium should be rechecked after the initial doses of vasopressin to insure some improvement of Na^+. If improvement is not adequate, it may be necessary to increase the fluids in addition to treating with vasopressin. Once the serum sodium has been corrected, input and output can be balanced by fluid and vasopressin. It is important to be careful because excess water during administration of vasopressin can create a syndrome of inappropriate antidiuretic hormone and potentially severe hyponatremia. During the first several days after surgery vasopressin or DDAVP should be withheld once daily until polyuria is reestablished and the persistence of diabetes insipidus is confirmed [*Pliška*, this volume].

If parenteral DDAVP is not available one may administer Pitressin tannate in oil. This will allow for a relative steady control of polyuria for 24–48 h and with intact thirst and ad libitum water intake there is little danger of excess water and hyponatremia. Further therapy need not be given until persistent diabetes insipidus is demonstrated.

References

1 Anderson, K.E.; Arner, B.: Effects of DDAVP, a synthetic analogue of vasopressin, in patients with cranial diabetes insipidus. Acta med. scand. *192:* 21–27 (1972).
2 Arduino, F.; Ferraz, F.P.J.; Rodrigues, J.: Antidiuretic action of chlorpropamide in idiopathic diabetes insipidus. J. clin. Endocr. Metab. *26:* 1325–1328 (1966).
3 Aronsson, A.S.; Andersson, K.E.; Bergstrand, C.G.; Mulder, J.L.: Treatment of diabetes insipidus in children with DDAVP, a synthetic analogue of vasopressin. Acta paediat. scand. *62:* 133–140 (1973).
4 Bode, H.H.; Harley, B.M.; Crawford, J.D.: Restoration of normal drinking behavior by chlorpropamide in patients with hypodipsia and diabetes insipidus. Am. J. Med. *51:* 304 (1971).
5 Brazuea, P.: Agents effecting the renal conservation of water; in Goodman, Gilman The pharmacologic basis of therapeutics; 4th ed., pp. 204–225 (Macmillan, New York, 1970).
6 Chan, W.Y.; Vignead, V. du: Comparison of the pharmacologic properties of oxytocin and its highly potent analogue, desamino-oxytocin. Endocrinology *71:* 977–982 (1962).
7 Cobb, W.E.; Spare, S.; Reichlin, S.: Neurogenic diabetes insipidus: management with dDAVP (1-desamino-8-*D*-arginine vasopressin). Ann. intern Med. *88:* 183–188 (1978).
8 Crawford, J.D.; Kennedy, G.C.; Hill, L.E.: Clinical results of treatment of diabetes insipidus with drugs of the chlorothiazide series. New Engl. J. Med. *262:* 737–743 (1960).
9 Dashe, A.M.; Kleeman, C.R.; Czaczkes, J.W.; Rubinoff, H.; Spears, J.: Synthetic vasopressin nasal spray in the treatment of diabetes insipidus. J. Am. med. Ass. *190:* 1069–1071 (1964).
10 Edwards, C.R.W.; Kitau, M.J.; Chard, T.; Besser, G.M.: Vasopressin analogue DDAVP in diabetes insipidus clinical and laboratory studies. Br. med. J. *iii:* 375–378 (1973).
11 Havard, C.W.H.: Thiazide-induced antidiuresis in diabetes insipidus. Proc. R. Soc. Med. *58:* 1005–1007 (1965).
12 Kauli, R.; Laron, A.: A vasopressin analogue in treatment of diabetes insipidus. Archs Dis. Childh. *49:* 482–584 (1974).
13 Kimbrough, R.D., Jr.; Cash, W.D.; Branda, L.A.; Chan, W.Y.; Vigneaud, V., du: Synthesis and biological properties of 1-desamino-8-lysin-vasopressin. J. biol. Chem. *238:* 1411–1414 (1963).
14 Kimura, T.; Matsui, K.; Sato, T.; Yoshinaga, K.: Mechanisms of carbamazepine

(Tegretol)-induced antidiuresis: evidence for release of antidiuretic hormone and impaired excretion of a water load. J. clin. Endocr. Metab. *38:* 356–362 (1974).

15 Kumar, R.S.; Sutow, W.W.; Cole, V.W.; Chlorpropamide in diabetes insipidus. Lancet *1:* 577–578 (1969).

16 Miller, M.; Dalakos, T.; Moses, A.M.; Fellerman, H.; Streeten, D.H.P.: Recognition of partial defects in antidiuretic hormone secretion. Ann. intern. Med. *73:* 721–729 (1970).

17 Miller, M.; Moses, A.M.: Mechanisms of chlorpromide action in diabetes insipidus. J. clin. Endocrinol. *86:* 1014 (1970).

18 Miller, M.; Moses, A.: Potentiation of vasopressin action by chlorpropamide in vivo. Endocrinology *86:* 1014 (1970).

19 Moses, A.M.: Synthetic lysine vasopressin nasal spray in the treatment of diabetes insipidus. Clin. Pharmacol. Ther. *5:* 422–427 (1964).

20 Moses, A.M.; Howanitz, J.; Gemert, M. van; Miller, M.: Clofibrate-induced antidiuresis. J. clin. Invest. *52:* 535–542 (1973).

21 Moses, A.M.; Moses, L.K.; Notman, D.D.; Springer, J.: Antidiuretic responses to injected desmopressin, alone and with indomethacin. J. clin. Endocr. Metab. *52:* 910–913 (1981).

22 Moses, A.M.; Numann, P.; Miller, M.: Mechanism of chlorpromide-induced antidiuretic in man: evidence for release of ADH and enhancement of peripheral action. Metabolism: *22:* 59 (1973).

23 Mulder, J.; Andersson, K.E.; Arner, B.; Aronsson, S.; Bachmann, R.; Hokfelt, B.: Pharmacology of DDAVP and its clinical trial in crancial diabetes insipidus. International Congress of Endocrinology, Washington 1972. Excerpta Med. Int. Congr. Ser., No. 256, p. 49 (1972).

24 Robinson, A.G.: DDAVP in the treatment of central diabetes insipidus. New Engl. J. Med. *294:* 507–511 (1976).

25 Sawyer, W.H.; Acosta, M.; Balaspiri, L.; Judd, J.; Manning, M.: Structural changes in the arginine vasopressin molecule that enhance antidiuretic activity and specificity. Endocrinology *94:* 1106–2225 (1974).

26 Sawyer, W.H.; Acosta, M.; Manning, M.: Structural changes in the arginine vasopressin molecule that prolong its antidiuretic action. Endocrinology *95:* 140–149 (1974).

27 Seif, S.M.; Zenser, T.V.; Ciarochi, F.F.; David, B.B.; Robinson, A.G.: DDAVP (1-desamino-8-*d*-arginine-vasopressin) treatment of central diabetes insipidus – mechanism of prolonged antidiuresis. J. clin. Endocr. Metab. *46:* 381–388 (1978).

28 Toman, J.E.P.: Drugs effective in convulsive disorders; in Goodman, Gilman, The pharmacologic basis of therapeutics; 4th ed. L.S. Goodman, pp. 204–225 (Macmillan, New York, 1970).

29 Walter, R.; Rundinger, J.; Schwartz, I.L.: Chemistry and structure-activity relations of the antidiuretic hormones. Am. J. Med. *42:* 653–677 (1967).

30 Ward, M.K.; Fraser, T.R.: DDAVP in treatment of vasopressin-sensitive diabetes insipidus. Br. med. J. *iii:* 86–89 (1974).

31 Webster, B.; Bain, J.: Antidiuretic effect and complications of chlorpropamide therapy in diabetes insipidus. J. clin. Endocr. Metab. *30:* 215–227 (1970).

32 Zaoral, M.; Kolc, J.; Sorm, F.: Amino acids and peptides. LXXI. Synthesis of 1-de-

amino-8-*d*-γ-aminobutyrine-vasopressin, 1-deamino-8-*D*-lysine-vasopressin, and 1-deamino-8-*D*-arginine-vasopressin. Collect. Czech. Chem. Commun. *32:* 1250–1257 (1967).

33 Zaoral, M.; Sorm, F.: Amino acids and peptides. LX. Synthesis of *D*-DAB^8-vasopressin. Collect. Czech. Chem. Commun. *31:* 310–314 (1966).

34 Zial, F.; Roderich, W..; Rosenthal, I.M.: Treatment of central diabetes insipidus in adults and children with desmopressin – a synthetic analogue of vasopressin. Archs intern. Med. *138:* 1382– 1383 (1978).

35 Geizerova, H.; Gyarfas, I.; Green, K.G.; Heady, J.A.; Oliver, M.F.; Strasser, T.: A co-operative trial in the primary prevention of ischaemic heart disease using clofibrate. Br. Heart J. *40:* 1069–1118 (1978).

Alan G. Robinson, MD, Department of Medicine, School of Medicine, University of Pittsburgh, Pittsburgh, PA 15261 (USA)

Treatment of Diabetes insipidus in Children and Adolescents

R. Kauli, A. Galatzer, Z. Laron

Institute of Pediatric and Adolescent Endocrinology, Beilinson Medical Center, Petah Tikva and Sackler School of Medicine, Tel Aviv University, Israel

Good control of diabetes insipidus (DI) is of the utmost importance in children and adolescents as fluid and electrolyte imbalance has a negative effect on growth and psychosocial adjustment [4, 10, 11]. Adequate water replacement to avoid hyperosmolarity is of particular importance in infants and young children, since they depend on adults for their fluid supply. It is, however, equally important to bear in mind the possible danger of water excess in infants and children receiving antidiuretic hormone (ADH) replacement therapy since self-control of fluid intake is limited at this young age [3, 9]. The hazard of water intoxication following ADH replacement exists also in patients with coexisting untreated adrenocorticotropic hormone (ACTH) deficiency which should be diagnosed and corrected first [4, 8, 9].

During the period from 1958 to 1983, we treated and followed in our institute 36 juvenile patients with cranial DI. A retrospective analysis on the course of illness with particular regard to the problems and effects of therapy is hereby presented.

Subjects and Methods

36 patients were all proven to have central DI with etiology as presented in table I. The age at diagnosis ranged from $2^{7}/_{12}$ to $15^{4}/_{12}$ years and the follow-up from 3 to 22 years. All patients found to have multiple pituitary hormone deficiencies (MPHD) have been receiving replacement therapy including L-thyroxine, hydrocortisone, sex steroids at the appropriate age and growth hormone when indicated. Up to 1972, the medications used included Pitressin tannate in oil (PTO) i.m., Pitressin powder (PP) nasal snuff, lysine-8-vaso-

Table I. Etiology of cranial DI in 36 juvenile patients

Idiopathic	
Isolated DI	13
With multiple pituitary hormone deficiencies	3
Hystiocytosis	
Isolated DI	4
With multiple pituitary hormone deficiencies	2
CNS tumors (with multiple pituitary hormone deficiencies)	
Craniopharyngiomas	6
Other tumors	5
Posttraumatic (isolated)	2
Postinfectious (isolated)	1
Total	36

pressin (LVP) nasal spray and, rarely, carbamazepine and chlorpropamide orally [7]. Since 1972, all patients have been switched to 1-desamino-8-D-arginine-vasopressin (DDAVP) intranasally (i.n.) [1, 5]. All newly diagnosed or newly referred patients have been treated exclusively with DDAVP i.n. Once the dose of DDAVP has been adjusted individually the patient is seen once every 4–6 months. At each visit the degree of control is evaluated on the basis of the information received on urinary habits and fluid requirements, 24-hour urine volume (measured between visits by the patients), continuity of night sleep and general health and daily functioning. Urinary specific gravity, urine and blood osmolarity and blood electrolytes are also determined.

At each visit all patients underwent a full clinical examination including assessment of growth and pubertal development. The effect of the degree of therapeutic control of DI on growth and puberty was noted particularly in patients who had isolated ADH deficiency. At the time of the study 14 patients had reached their final height; this was compared with their expected mid-parental height. Data on school achievement, psychosocial activities and family life were collected from the patients and their parents as well as from teachers, social counsellors, school doctors and nurses and psychologists.

Results

Control of Water Balance. The superiority of DDAVP over the previously used medications [2, 4–6] in achieving a good water balance is shown in table II, which presents the ranges of urine volumes and concentration on different therapeutic regimens. In addition, the prolonged antidiuretic effect of DDAVP gave patients a feeling of security and freedom during the day and an uninterrupted night's rest, contributing to marked improvement in school achievements and psychosocial re-

Table II. Ranges of urine volume and concentration in 36 patients with cranial DI on different therapeutic regimens

Therapy	n	Dose range per 24 h	Urine volume l/24 h	Urinary specific gravity[1]	Urine osmolarity mosm/kg water[1]
Before	36	–	4.0–12.0	1,000–1,003	44–220
PTO	8	2–5 IU/24 h	2.0–3.5	1,007–1,018	215–390
PP	14	40–80 mg	1.8–3.6	1,004–1,018	190–390
LVP	3	8–20 IU	2.4–3.8	1,008–1,016	140–289
Chlorpropamide	2	200–400 mg	4.0–6.0	1,006–1,008	156–203
DDAVP	34	2.5–30 µg	0.9–1.7	1,012–1,025	420–1,005

[1] Values were obtained during follow-up visits.

habilitation. The superiority of effect and the ease of administration of DDAVP helped to regain the confidence and cooperation of a number of patients who had previously neglected their treatment.

Dose Requirements. We did not find a correlation between the dose of DDAVP which was used for effective control and age or body size (fig. 1) as reported by others [1, 4, 8]. The dose ranges of the various medications used in our patients are indicated in table II. For DDAVP i.n. the dose range was from 2.5 to 30 µg per 24 h, in most patients given in two divided doses. In a few patients a single dose of 5 µg, taken at bedtime, kept them symptom-free for approximately 20 h. Only 3 patients had to take medication 3 times daily. It has been our practice to start treatment with a dose of 2.5 µg and to increase it gradually until normal urine output and osmolarity are achieved. Once established, the dose need not be changed with age and growth. In a few cases, however, dose requirements changed with time in both directions (fig. 3). In 3 patients, the DI resolved within 2–18 months after onset.

Throughout the years of our experience with DDAVP we have observed no side effects of this drug. Since lack of cooperation and inconsistency of treatment are major problems in treating children with DI, the introduction of DDAVP constitutes a major advance in this field [2, 5, 7, 8]. It should be noted, however, that we did have 2 cases of water intoxication in patients who had intentionally taken overdoses of DDAVP to achieve prolonged antidiuresis. Both were brought to our clinic with complaints of blurred vision, headaches and drowsiness and both were

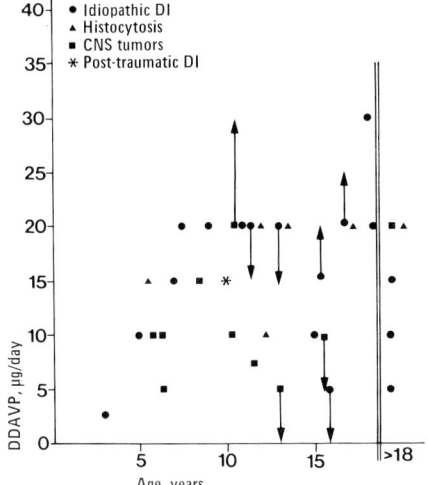

Fig. 1. Dose requirements of DDAVP intranasally in patients with cranial DI commencing at different ages. Arrows indicate changes of dose with time.

found to have hyponatremia and inappropriately high osmolarity of the urine. These findings returned to normal and the symptoms disappeared after less than 24 h fluid restriction and discontinuation of DDAVP.

Effects of Therapy on Growth. We have observed that growth is retarded in children with untreated or poorly controlled isolated cranial DI and on the other hand that growth is normal with adequate therapy. Figure 2 shows the growth chart of a patient in whom diagnosis was made and treatment started at age $3^{2}/_{12}$ years, with symptoms dating back at least 1 year before; without therapy growth showed progressive retardation, whereas after institution of therapy there was a remarkable catch-up and normalization. Figure 3 shows the growth curve of a boy diagnosed to have DI at age $6^{4}/_{12}$ years in whom the initial treatment with PP presented considerable difficulties due to side effects; he refused PTO injections and LVP proved ineffective. Up to the age of 11 years he was treated with PP on an irregular basis, and had inadequate control. The growth curve shows an obvious progressive slowing of growth rate. When therapy was changed to DDAVP at age 11 his attitude towards therapy changed; there was a remarkable improvement of control, and growth normalized. Figure 4 shows the normal growth of a girl who de-

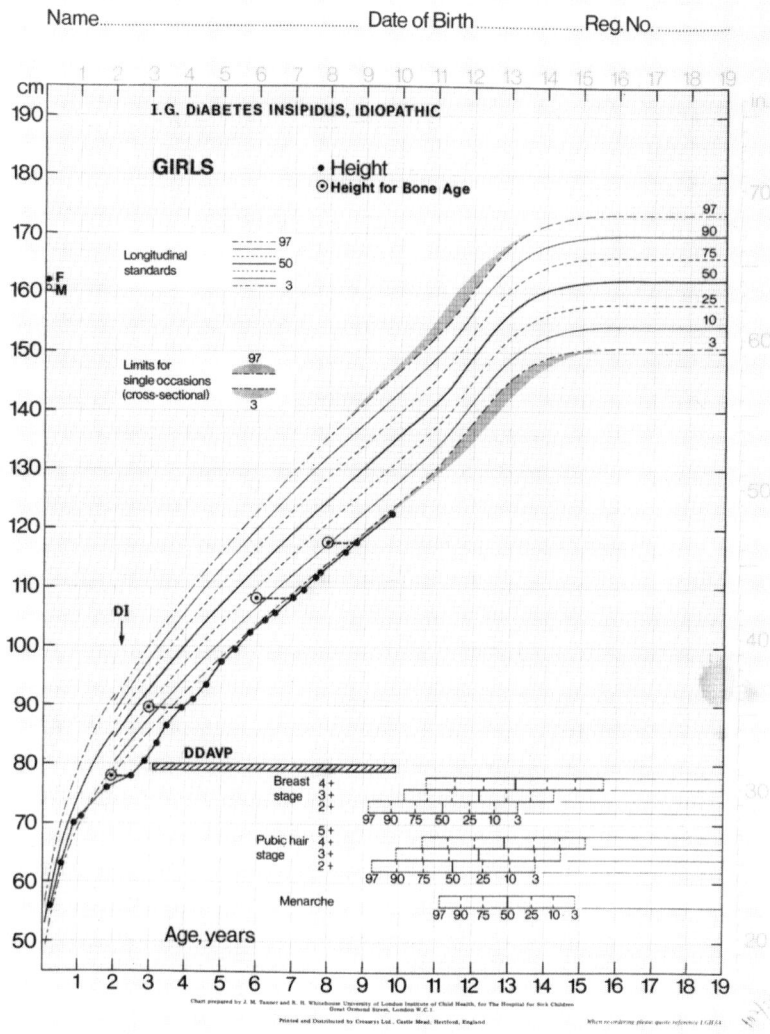

Fig. 2. Linear growth of girl with idiopathic cranial DI diagnosed and treated with DDAVP since age 3²/₁₂ years. Symptoms started at least 1 year before. Note slowing of growth before therapy and catch-up with subsequent normal growth after institution of therapy. Her short stature is familial.

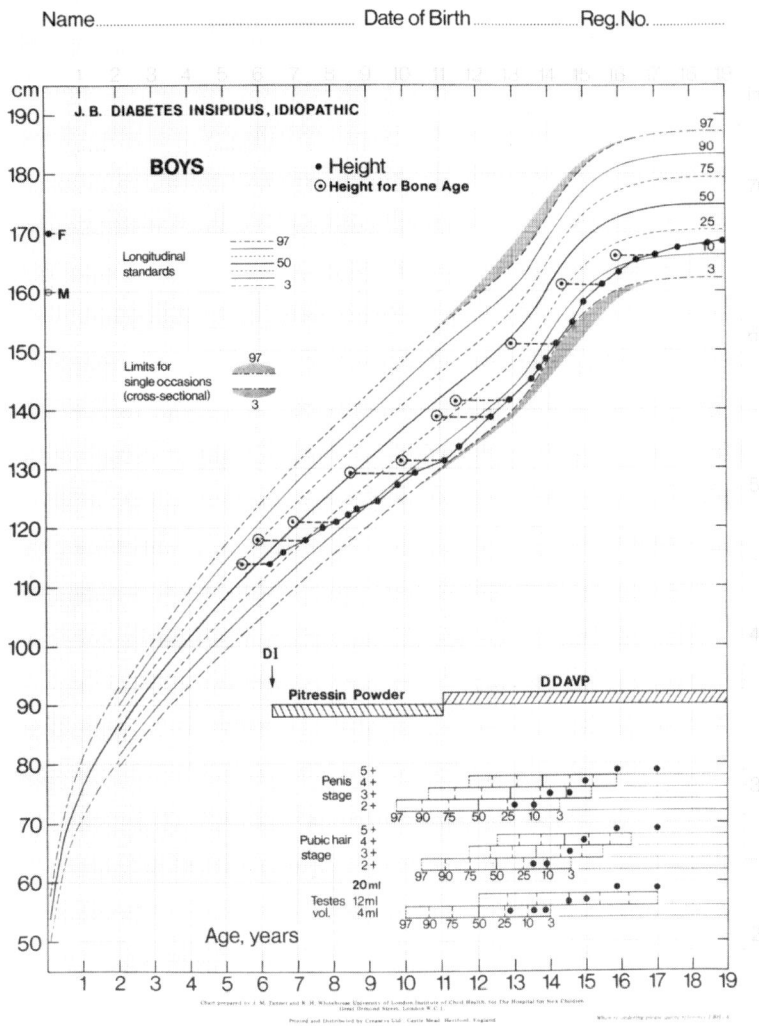

Fig. 3. Linear growth of boy with idiopathic cranial DI since age 6⁴/₁₂ years treated inconsistently with PP until age 11 when therapy was switched to DDAVP. Note progressive slowing of growth rate during therapy with PP and normalization of growth with catch-up after changing treatment to DDAVP.

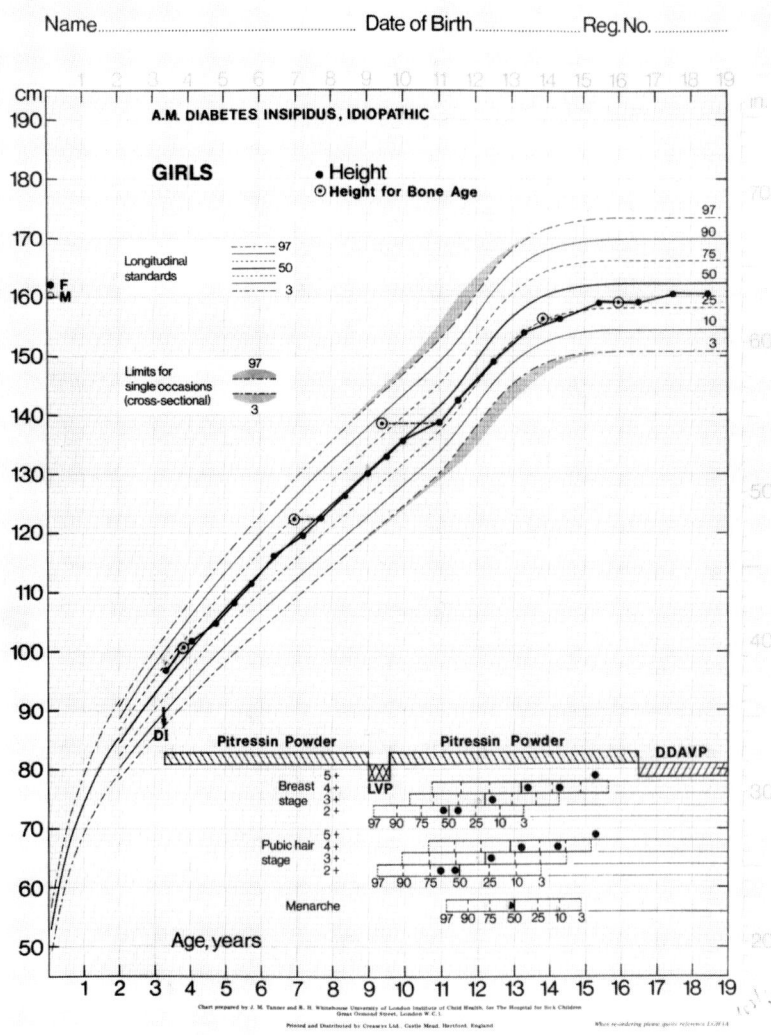

Fig. 4. Normal growth and puberty in a girl with idiopathic cranial DI diagnosed at age 3, treated with PP on a regular basis (in spite of side effects).

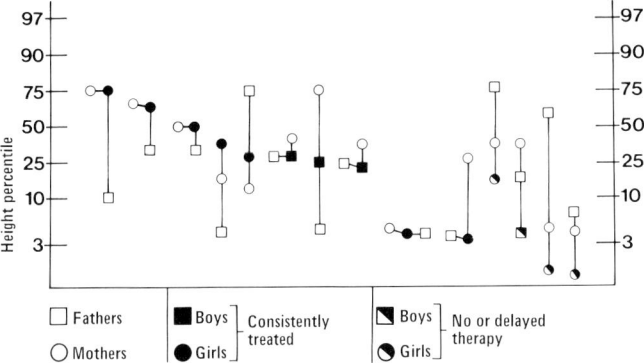

Fig. 5. Final height of 14 juvenile patients with isolated cranial DI in relation to that of their parents. Note that final height of untreated patients is below expected mid-parental height, in contrast to consistently treated patients whose final height matches their genetic growth potential.

veloped DI at age 3 and was treated with PP; despite marked chemical rhinitis and abdominal pain she was a cooperative patient and her DI was well controlled. By the time therapy was changed to DDAVP at age $16^{6}/_{12}$, she had surpassed the expected mid-parental height. Four of our patients with isolated DI (2 girls and 1 boy with idiopathic DI and 1 girl with post-traumatic DI) had grossly inadequate treatment during childhood, 1 because of a significant delay (2 years) in diagnosis and initiation of treatment and the other 3 because of total lack of cooperation, due to the side effects and limited effectiveness of the previous medications they used; 1 of the latter resumed therapy (with DDAVP) at the age of 22 and the other 2 remained untreated. All 4 patients showed marked growth retardation which could only be explained by the uncontrolled DI since they had no other hormone deficiencies and no other nutritional or metabolic problems aside from the chronically deranged water and electrolyte balance.

The effect of therapy on the growth in our patients is further illustrated in figure 5 which shows the final heights of 14 patients with isolated DI (11 with idiopathic DI) and 3 with histiocytosis (treated by chemotherapy) in relation to that of their parents. It is evident that the patients with long-standing lack of treatment suffered stunted growth.

Effect of Therapy on Pubertal Development. The patients with cranial DI who had intact gonadotrophin secretion developed normal puberty even when the growth rate was slowed by inadequate treatment. It would appear therefore that the process of puberty is less affected than growth by the metabolic disturbance associated with uncontrolled DI. However, 3 girls with poor control at the time of pubertal development showed delayed puberty and delayed menarche at ages $14^{9}/_{12}$, $15^{4}/_{12}$ and 16 years. It is of interest that 2 girls (1 with idiopathic DI consistently treated and the other with transient postinfectious DI, requiring no treatment), developed precocious puberty.

School Performance and Psychosocial Rehabilitation. Children with uncontrolled DI often experience severe difficulty in learning and in their psychosocial adjustment due to the chronic state of fatigue resulting from lack of sleep, and the frequent interruptions of their daily routine due to the polyuria and polydipsia and the enuresis. In the past, the side effects and the inconsistent effectiveness of the drug therapy then available aggravated these problems. With the introduction of DDAVP children have shown an improvement in psychological status [11]. In our patients there was a significant improvement in school achievements and the quality of life. Several patients participated in school excursions and sport activities for the first time in their lives because of the confidence gained from the DDAVP therapy. Patients with CNS tumors and other pituitary hormone deficiencies, who required other substitutional therapy and were limited in other ways, also stressed the importance of DDAVP in their rehabilitation and in enhancing the quality of life. Adolescents with MPHD ranked the importance to their antidiuretic therapy before treatment with growth hormone and with sexual hormones. The patients with isolated ADH deficiency who have reached adulthood are adequately treated with DDAVP and lead a normal life.

In conclusion, our experience confirms the recognized need for good control of DI to assure normal life in children and adolescents.

References

1 Aronson, A.S.; Andersson, K.E.; Bergstrand, C.G.; Mulder, J.L.: Treatment of diabetes insipidus in children with DDAVP, a synthetic analogue of vasopressin. Acta paediat. scand. *62:* 133–140 (1973).

2   Becker, D.J.; Foley, T.P., Jr.: 1-Deamino-8-*D*-arginine-vasopressin in the treatment of central diabetes insipidus in childhood. J. Pediat. *92:* 1011–1015 (1978).
3   Becker, D.J.; Foley, T.P., Jr.: The effect of water deprivation and water loading during treatment with 1-deamino-*D*-arginine-vasopressin in central diabetes insipidus in childhood. Acta endocr., Copenh. *97:* 358–360 (1981).
4   Czernichow, P.: Traitements du diabete insipide; in Bertrand, Rappaport, Sizonenko, Endocrinologie pédiatrique, pp. 541–543 (Payot, Lausanne 1982).
5   Kauli, R.; Laron, Z.: A vasopressin analogue in treatment of diabetes insipidus. Archs, Dis. Childh. *49:* 482–485 (1974).
6   Lee, W.P.; Lippe, B.M.; LaFranchi, S.H.; Kaplan, S.A.: Vasopressin analog DDAVP in the treatment of diabetes insipidus. Am. J. Dis. Child. *130:* 166–169 (1976).
7   Laron, Z.; Kauli, R.: Diabetes insipidus; in Gellis, Kagan, Current pediatric therapy; 8th ed., pp. 342–344 (Sauders, Philadelphia 1978).
8   Limal, J.M.; Mugner, E.; Czernichow, P.: Traitement du diabète insipide pitressosensible de l'enfant par le DDAVP. Archs. fr. Pédiat. *34:* 965–972 (1977).
9   Perheentupa, J.: Deficient AVP and thirst functions, diabetes insipidus (DI) and hypernatremic dehydration, treatment; in Brook, Clinical pediatric endocrinology, pp. 317–319 (Blackwell, Oxford 1981).
10 Vest, M.; Talbot, N.B.; Crawford, J.D.: Hypocaloric dwarfism and hydronephrosis in diabetes insipidus. Am. J. Dis. Child. *105:* 175–181 (1963).
11 Waggoner, R.W., Jr.; Slonim, A.E.; Armstrong, S.H.: Improved psychological status of children under DDAVP therapy for central diabetes insipidus. Am. J. Psychiat. *135:* 361–362 (1978).

Dr. R. Kauli, Institute of Pediatric and Adolescent Endocrinology,
Beilinson Medical Center, IL-49100 Petah-Tikva (Israel)

Peroral Administration of Antidiuretic Peptides to Conscious Dogs, Normal Humans and Patients with Diabetes insipidus

Hans Vilhardt, Mogens Hammer, Peter Bie[1]

Department of Medical Physiology C, The Panum Institute, University of Copenhagen and Department of Medicine P, Rigshospitalet, Copenhagen, Denmark

In 1913 *von den Velden* demonstrated an antidiuretic effect of subcutaneous injections of pituitary extracts to a patient with diabetes insipidus. This was later followed up by showing that the same effect could be obtained by intranasal administration of the pituitary extract [*Blumgart*, 1922] and this route of application of vasopressin and its analogues in the management of diabetes insipidus has been prevailing ever since. In a case report published in *Endocrinology* in 1918 the Norwegian physician *Motzfeldt* described a female diabetes insipidus patient who could reduce her nightly polyuria from 2,500 to 300 ml by eating 2–7 fresh pituitary bodies from cattle every evening. This observation apparently did not attract much attention at the time and it took more than 60 years to rediscover it. The present presentation can therefore be regarded as a direct continuation of the work of *Motzfeldt* as published in 1918.

Experiments in Dogs

In previous experiments in trained conscious dogs given a sustained water load we demonstrated that peroral administration of arginine vasopressin (AVP) and its structural analogues 1-deamino-8-*D*-AVP

---

[1] We are grateful to Ferring Pharmaceuticals for providing the peptides and tablets used in this study. Thanks are due to Dr. *J. Rask Madsen,* Department of Gastroenterology, University Hospital of Odense, Denmark, for assistance in introducing the duodenal tubes to their correct position in the gut.

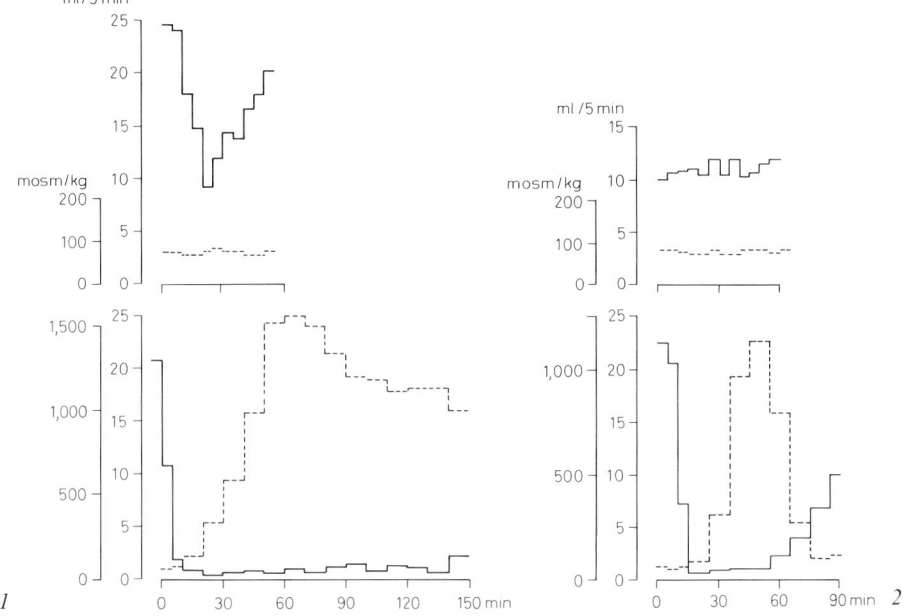

Fig. 1. The effect of 5 μg (upper panel) and 50 μg (lower panel) of perorally administered 1-deamino-AVP on urine volume (solid lines) and osmolality (broken lines) in a hydrated dog.

Fig. 2. The effect of 5 μg (upper panel) and 50 μg (lower panel) of perorally administered 8-D-AVP on urine volume (solid lines) and osmolality (broken lines) in a hydrated dog.

(DDAVP) and 4-asparagine-DDAVP resulted in an antidiuretic response with a concomitant increase in urine osmolality [*Vilhardt and Bie,* 1983]. On an equimolar base DDAVP was about 10 times as potent as AVP, the lowest active doses of DDAVP being 3–5 μg. This indicates that DDAVP is either more resistant to enzymic degradation in the gastrointestinal tract and/or it is absorbed by the intestinal mucosa to a higher degree than AVP. It should be pointed out, however, that the antidiuretic activity of the two substances has not been properly established in the conscious dog.

DDAVP is modified in position 1 and position 8 of the AVP molecule and to investigate which of these two modifications was responsible for the higher bioavailability of DDAVP we administered 1-deamino-AVP and 8-*D*-AVP intragastrically to the hydrated dogs. As seen from figures 1 and 2, the antidiuretic response of 1-deamino-AVP is

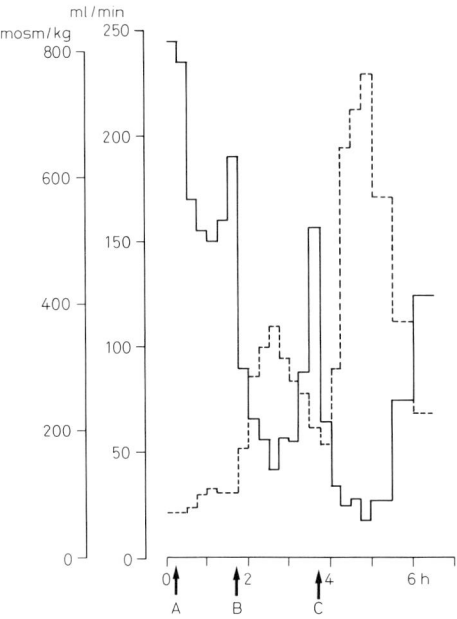

Fig. 3. The effect on urine volume (solid line) and osmolality (broken line) in a hydrated human volunteer (22-year-old male, 64 kg) after peroral administration of 20 µg (A), 40 µg (B) and 60 µg (C) of DDAVP.

more profound and in particular more prolonged than the effect caused by 8-D-AVP, suggesting that it is the deamination at position 1 of AVP which leads to a higher bioavailability after peroral administration.

Experiments in Normal Humans

Healthy volunteers of both sexes aged 18–42 years were given a peroral water load corresponding to 1.5% of the body weight. Subjects were not fasted and were allowed to eat freely during the experiment. Urine was voided every 15 min and volume and osmotic concentration were measured. Fluid loss with the urine was currently replaced by drinking equal amounts of tap water. When urine output exceeded 150 ml/15 min DDAVP was administered perorally in 50 ml of water. The threshold dose for an antidiuretic effect was in most cases found to be 50 µg of peptide but some persons responded even to 20 µg (fig. 3). Doses of 100 µg or more caused an antidiuresis lasting approximately 6 h.

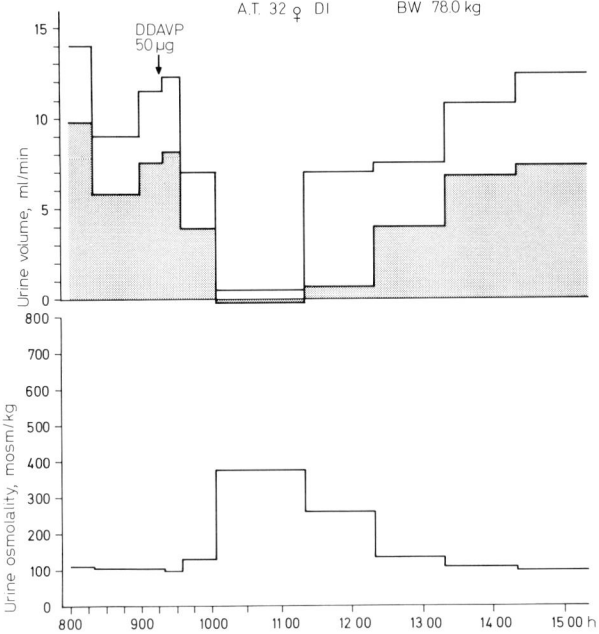

Fig. 4. The effect on urine volume (upper panel), free water clearance (shaded area) and urine osmolality (lower panel) of a tablet containing 50 μg of DDAVP in a patient with diabetes insipidus, who had received no treatment for the preceding 36 h.

In 2 subjects a duodenal tube was inserted through the nose and under X-ray guidance the tip was placed at the junction of duodenum and the jejunum. After hydration DDAVP (200 μg) was injected into the intestinal lumen through the tube. In both cases a profound and long-lasting antidiuretic effect was obtained [*Vilhardt and Bie,* 1984].

Experiments in Patients with Diabetes insipidus

The results of the studies performed in normal volunteers clearly indicated that DDAVP can be absorbed from the human gastrointestinal tract. The duration of the antidiuretic response with the doses used, however, appeared not of sufficient duration, for example, to ensure a full night's sleep for a patient with diabetes insipidus. To prolong the effect of perorally administered DDAVP, specially formulated tablets

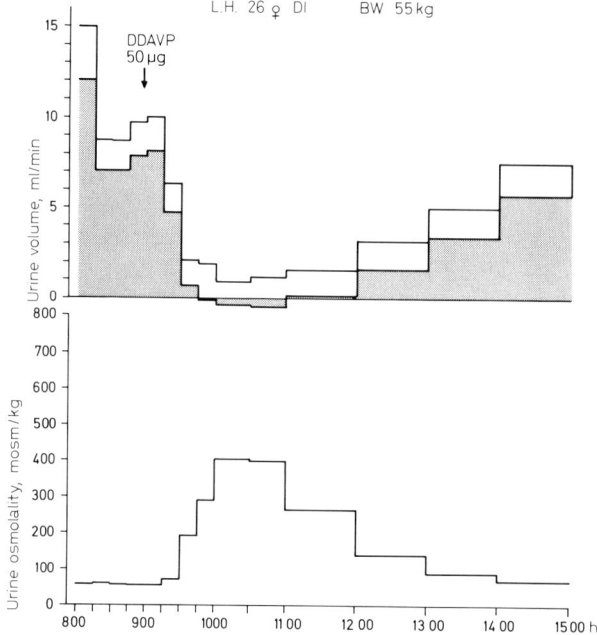

Fig. 5. The effect on urine volume (upper panel), free water clearance (shaded area) and urine osmolality (lower panel) of a tablet containing 50 μg of DDAVP in a patient with diabetes insipidus, who had received no treatment for the preceding 36 h.

were manufactured (Ferring Pharmaceuticals, Malmö) each containing 50 or 100 μg of the peptide.

Patients of both sexes suffering from diabetes insipidus of several years duration were included in the study. Their polyuria was well controlled by intranasal application of DDAVP. All medication was discontinued 36 h before initiation of treatment with DDAVP tablets. Urine volume and osmolality were then measured before and for 6 h after administration of a tablet containing 50 μg of DDAVP. In all cases antidiuresis occurred with a concomitant increase in urine osmolality (not always pronounced, probably due to decreased renal papillary osmolality caused by the preceding polyuria). Antidiuresis usually lasted for several hours (fig. 4, 5). The treatment was then continued on an outpatient basis initially with weekly control visits to the hospital. Most patients could satisfactorily control their polyuria on 2×100 μg DDAVP/24 h, a few required an additional 100 μg. 3 patients have at present been treated in this way for a period of more than 6 months. No side effects have been observed and all patients preferred the tablets to the intranasal instillation.

Discussion

It is generally believed and in accordance with most text books in gastrointestinal physiology that perorally ingested proteins and peptides are degraded to amino acids by pancreatic and intestinal enzymes. Yet it has been demonstrated that di- and tripeptides undergo intestinal absorption in animals [*Matthews* et al., 1969] and in man [*Adibi and Morse*, 1971]. The present experiments show that vasopressin and its analogues can be absorbed after peroral administration both in animal and man. The observed antidiuretic effects indicate that the peptides are absorbed in an intact form since any cleavage of peptide bonds or reduction of the disulfide bridge invariably lead to complete inactivation of the biological activity of the molecules.

The results of the experiments in dogs suggest that when given in equimolar doses DDAVP enters the blood stream more readily than AVP and that 1-deamino-AVP is absorbed to a higher degree than 8-*D*-AVP. Only the 8-*D* substituted analogues are resistant to tryptic cleavage and the results therefore suggest that trypsin plays a minor role in degradation of these small peptides in the gut.

The present data do not allow any conclusions as to the absorption kinetics of the analogues. The amount of peptide entering the blood is the net result of metabolic stability and absorption rate of the peptides and neither of these parameters is known. The different half-lives in plasma and the different antidiuretic activity of the analogues add to the difficulties. Apparent absorption kinetics can probably only be obtained through in vitro studies.

The demonstration that DDAVP given perorally exerts an antidiuretic activity in patients with diabetes insipidus is of therapeutic interest. The doses needed to control the polyuria are somewhat higher than those used for intranasal administration, but this may be outweighed by several merits of the tablet form. Some patients find it difficult to handle the intranasal application system, others complain of constant mucosal irritation in the nose, and in patients with upper respiratory infections (common cold) the antidiuretic effect following intranasal administration may be erratic. These disadvantages are not encountered in peroral therapy and it is characteristic that all the patients we have treated prefer the tablets to the intranasal form. We therefore believe that DDAVP in a properly formulated tablet will become a therapeutic option for the future treatment of patients with diabetes insipidus.

References

Adibi, S.A.; Morse, E.-L.: Intestinal transport of dipeptides in man: relative importance of hydrolysis and intact absorption. J. clin. Invest. *50:* 2266 (1971).
Blumgart, H.L.: The antidiuretic effect of pituitary extract applied intranasally in a case of diabetes insipidus. Archs intern. Med. *29:* 508 (1922).
Matthews, D.M.; Lis, M.T.; Cheng, B.; Crampton, R.F.: Observations on the intestinal absorption of some oligopeptides of methionine and glycine in the rat. Clin. Sci. *37:* 751 (1969).
Motzfeldt, K.: Diabetes insipidus. Endocrinology *2:* 112 (1918).
Velden, R. von den: Die Nierenwirkung von Hypophysenextrakten beim Menschen. Berl. klin. Wschr. *2:* 2083 (1913).
Vilhardt, H.; Bie, P.: Antidiuretic response in conscious dogs following peroral administration of arginine vasopressin and its analogues. Eur. J. Pharmacol. *93:* 201 (1983).
Vilhardt, H.; Bie, P.: Antidiuretic effect of perorally administered DDAVP in hydrated humans. Acta endocr., Copenh. (1984).

Hans Vilhardt, MD, Department of Medical Physiology C, Panum Instituttet, Blegdamsvej 3c, DK-2200 Copenhagen N (Denmark)

Subject Index

Accessory nucleus 4
Acetylcholine 31
Adipsia
   in adults 161
   in children 203
   treatment 299
Adrenal insufficiency and DI 196, 257
Angina 293, 294
Angiotensin and thirst 80
Angiotensin II 31
Anterior pituitary function 160
Apomorphine 186
Autoimmunity and DI 237

Brain stem 14
Brattleboro rats
   animal model of DI 110
   genetic defect 37
   immunohistochemistry 2
   receptors 100

Calcium 30
   and secretion 59
Carbamazepine 141, 295
Catecholamine 14, 30
Clofibrate 296
Chlorpropamide
   and hypoglycemia 295
   treatment of DI 295
Chronic hypernatremia, see Adipsia
Cord
   arteriovenous gradient 47

   during hypoxia 47
   plasma levels 47
Craniopharyngioma 199, 215
CRH
   and vasopressin 9
   colocalization with AVP 9

DDAVP
   binding to receptors 99
   during pregnancy 272
   hemostatic effect 285
   inactivation 287
   memory 286
   pharmacokinetics 286
   pharmacology 284
   plasma value 286
   treatment
      in adults 297
      in children 305
   water intoxication
      in adults 298
      in children 306
      postoperative period
         259
Dehydration test
   in adults 165, 179
   in children 192
Dopamine 14
Dysgerminoma
   children 199, 240
   tumoral markers 243
   X-ray 216, 242

Encephalomalacia 159

Familial DI 202
Fertility with DI 268
Fetal AVP
 in amniotic fluid 46
 in CSF 46
 kidney receptors 43, 45
 in plasma 43
Flush 294

GABA 31
Gene 25, 37
Genetics
 familial central DI 202
 nephrogenic DI 228
Gestation 268
Glycoprotein 26
Gomori staining 2
Granules 23, 24
Growth
 in DI 307
 in nephrogenic DI 239

Histiocytosis 158, 199, 218, 235
Hypertonic NaCl infusion 128

Idiopathic DI
 in adults 158
 anatomy 158, 202
 and autoimmunity 232
 in children 200
Immunofluorescence 2
Immunoperoxidase 2
Indomethacin 227
Infant AVP physiology 47

Lactation, with DI 270
Locus ceruleus 14
Lung fluid production and vasopressin 44

Magnocellular system 3

Neurophysins 22, 28, 30

Oat cell carcinoma 28
Opiates 31
Organum vasculosum 12

Paraventricular nucleus 4
Parturition
 cesarian section 268
 hypocontractile labor 268
Parvocellular nucleus 9, 14
Phasic pattern 53, 54
Pitressin tannate in oil 293
Pituicytes 31
Placenta
 transfer of AVP 43
 transfer of DDAVP 272
 water transfer across 44
Post-surgical DI 158, 197
Precursor 24
Pregnancy
 effect on DI 266
 nephrogenic DI 271
 and osmotic threshold 267
 and oxytocin 267, 270
Primary empty sella 217
Primary polydipsia
 animal model 105
 in children 203
 in human adults 162, 170, 178
 differentiation from neurogenic DI 179
 hydronephrosis in humans 156
 maximum urinary concentration 106
 treatment with DDAVP 162, 186
Prostaglandins 269
 AVP action and nephrogenic DI 143, 229
 and satiety 81
Proteolytic enzymes 27, 29
Psychosocial patterns 312
Pubertal development 312

Receptors
 affinity constant 92
 and antagonists 98
 binding to kidney membrane 91
 hepatocyte 93
 and inositol lipids 96
 pituitary cells 94
 regulation 100
 spare 93
 for thirst
  exteroceptive receptors 74

Subject Index

and extracellular dehydration 74
intracellular dehydration 72
and sodium 73
vascular bed 94
vasopressin-sensitive cyclase 95
Refractory period 62
Renal refractoriness to vasopressin 268

Sarcoidosis 161
Satiety 77
Sheehan's syndrome 272
Spinal cord 14
Stalk enlargement 158, 202, 217
Stria terminalis 14
Substance P 31
Suprachiasmatic nucleus 12
Supraoptic nucleus 4

Thiazide diuretics 296
Thirst
  and extracellular dehydration 70, 71
  and intracellular dehydration 70
  sodium-sensitive 70
Transport 28
Treatment
  and adipsia 299
  central DI
    adults 292
    children 304

during pregnancy 272, 273
nephrogenic DI 225
oral administration 314
postoperative 259, 300
test
  in central DI 191
  in nephrogenic DI 229
Triphasic DI 251
Tumors
  metastases 160
  primary brain 157, 198

Urinary AVP 169
Uterine contraction 269

Vasopressin
  analogues 280
  snuff 293
  and thirst 82
Vasotocin in fetus 42

Water intoxication, during treatment with DDAVP
  in adults 308
  in children 306

Zona externa of median eminence 8

NO LONGER THE PROPERTY
OF THE
UNIVERSITY OF R.I. LIBRARY